D1825813

Pragmatics in Neurogenic
Communication Disorders

Related Pergamon Books

MICHEL PARADIS *Foundations of Aphasia Rehabilitation*
MICHEL PARADIS *Apects of Bilingual Aphasia*
FRANCO FABBRO *Concise Encyclopedia of Language Pathology*
K. JASZCZOLT and
K. TURNER *Contrastive Semantics & Pragmatics, 2-Volume Set*
J. L. MEY *Concise Encyclopedia of Pragmatics*

Related Pergamon Journals

Journal of Communcation Disorders
Editor: Theodore J. Glattke

Journal of Fluency Disorders
Editor: G. J. Brutten

Journal of Neurolinguistics
Editor: J. C. Marshall

Journal of Pragmatics
Editor: J. L. Mey

Language Sciences
Editor: Nigel Love

Neuropsychologia
Editor: S. D. Iversen

Free specimen copies of journals available on request

PRAGMATICS IN NEUROGENIC COMMUNICATION DISORDERS

Edited by

MICHEL PARADIS
McGill University

INTERNATIONAL ASSOCIATION OF
LOGOPEDICS AND PHONIATRICS

PERGAMON

U.K.	Elsevier Science Ltd, The Boulevard, Langford Lane, Kidlington, Oxford OX5 1GB, U.K.
U.S.A.	Elsevier Science Inc., 660 White Plains Road, Tarrytown, New York 10591-5153, U.S.A.
JAPAN	Elsevier Science Japan, Tsunashima Building Annex, 3-20-12 Yushima, Bunkyo-ku, Tokyo 113, Japan

First edition 1998

Library of Congress Cataloging in Publication Data

Pragmatics in neurogenic communication disorders/edited by Michel Paradis.
    p. cm.
"International Association of Logopedics and Phoniatrics."
Includes bibliographical references and indexes.
ISBN 0-08-043065-1
 1. Aphasia. 2. Pragmatics. I. Paradis, Michel.
II. International Association of Logopedics and Phoniatrics.
    [DNLM: 1. Aphasia—etiology. 2. Aphasia—physiopathology.
3. Brain Diseases—complications. 4. Communication. WL 340.5 P898 1998]
RC425.P726   1998
616.85'52—dc21
DNLM/DLC
for Library of Congress   98-6486
                                                                         CIP

ISBN 0-08-043065-1

Printed and bound in Great Britain by Galliard (Printers) Ltd, Great Yarmouth

PII: S0911-6044(98)00022-0

Publishers' Preface

In recognition of the current importance of this topic, this collection was also published as Volume 11 (1–2) of the *Journal of Neurolinguistics*

Contents

III. Compensatory strategies

 Pergamon

List of Contributors

John S. Ashford, Veterans Administration Medical Center, Nashville, Tennessee, U.S.A.

Linda L. Auther, Veterans Administration Medical Center, Nashville, Tennessee, U.S.A.

Jan R. Avent, Department of Communicative Sciences and Disorders, California State University, Hayward, California, U.S.A.

Ronald Bloom, Speech-Language-Hearing Sciences, Hofstra University, Hempstead, New York, New York, U.S.A.

Joan C. Borod, Department of Psychology, Queens College, and Department of Neurology, Mount Sinai Medical Center, CUNY, New York, New York, U.S.A.

Frauke Bürk, Speech-Language Therapy, Sanderson Hospital, Newcastle-upon-Tyne, Great Britain.

Sandra Bond Chapman, Callier Center for Communication Disorders, University of Texas, Dallas, Texas, U.S.A.

Karen L. Chobor, Department of Neurology, New York University Medical Center, New York, New York, U.S.A.

Nina Dronkers, Veterans Administration Northern California Health Care System and University of California, Davis, California, U.S.A.

Bibi Fex, Logopedic Unit, ENT Department, Helsingborg Hospital, Helsingborg, Sweden.

Constance R. Henschel, Department of Hearing and Speech Sciences, Vanderbilt University School of Medicine, Nashville, Tennessee, U.S.A.

Amy Peterson Highley, Callier Center for Communication Disorders, University of Texas, Dallas, Texas, U.S.A.

Judy Kegl, Center for Molecular and Behavioral Neuroscience, Rutgers University, Newark, NJ, U.S.A

Andrew Kertesz, Department of Clinical Neurological Sciences and Disorders, St. Joseph's Health Centre, University of Western Ontario, London, Ontario, Canada.

Howard S. Kirshner, Department of Neurology, Vanderbilt University School of Medicine, Nashville, Tennessee, U.S.A.

Elissa Koff, Department of Psychology, Wellesley College, Wellesley, Massachusetts, U.S.A.

Minna Laakso, Department of Phonetics, University of Helsinki, Helsinki, Finland.

Matti Laine, Department of Neurology, University of Turku, Turku, Finland.

Ruth Lesser, Department of Speech, University of Newcastle-upon-Tyne, United Kingdom.

Carl A. Ludy, Veterans Administration Northern California Health Care System, Davis, California, U.S.A.

Jennifer E. Peacock, School of Communication Sciences and Disorders, University of Western Ontario, London, Ontario, Canada.

Marjorie Perlman Lorch, Applied Linguistics, Birkbeck College, University of London, London, Great Britain.

Ann-Christin Månsson, Logopedic Unit, ENT Department, Helsingborg Hospital, Helsingborg, Sweden.

Nick Miller, Department of Speech, University of Newcastle-upon-Tyne, United Kingdom.

Loraine K. Obler, Program in Speech and Hearing Sciences, CUNY Graduate School, New York, New York, U.S.A.

Joseph B. Orange, School of Communication Sciences and Disorders, University of Western Ontario, London, Ontario, Canada.

Michel Paradis, Department of Linguistics, McGill University, and Cognitive Neuroscience Laboratory, Université du Québec à Montréal, Montreal, Canada.

Lisa Perkins, Department of Speech, University of Newcastle-upon-Tyne, United Kingdom.

Howard Poizner, Center for Molecular and Behavioral Neuroscience, Rutgers University, Newark, NJ, U.S.A

Brenda B. Redfern, Veterans Administration Northern California Health Care System, Davis, California, U.S.A.

Juha Rinne, Department of Neurology, University of Turku, Turku, Finland.

Avraham Schweiger, Department of Speech and Hearing Sciences, CUNY Graduate Center, New York, New York, U.S.A.

Luise Springer, Lehranstalt für Logopädie, RWTH Aachen, Germany.

Jennifer L. Thompson, Callier Center for Communication Disorders, University of Texas, Dallas, Texas, U.S.A.

Elina Vuorinen, Pulsin Neurocenter, Turku, Finland.

Robert T. Wertz, Veterans Administration Medical Center, and Department of Hearing and Speech Sciences, Vanderbilt University School of Medicine, Nashville, Tennessee, U.S.A.

Anne Whitworth, Department of Speech, University of Newcastle-upon-Tyne, United Kingdom.

Pergamon

J. Neurolinguistics, Vol. 11, Nos 1–2, p. ix-x, 1998
Published by Elsevier Science Ltd. Printed in Great Britain
0911-6044/98 $19.00 + 0.00

PII: S0911-6044(98)00018-9

Preface

At its August 1995 meeting in Cairo, Egypt, on the occasion of the XXIIIrd Congress of the International Association of Logopedics and Phoniatrics (IALP), the Aphasia Committee undertook to prepare a report on pragmatics in neurogenic communication disorders to be presented at the XXIVth IALP congress in Amsterdam in August 1997.

The *Longman Dictionary of Applied Linguistics* defines pragmatics as the study of the use of language in communication, particularly the relationships between sentences and the contexts and situations in which they are used (Richards, Platt, & Weber, 1985). Each member of the Aphasia Committee set out to investigate pragmatics from one of two angles: (1) as a compensatory strategy for individuals whose implicit linguistic competence has been impaired, and (2) as an element affected by a variety of neurogenic conditions, from focal damage to the right or the left hemisphere to various types of progressive dementias. A first draft of these investigations was discussed by the members of the Aphasia Committee at a meeting held in Montreal, Canada, on 1 and 2 July 1997. After a thorough discussion of each contribution, each author wrote a final report which is published in the present volume. Hence, all contributions report on research conducted specifically for this publication over the past three years by members of the IALP Aphasia Committee and their collaborators.

The purpose of this work is to focus the attention of language pathologists and neurolinguists around the world on the importance of pragmatics in verbal communication and its disorders. The various contributions illustrate why pragmatics might in fact be considered more critical than morphosyntax for verbal communication. On the one hand, it can always be used by speakers to overcome gaps in their linguistic competence, as is routinely done by children, second language learners, and aphasic patients. On the other, whereas languages with the relevant structural information can always take pragmatic cues into account, languages with no morphosyntactic marking for certain concepts, such as gender in pronominal reference, for instance (e.g., spoken Chinese), must rely solely on discourse context and other pragmatic cues. In both cases, neither grammar nor pragmatics is entirely sufficient but both are necessary for arriving at an interpretation for every utterance in normal verbal communication. Chomsky (1980) explicitly recognized that "the system of language [i.e., implicit linguistic competence, the grammar] is only one of a number of cognitive systems that interact in the most intimate way in the actual use of language" (p. 188).

Pragmatic competence comprises two major aspects that will be considered here: (a) the organization of discourse, that is, when and how utterances are appropriately used in verbal interactions, and (b) the meaning derived from aspects of language use other than the grammar, such as inferences from the situational context, paralinguistic phenomena (affective prosody, gestures, facial expressions) or general knowledge. Such inferences, it is reckoned, are necessary to interpret not only jokes, sarcasm, indirect speech acts and non-stereotyped metaphors, but every single utterance, if only to determine whether it is to be taken literally or not.

After a discussion of the historical and theoretical background for the study of pragmatics in neurogenic communication disorders, a first set of five papers examine ways in which the pragmatic aspects of discourse are disordered by neurogenic pathologies; a second group of five papers investigate pragmatic deficits in communication disorders of varying etiology; and a third and last group of four papers consider the use of pragmatics as a compensatory strategy for patients with aphasia.

These contributions, which appear in the third volume published since 1993 by the IALP Aphasia Committee, are simultaneously published in a special issue of the *Journal of Neurolinguistics* (1998), volume 11, numbers 1 and 2.

Acknowledgments

Thanks are due to Lara Riente for her careful copy-editing of the text and the production of the camera-ready copy. Dr. Zofia Laubitz revised the texts contributed by the nonnative speakers of English. The research was supported in part by the Quebec Ministry of Education FCAR grants 92ER1050 and 96ER0404 and the Social Sciences and Humanities Research Council of Canada grant 410-94-0770.

References

Chomsky, N. (1980). *Rules and representations*. Oxford: Basil Blackwell.
Richards, J., Platt, J., & Weber, H. (1985) *Longman Dictionary of Applied Linguistics*. Harlow, UK: Longmans.

Pergamon

J. Neurolinguistics, Vol. 11, Nos 1–2, p. 1–10, 1998
© 1998 Published by Elsevier Science Ltd. All rights reserved
Printed in Great Britain
0911-6044/98 $19.00 + 0.00

PII: S0911-6044(98)00001-3

The other side of language: Pragmatic competence

Michel Paradis

Department of Linguistics, McGill University

Abstract—Language pathology has traditionally been concerned with deficits in left-hemisphere-based linguistic competence, namely, in phonology, morphology, syntax and semantics. It has become increasingly apparent over the past twenty years that linguistic competence is not sufficient for normal verbal communication. Right-hemisphere-based pragmatic competence is at least equally necessary. As a result, on the one hand, neuropsychologists have been investigating pragmatic deficits, and on the other, language pathologists have been using aphasic patients' preserved pragmatic abilities to help them compensate for their deficits in linguistic competence. From the viewpoint of linguistic theory, there is now an external justification for treating sentence grammar independently of pragmatics.

As defined by Chomsky (1980), grammar characterizes the properties of sentences, but the production and interpretation of sentences by a speaker makes use of much else including, in particular, "pragmatic competence" (p. 206). Generative grammars, through their successive mutations over the past 40 years (e.g., from transformational to government and binding to the minimalist program), have been concerned with (extralinguistic context-independent) sentence grammar. Linguistic competence, as described by Chomsky (1965) is the implicit knowledge of the speaker-hearer's [sentence] grammar. The automatic, unconscious use of this linguistic competence is what allows speaker-hearers to produce and understand sentences.

In the normal use of language, various additional cognitive entities have been recognized either as necessary (e.g., conceptual and episodic memory, the speaker's "encyclopædia", i.e., that which the speaker speaks *about*) or as placing constraints on the use of linguistic competence (e.g., short-term memory constraints, attention span limitations, etc.). Moreover, in the normal use of language, in addition to the interpretation of the literal meaning of sentences, a discourse grammar, including rules of presuppositions and inference, and in general any extra-sentential context-dependent phenomena, is required. Sociolinguistic rules, which determine the appropriate choice among the various possible structures available in linguistic competence, are equally necessary. Paralinguistic competence is likewise required, comprising the use of intonation, gestures, and anything that serves to specify the meaning of the sentence— such as whether it is meant as a sarcastic remark or a compliment, an indirect request or a factual question, whether it is to be taken with a figurative, metaphoric, idiomatic meaning or at face value.

Indeed, we may estimate that more than half of what we say is not literally what we mean—at least not entirely. Most of the time, we mean more than what we say, or something different than what we actually say, or even the opposite of what we say. Not that we are particularly devious or deliberately deceptive, but this is simply the way language works in communication. We rarely state everything that we mean, leaving out whatever we consider to be obvious to the listener, often in the form of an inference (e.g., in the statement, "Bill drove John to the hospital—He had a broken leg", there is no need to specify that John, not Bill, had the broken leg for reasons of inference from general knowledge: people with broken legs do not usually drive friends to the hospital, though the opposite is not uncommon). Often we say something different from what we mean, as in the use of figurative speech (*a dirty look*), idiomatic expressions (*she's got him eating out of her hand*), metaphors (*the customs officer was a brick wall*), and indirect speech acts ("Is there any salt on the table?" uttered at the dinner table after sipping the first spoonfuls of one's soup, is not interpreted as a question, but as an obvious request to pass the salt).

The illocutionary force of indirect speech acts and the meaning of idiomatic expressions are so immediately obvious that their literal interpretation is a recurrent theme in newspaper comics (Dagwood to Blondie on her way to the kitchen: "If you happen to see a turkey sandwich in there, with lettuce, pickles and mayonnaise on it, would you bring it to me?" Blondie, coming back: "I didn't see one". Idiomatic expressions: Hägar: "I'm so hungry I could eat a horse." Waiter: "Rare, medium or well done?"). Sometimes we mean the opposite of what we say (irony, sarcasm) although what we mean is obvious to our interlocutor because of the tone of voice and/or concurrent circumstances (e.g., "That's a fine job!" uttered in the presence of a spectacularly bad performance). There is no doubt that the statement "the cat is on the mat" will be unambiguously interpreted as a warning when shouted at someone who is about to open the door; as a scolding if the cat's owner was told in no uncertain terms that it was all right to let the cat in the house, as long as it did not shed its hair on the carpet; and as a congratulatory remark if the cat's owner has been training the cat for days to lie on the mat, not on the sofa.

In the literature on linguistics and the pathology of communication, there are at least two clearly distinct domains subsumed under the term "pragmatics": discourse structure and nonliteral meanings. The two domains may or may not share a single source, but both have traditionally been considered to constitute pragmatics and, more interestingly, both have been reported to be vulnerable to right hemisphere damage, frontal lobe lesions and schizophrenia, while relatively preserved in the context of dysphasia (Pierce & Wagner, 1985). Whether discourse analysis and the respect of conversational maxims are to be subsumed under the term "pragmatics", as two of its major components, or whether the two areas may be considered distinct, albeit with some possible overlap, is a decision contingent upon the level of abstraction of the analyst's perspective and theoretical framework. The common denominator seems to be the necessity to rely on context in order to derive an interpretation. This context can be situational (including paralinguistic cues), but also discursive (including structure and contents, as well as turn-taking and the like, from which inferences and implications must be made).

Over the past century, damage to specific areas of the left cerebral hemisphere (LH) has been reported to disrupt the comprehension and/or production of various aspects of phonology, morphology, syntax, and the lexicon. In addition to mild symptoms of the same type subsequent to right hemisphere (RH) lesions in sites homologous to the LH classical language areas in some patients (Joanette, 1980; Joanette, Lecours, Lepage, & Lamoureux, 1983; Joanette, Goulet, & Hannequin, 1990), clear deficits of a different nature, affecting the comprehension and production of humor, affect, and various aspects of the nonliteral interpretation of utterances, have been reported over the past 20 years or so (Hier & Kaplan, 1980; Dwyer & Rinn, 1981; Ross, 1981, 1984, 1993; Brownell, Michel, Powelson, & Gardner, 1983; Foldi, Cicone, & Gardner, 1983; Gardner, Brownell, Wapner, & Michelow, 1983; Brookshire & Nicholas, 1984; Heilman, Bowers, Speedie, & Costlett, 1984; Hirst, LeDoux, & Stein, 1984; Tompkins & Mateer, 1985;

Bihrle, Brownell, Powelson, & Gardner, 1986; Brownell, Potter, Bihrle, & Gardner, 1986; McDonald & Wales, 1986; Foldi, 1987; Roman, Brownell, Potter, Seibold, & Gardner, 1987; Bihrle, Brownell & Gardner, 1988; Brownell, 1988; Weylman, Brownell, Roman, & Gardner, 1989; Brownell, Simpson, Birhle, Potter, & Gardner, 1990; Joanette & Brownell, 1990, 1993; Joanette, Goulet, & Hannequin, 1990, 1993; Kaplan, Brownell, Jacobs, & Gardner, 1990; Molloy, Brownell, & Gardner, 1990; Brownell, Carroll, Rehak, & Wingfield, 1992; Lalande, Braun, Charlebois, & Whitaker, 1992; Rehak, Kaplan, & Gardner, 1992; Rehak, Kaplan, Weylman, Kelly, Brownell, & Gardner, 1992; Brownell, Gardner, Prather, & Martino, 1995; Bara, Tirassa, & Zettin, 1997; Dipper, Bryan, & Tyson, 1997). Deficits secondary to RH damage thus typically involve those aspects of language use other than the ones involved in the literal interpretation of (context-independent) sentences.

More specifically, patients with RH damage have been variously shown to be insensitive to: connotative meaning (Brownell, Potter, Michelow, & Gardner, 1984); figurative speech, even when supportive contextual cues are available (Myers & Linebaugh, 1981); metaphors (Gardner, Ling, Flamm, & Silverman, 1975; Winner & Gardner, 1977; Brownell, Potter, Michelow, & Gardner, 1984; Bryan, 1988; Brownell, Simpson, Bihrle, Potter, & Gardner, 1990; Tompkins, 1990); idioms, even familiar ones (Weylman et al., 1989; Tompkins, Boada, & McGarry, 1992); the emotive meaning of words and emotions that have to be inferred from context (Cicone, Wapner, & Gardner, 1980; Brownell et al., 1984; Ross, 1984; Cancelliere & Kertesz, 1990; Ostrove, Simpson, & Gardner, 1990; Tompkins, 1991; Bloom, Borod, Obler, & Gerstman, 1992; Borod, Andelman, Obler, Tweedy, & Welkowitz, 1992); and indirect speech acts (Foldi, 1987; Stemmer, 1994). Many have been reported not to be able to use prosody to interpret (or convey) emotional content (Weintraub, Mesulam, & Kramer, 1981; Heilman, Bowers, Speedie, & Coslett, 1984; Shapiro & Danly, 1985; Tompkins & Flowers, 1985; Tompkins & Mateer, 1985; Schlanger, Schlanger, & Gerstman, 1986; Ross, Edmondson, Seibert, & Homan, 1988; Alexander, Benson, & Stuss, 1989; Behrens, 1989; Bryan, 1989; Bowers, Coslett, Bauer, Speedie, & Heilman, 1987; Cancelliere & Kertesz, 1990; Cohen, Prather, Town, & Hynd, 1990). Patients with RH damage also fail to understand the moral, punchline, theme, or main point of a story (Gardner et al., 1975; Brownell et al., 1986; Bryan, 1988; Hough, 1990; Schneiderman, Murasugi, & Saddy, 1992) and have problems in the organization of discourse (Delis, Wapner, Moses, & Gardner, 1983; Rehak, Kaplan, & Gardner, 1992; Rehak, Kaplan, Weylman, Kelly, Brownell, & Gardner, 1992). Overall, these patients seem to have difficulty in using contextual information to interpret discourse.

However, the role of the RH in language processing has not been investigated as thoroughly as that of the LH. Thus, whereas deficits in implicit linguistic competence have long been commonly referred to as *aphasia* (or, more etymologically accurate, and still current in British English, *dysphasia*) there is no label to refer to impairments in the ability to infer what is meant from the context in which something is said. I propose to use the term *dyshyponoia*, from the Greek ὑπονοώ (to grasp what is "understood" (in the sense of the French *sous-entendu*), albeit unsaid in an utterance) to refer to a difficulty in drawing appropriate inferences from extra-sentential information, leading to problems in the interpretation of the unspoken component of an utterance (its illocutionary force or pragmatic component), with preserved comprehension of the literal meaning of a sentence (its semantics, derivable solely from the lexical meaning of words and morphosyntactic structure of the sentence).

Whereas communication disorders subsequent to LH lesions (dysphasias) have been studied for over a hundred years, communication disorders following RH lesions (dyshyponoia) have been systematically investigated only over the past one or two decades. Consequently, whereas detailed information ia available concerning the correlation between lesion site and size and specific linguistic deficits, right hemisphere lesion sites have not been correlated with specific pragmatic deficits. Nevertheless, as with the dysphasias, dyshyponoiac deficits are varied in their manifestation, both quantitatively and qualitatively, though all seem to result in some way from a problem with grawing inferences from the unsaid. A first approximation would suggest that, while LH lesions cause deficits in implicit linguistic competence, RH lesions cause deficits in various aspects of pragmatic competence. From the fragmentary data available to date, it may be reasonably assumed that context-independent sentence grammar is separable from other aspects of sentence interpretation (discourse context-dependent rules and nonliteral meanings) and is indeed separated neurofunctionally (i.e., whereas focal LH damage causes context-independent sentence grammar deficits, focal RH damage causes deficits in context-dependent interpretations, nonliteral interpretations, and affect-related aspects of language processing).

The legitimacy of (context-independent) sentence grammar has not gone unchallenged. It has been argued increasingly often over the years that generative grammars do not account for many of the phenomena that are involved in sentence interpretation (Lakoff, 1974; Grace, 1987: 27). Many linguists have pointed out that, in normal language use, in addition to a sentence grammar to interpret the literal meaning of each sentence, a discourse grammar, comprising presuppositions and inferences and in general all phenomena that are dependent on discourse context (anaphora, cataphora, sequence of tenses, cohesion, etc.) is just as necessary for the interpretation of utterances. Some have stressed the importance of situational context (the physical environment in which the sentence is uttered, the people present, sociolinguistic considerations, paralinguistic phenomena (intonation, stress, facial expressions and gestures, etc.). Since the normal use of language involves sentences uttered in context and deriving their meaning in part from that context, the question has arisen as to whether there is any justification for positing a linguistics (i.e., the study of sentence grammar) outside of pragmatics (i.e., the study of utterances in context). Some linguists have attempted to incorporate aspects of pragmatics into structural trees (Ross, 1970; McCawley, 1973; Parisi & Antinucci, 1976; Sadock, 1974), but without much success. The tendency among many European linguists (van Dijk, 1981; Stickel, 1983; Contini-Morava, 1989; Davis & Taylor, 1990) has been to abandon linguistics for pragmatics, in particular the discourse rules, speech acts, and paralinguistic phenomena on which the interpretation of sentences is based.

Not only is (context-independent) sentence grammar theoretically separable from other aspects of sentence interpretation in normal language use, but it has in fact been shown to be clearly separated neurofunctionally. Even though sentence grammars are not sufficient to account for the interpretation of sentences in natural language use, there nevertheless appears to be a good theory-external justification for the study of the structure of words, phrases and sentences in isolation (given that they are selectively impaired in dysphasia).

There are also good theory-external reasons to distinguish between the linguistic construct 'sentence' and the pragmatic construct 'utterance'. The former is independent of context and its semantic interpretation can be derived from the meaning of its words and its underlying grammatical structure; the meaning of the latter requires, in addition, inferences from its various contexts (situational, discourse, general knowledge). The sentence component of the utterance is compromised in dysphasia (as is, in some forms of dysphasia, the literal meaning of words). The inferential component is compromised in dyshyponoia (as is, in some forms of dyshyponoia, the nonliteral meaning of words or expressions—note that a selection of the figurative word meaning rests on inferences from the context).

It has been amply demonstrated that (context-independent) sentence grammar (linguistic competence, as narrowly defined in theoretical linguistics) is subserved by areas of the LH, and that consequently damage to specific areas of the LH results in language structure deficits (i.e., dysphasia: deficits in phonology, morphology, syntax, and the lexicon, in both comprehension and production). There is increasing evidence that the pragmatic aspects of language are subserved by the RH, and that consequently damage to specific areas of the RH results in context-dependent language deficits (i.e., dyshyponoia: deficits in the use of presupposition, inference, anaphora, cataphora, sarcasm, indirect speech acts, idioms, metaphors, and any aspect of the non-literal interpretation of sentences, in both comprehension and production, including the recognition and production of affective prosody, as well as disruption of discourse structure and conversational conventions). Patients with RH damage have been shown to rely on their largely intact linguistic abilities when understanding discourse (Brownell, Carroll, Rehak, & Wingfield, 1992). Not only are individuals with dyshyponoia able to use sentence grammar, but individuals with dysphasia are able to use pragmatic elements to understand utterances or to convey their needs (Schlanger et al., 1976; Wilcox, Albyn Davis, & Leonard, 1978; Prinz, 1980; Deloche & Seron, 1981; Feyereisen & Seron, 1982; Williams & Canter, 1982; Schnitzer, 1985; Hupet, Seron, & Frederix, 1986; Hough, Pierce, & Cannito, 1989). A clear double dissociation is thus evident. Both right and left cerebral hemispheres supply a specific contribution to the microgenetic processing of language; neither is sufficient but both are necessary. Just as not all patients with dysphasia show evidence of agrammatism or phonemic paraphasias, not all patients with dyshyponoia show deficits in all aspects of pragmatics. Indeed, different individuals seem to show different deficits (Brownell et al., 1990), suggesting that there may be a dissociation between various types of pragmatic phenomena, possibly, but not necessarily, related to lesion site.

It is thus feasible to posit a distinction between, on the one hand, the grammar necessary to interpret the literal meaning of sentences in the absence of context (linguistic competence) and, on the other hand, the inferences from the discursive and situational context at the time of the utterance on the basis of which elements are selected from among the available linguistic choices (pragmatic competence). As we have seen, linguistic competence is subserved by specific areas of the LH, and there is increasing evidence that pragmatic competence is subserved by specific areas of the RH. Thus, if it is confirmed that semantics (meaning derived entirely from linguistic competence) is subserved by areas of the LH and pragmatics (meaning inferred from nonlinguistic phenomena) is subserved by areas of the RH, it may be possible to determine whether a particular type of problematic meaning, in the grey area between the two, is treated neurolinguistically as semantic or pragmatic, depending on its vulnerability to a right or left hemispheric cerebrovascular lesion.

Nevertheless, many questions remain unanswered. For example, what is/are the specific contribution(s) of the RH to language use? Are all nonliteral meanings equally susceptible to damage, or do the various types of nonliteral interpretation form a hierarchy of vulnerability? Do different patients show preferential deterioration of different types of nonliteral meanings, thus pointing to the neurofunctional independence (or modularity) of these various aspects? If these aspects are neurofunctionally modular, are they correlated with specific lesion sites? Do patients with difficulties in interpreting nonliteral utterances also show impairment in their use of discourse strategies?

Researchers still need to tease out whether RH deficits affecting nonliteral meanings are due to (1) an inability to take context into account (in which case patients with RH damage would have problems with homophones and polysemic words when only context can disambiguate them); or (2) an inability to access figurative, nonliteral meanings *per se*; or (3) a specific inability to use context to derive nonliteral meanings (while knowledge of the nonliteral meaning *per se* is preserved) combined with an ability to use context to disambiguate literal (concrete) meanings (e.g., *He took his money to the bank/ They sat on the river bank*), and of course grammatical ambiguity (since the grammar is generally intact).

There is, however, some evidence pointing in the direction of modularity: Various patients with RH damage have been reported as manifesting different deficits (Joanette, Goulet, & Daoust, 1991; Chantraine, Joanette, & Ska, 1998). It is therefore plausible to hypothesize that double dissociations may emerge between the various types of deficits reported so far when a sufficient number of patients with RH damage have been assessed on various tasks. It is also not implausible to speculate that a correlation of specific linguistic pragmatic deficits with specific lesion sites will then reveal itself. There is recent evidence (McDonald & Pearce, 1996; Bara et al., 1997) suggesting that pragmatic inference is applied to the output of a prior semantic analysis. In other words, the output of semantic analysis serves as the input to pragmatic analysis. Thus, from a theoretical standpoint, sentence grammar and pragmatics are independent, and hence the speaker's implicit intentions (the illocutionary force of the utterance) are not part of the grammar. Some of the speaker's intentions may be expressed by explicit grammatical markers (morphological, syntactic or lexical) while others are encoded by markers external to the grammar (affective prosody, facial expressions, gestures, or simply reference to the context). In any case, it is pragmatic competence that determines the appropriate choice as a function of the context.

Some studies have failed to find certain expected deficits in their patient population. Conflicting evidence of patients with RH damage exhibiting or not exhibiting certain deficits may result from (1) mixing patients with different lesion sites (and, in the case of closed head injury, multiple sites); (2) lack of control for the conventional/novel dimension of the stimuli; and (3) severity of the deficit. With reference to (1), closed head injury populations may not constitute the best populations for studying the role of the RH in processing pragmatic aspects of language use (even if they do yield evidence for certain dissociations). Firstly, lesions in closed head injuries may not affect the relevant RH areas (which is true of all RH damage until the pragmatic areas—if circumscribed in the same way as the classical language areas—have been identified). Secondly, patients with closed head injuries constitute an extremely heterogeneous group as far as lesion site is concerned. In Bara, Tirassa, and Zettin (1997), for example, of the 13 patients examined, only 3 have mainly RH damage (2 temporal, 1 fronto-parietal) while 5 others have some RH involvement (including 2 bilateral occipital); there are also 3 patients with left fronto-temporal damage. Thirdly, even though patients with closed head injuries often exhibit RH-type deficits, the absence of a certain kind of deficit in a number of patients cannot serve as evidence against the role of the RH in subserving pragmatic competence. Pragmatics will be impaired to the extent that its neural substrate is affected. The study of the possible lateralization of pragmatic phenomena and then their more precise localization within one hemisphere will require patients with well-documented circumscribed lesions so as to be able to test possible dissociations and their anatomical correlations, if any (in fact, other areas, including the left frontal lobe, subcortical structures and the cerebellum may play some role as well). With reference to (2), the "can/could/will/would you" requests, for example, are so common that they may have been lexicalized; e.g., "Please pass the salt" may have become the most frequent meaning of "can/could/will/would you pass the salt" and may thus be represented in the mental lexicon as a request marker, rather than a question about ability. Its pragmatic ambiguity may have been reduced to plain semantic ambiguity, as in *bank* or *bat*. Only individuals with severe dyshyponoia are likely to misunderstand highly conventional words, phrases, or idiomatic expressions. The patients with RH damage studied by Stemmer, Giroux, & Joanette (1994) had problems with nonconventional indirect speech acts, but not with conventional ones. With reference to (3), it remains to be determined whether the various pragmatic tasks constitute a hierarchy of difficulty or are neurofunctionally modular. If the former is true, then one might expect certain tasks to be consistently affected before others. In the latter case, one might expect deficits to be distributed randomly across patients (contingent upon which substrate is selectively affected), though the number of deficits may increase with the severity of the pathological condition.

Interestingly, many of the aspects of verbal communication unavailable to patients with dyshyponoia are the very ones retained by patients with dysphasia. Indeed, most patients with dysphasia appear to have intact knowledge of how to use language. What is deficient is their linguistic code. The pragmatics of reference are preserved in patients with both Broca's and Wernicke's aphasia, despite syndrome-specific problems in retrieving content words and/or closed class grammatical elements (Wulfeck, Bates, Juarez, Opie, Federici, McWhinney, & Zurif, 1989). Therapy may try to capitalize on the dysphasic patient's preserved RH-based pragmatic aspects of verbal communication by using paralinguistic features such as intonation, gestures, and facial expressions and a greater reliance on inference to aid the comprehension and production of verbal messages and thus circumvent the loss of linguistic structure (Gallagher & Prutting, 1983; Green, 1984; Penn, 1984, 1985; Davis & Wilcox, 1985; Garrett, Beukelman, & Low-Morrow, 1989; Holland, 1991; Gallagher, 1991, Lesser & Milroy, 1993; Perkins & Lesser, 1993) which, especially in the elderly patient, may be difficult to restore.

To conclude, it has become apparent that language is not just the language system, which has been the focus of attention of linguists and aphasiologists over the years, but linguistic competence *plus* pragmatic competence. A generative grammar is not only constrained by short-term memory and attention span and interaction with the speaker's encyclopaedia, but is also integrated with pragmatic competence in the normal use of language. The study of sentence grammar (in whatever form—see Paradis, 1994) is a legitimate enterprise, as it constitutes a component of the neurofunctional processing of normal verbal communication. However, the study of pragmatics is just as necessary. It is important, not just for clinicians but also for the general public, to become aware of dyshyponoia as a social handicap at least as significant as aphasia. On the other hand, the use of pragmatics in compensating for linguistic lacunae (as is spontaneously done by infants and second language learners—see Paradis, 1998) is likely to continue to be of benefit in general language therapy practice.

The division of labour between the left and right hemispheres in subserving linguistic and pragmatic competence seems well established. However, the RH's contributions to pragmatics in language use and the integration of each hemisphere's contribution to discourse will require further investigation.

References

Alexander, M. P., Benson, D. F., & Stuss, D. T. (1989). Frontal lobes and language. *Brain and Language, 37,* 656-691.

Bara, B. G., Tirassa, M., & Zettin, M. (1997). Neuropragmatics: Neuropsychological constraints on formal theories of dialogue. *Brain and Language, 59,* 7-49.

Behrens, S. J. (1989) Characterizing sentence intonation in a right hemisphere damaged population. *Brain and Language, 37,* 181-200.

Bihrle, A. M., Brownell, H. H., & Gardner, H. (1988). Humor and the right hemisphere: A narrative perspective. In H.A. Whitaker (Ed.), *Contemporary reviews in neuropsychology* (pp. 109-126). New York: Springer-Verlag.

Bihrle, A. M., Brownell, H. H., Powelson, J. A., & Gardner, H. (1986). Comprehension of humorous and non-humorous materials by left and right brain-damaged patients. *Brain and Cognition, 5,* 399-411.

Bloom, R. L., Borod, J. C., Obler, L. K., & Gerstman, L. J. (1992). Impact of emotional content on discourse production in patients with unilateral brain damage. *Brain and Language, 42,* 153-164.

Borod, J. C., Andelman, F., Obler, L. K., Tweedy, J. R., & Welkowitz, J. (1992). Right hemisphere specialization for the identification of emotion words and sentences: Evidence from stroke patients. *Neuropsychologia, 30,* 827-844.

Bowers, D., Coslett, H. B., Bauer, R. M., Speedie, L. J., & Heilman, K. M. (1987). Comprehension of emotional prosody following unilateral hemispheric lesions: Processing effect versus distraction effect. *Neuropsychologia, 25,* 317-328.

Brookshire, R. H., & Nicholas, L. E. (1984). Comprehension of directly and indirectly stated main ideas and details in discourse by brain-damaged listeners. *Brain and Language, 21,* 21-36.

Brownell, H. H. (1988) Appreciation of metaphoric and connotative word meaning by brain-damaged patients. In C. Chiarello (Ed.), *Right hemisphere contributions to lexical semantics* (pp. 19-32). New York: Springer.

Brownell, H. H., Carroll, J. J., Rehak, A., & Wingfield, A. (1992). The use of pronoun anaphora and speaker mood in the interpretation of conversational utterances by right hemisphere brain-damaged patients. *Brain and Language, 43,* 121-147.

Brownell, H.H., Gardner, H., Prather, P., & Martino, G. (1995). Language, communication, and the right hemisphere. In H. S. Kirshner (Ed.), *Handbook of neurological speech and language disorders* (pp. 325-349). New York: Dekker.

Brownell, H. H., Michel, D., Powelson, J., & Gardner, H. (1983). Surprise but not coherence: Sensitivity to verbal humor in right-hemisphere patients. *Brain and Language, 18,* 20-27.

Brownell, H. H., Potter, H. H., Bihrle, A. M., & Gardner, H. (1986). Inference deficits in right brain-damaged patients. *Brain and Language, 27,* 310-321.

Brownell, H. H., Potter, H. H., Michelow, D., & Gardner, H. (1984). Sensitivity to lexical denotation and connotation in brain-damaged patients: A double dissociation? *Brain and Language, 22,* 253-265.

Brownell, H. H., Simpson, T. L., Bihrle, A. M., Potter, H. H., & Gardner, H. (1990). Appreciation of meta-phoric alternative meanings by left and right brain-damaged patients. *Neuropsychologia, 28,* 375-383.

Bryan, K. L. (1988). Assessment of language disorders after right hemisphere damage. *British Journal of Disorders of Communication, 23,* 111-125.

Bryan, K. L. (1989). Language prosody and the right hemisphere. *Aphasiology, 3,* 285-299.

Cancelliere, A. E., & Kertesz, A. (1990). Lesion localization in acquired deficits of emotional expression and comprehension. *Brain and Cognition, 13,* 133-147.

Chantraine, Y., Joanette, Y., & Ska, B. (1998). Conversational abilities in patients with right hemisphere damage. *Journal of Neurolinguistics, 11,* 21-32.

Chomsky, N. (1965). *Aspects of the theory of syntax.* Cambridge, MA: MIT Press.

Chomsky, N. (1980). *Rules and representations.* Oxford: Basil Blackwell.

Cicone, M., Wapner, W., & Gardner, H. (1980). Sensitivity to emotional expressions and situations in organic patients. *Cortex, 16,* 145-158.

Cohen, M., Prather, A., Town, P., & Hynd, G. (1990) Neurodevelopmental differences in emotional prosody in normal children and children with left and right temporal lobe epilepsy. *Brain and Language, 38,* 122-134.

Contini-Morava, E. (1989). *Discourse pragmatics and semantic categorization.* Berlin: Mouton de Gruyter.

Davis, G. A., & Wilcox, M. J. (1985) *Adult aphasia rehabilitation: Applied pragmatics.* San Diego: College-Hill Press.

Davis, H., & Taylor, T. (Eds.) (1990). *Redefining linguistics.* London: Routledge.

Deloche, G., & Seron, X. (1981). Sentence understanding and knowledge of the world: Evidences from a sentence-picture matching task performed by aphasic patients. *Brain and Language, 14,* 57-69.

Dipper, L., Bryan, K., & Tyson J. (1997). Bridging inference and Relevance Theory: An account of right hemisphere inference. *Clinical Linguistics and Phonetics,* 11, 213-228.

Dwyer, J., & Rinn, W. (1981). The role of the right hemisphere in contextual inference. *Neuropsychologia, 19,* 479-482.

Feyereisen, P., & Seron, X. (1982). Non-verbal communication and aphasia: A review. *Brain and Language, 16,* 191-236.

Foldi, N. S. (1987). Appreciation of pragmatic interpretations of indirect commands: Comparisons on right and left hemisphere brain-damaged patients. *Brain and Language, 31,* 88-108.

Foldi, N. S., Cicone, M., & Gardner, H. (1983). Pragmatic aspects of communication in brain-damaged patients. In S. Segalowitz (Ed.), *Language functions and brain organization* (pp. 51-86). New York: Academic Press.

Gallagher, T. (Ed.) (1991). *Pragmatics of language: Clinical practice issues*. San Diego: Singular.

Gallagher, T., & Prutting, C. (1983). *Pragmatic assessment and intervention issues in language*. San Diego: College-Hill.

Gardner, H., Brownell, H., Wapner, W., & Michelow, D. (1983). Missing the point: The role of the right hemisphere in the processing of complex linguistic materials. In E. Perecman (Ed.), *Cognitive processing in the right hemisphere* (pp. 169-191). Orlando: Academic Press.

Gardner, H., Ling, P. K., Flamm, L., & Silverman, J. (1975). Comprehension and appreciation of humorous material following brain damage. *Brain, 98*, 394-412.

Garrett, K., Beukelman, D., & Low-Morrow, D. (1989). A comprehensive augmentative communication system for an adult with Broca's aphasia. *Augmentative and Alternative Communication, 5*, 55-61.

Grace, G. W. (1987). *The linguistic construction of reality*. London: Croom Helm.

Green, G. (1984). Communication in aphasia therapy: Some of the procedures and issues involved. *British Journal of Disorders of Communication, 19*, 35-46.

Heilman, K. M., Bowers, D., Speedie, L., & Costlett, H. B. (1984). Comprehension of affective and nonaffective prosody. *Neurology, 34*, 917-921.

Hier, D. B., & Kaplan, J. (1980). Verbal comprehension deficits after right hemisphere damage. *Applied Psycholinguistics, 1*, 279-294.

Hirst, W., LeDoux, J., & Stein, S. (1984). Constraints on the processing of indirect speech acts: Evidence from aphasiology. *Brain and Language, 23*, 26-33.

Holland, A. (1991). Pragmatic aspects of intervention in aphasia. *Journal of Neurolinguistics, 6*, 197-211.

Hough, M. S. (1990) Narrative comprehension in adults with right and left hemisphere brain-damage: Theme organization. *Brain and Language, 38*, 253-277.

Hough, M., Pierce, R., & Cannito M. (1989). Contextual influences in aphasia: Effects of predictive vs. nonpredictive narratives. *Brain and Language, 36*, 325-334.

Hupet, M., Seron, X., & Frederix, M. (1986). Sensitivity to contextual appropriateness conditions for pragmatic indicators. *Brain and Language, 28*, 126-140.

Joanette, Y. (1980). Contribution à l'étude anatomo-clinique des troubles du langage dans les lésions cérébrales droites du droitier. Ph. D. Dissertation. Université de Montréal.

Joanette, Y., & Brownell, H. H. (Eds.) (1990). *Discourse ability and brain damage*. New York: Springer

Joanette, Y., & Brownell, H. H. (Eds.) (1993). *Narrative discourse in normal aging and neurologically impaired adults*. San Diego: Singular.

Joanette, Y., Goulet, P., & Daoust, H. (1991). Incidence et profils des troubles de la communication verbale chez les cérébrolésés droits. *Revue de Neuropsychologie, 1*, 3-27.

Joanette, Y., Goulet, P., & Hannequin, D. (1990). *Right hemisphere and verbal communication*. New York: Springer.

Joanette, Y., Goulet, P., & Hannequin, D. 1993. Verbal communication deficits after right-hemisphere damage. In G. Blanken, J. Dittmann, H. Grimm, J. Marshall and C.-W. Wallesch (Eds.), *Linguistic disorders and pathologies* (pp. 383-388). Berlin: Walter de Gruyter.

Joanette, Y., Lecours, A. R., Lepage, Y., & Lamoureux, M. (1983). Language in right handers with right hemisphere lesions: A preliminary study including anatomical, genetic, and social factors. *Brain and Language, 20*, 217-248.

Kaplan, J., Brownell, H. H., Jacobs, J. R., & Gardner, H. (1990). The effects of right hemisphere damage on the pragmatic interpretation of conversational remarks. *Brain and Language, 38*, 315-333.

Lakoff, G. (1974). Dialogue with George Lakoff. In H. Parret, *Discussing Language*, The Hague: Mouton, 151-178.

Lalande, S., Braun, C., Charlebois, N., & Whitaker, H. A. (1992). Effects of right and left hemisphere cerebrovascular lesions on discrimination of prosodic and semantic aspects of affect in sentences. *Brain and Language, 42*, 165-186.

Lesser, R., & Milroy, L. (1993). *Linguistics and aphasia: Psycholinguistic and pragmatic aspects of intervention*. Harlow, UK: Longmans.

McCawley, J. (1973). Remarks on the lexicography of performative verbs. Paper presented at the Texas Conference on Performatives, Conversational Implications, and Presuppositions. Austin, Texas.

McDonald, S., & Pearce, S. (1996). Clinical insights into pragmatic theory: Frontal lobe deficits and sarcasm. *Brain and Language, 52*, 81-104

McDonald, S., & Wales, R. (1986). An investigation of the ability to process inferences in language following right hemisphere brain damage. *Brain and Language, 29*, 68-80.

Molloy, R., Brownell, H. H., & Gardner, H. (1990). Discourse comprehension by right hemisphere stroke patients: Deficits of prediction and revision. in Y. Joanette and H. Brownell (Eds.), *Discourse ability and brain damage: Theoretical and Empirical Perspectives* (pp. 113-130). New York: Springer.

Myers, J. L., & Linebaugh, C. (1981). Comprehension of idiomatic expressions by right hemisphere damaged adults. *Clinical Aphasiology—Proceedings of the Conference* (pp. 254-261). Minneapolis: BRK Publishing.

Ostrove, J. M., Simpson, T., & Gardner, H. (1990). Beyond scripts: A note on the capacity of right hemisphere damaged patients to process social and emotional content. *Brain and Cognition, 12*, 144-154.

Paradis, M. (1994). Neurolinguistic aspects of implicit and explicit memory: Implications for bilingualism and SLA. In N. Ellis (Ed.), *Implicit and explicit learning of languages* (pp. 393-419). London: Academic Press.

Paradis, M. (1998). Aphasia in bilinguals: How atypical is it? In P. Coppens, A. Basso, & Y. Lebrun (Eds.), *Aphsaia in atypical populations* (pp. 35-66). Mahwah, NJ: Lawrence Erlbaum Associates.

Parisi, D., & Antinucci, F. 1976. *Essentials of grammar*, New York: Academic Press.

Penn, C. (1984). Compensatory strategies in aphasia: Behavioural and neurological correlates. In K. Grieve and R. Griesel (Eds.), *Neuropsychology III*, Pretoria: Monicol.

Penn, C. (1985). The profile of communicative approaches: a clinical tool for the assessment of pragmatics. *South African Journal of Communication Disorders, 32*, 18-23.

Perkins, L., & Lesser, R. (1993). Pragmatics applied to aphasia rehabilitation. In M. Paradis (Ed.), *Foundation of aphasia rehabilitation* (pp. 211-246). Oxford: Pergamon Press.

Pierce, R.S., & Wagner, C.M. (1985). The role of context in facilitating syntactic decoding in aphasia. *Journal of Communication Disorders, 18*, 203-213.

Prinz, P. (1980). A note on requesting strategies in adult aphasics. *Journal of Communication Disorders, 13*, 65-73.

Rehak, A., Kaplan, J. A., & Gardner, H. (1992). Sensitivity to conversational deviance in right-hemisphere-damaged patients. *Brain and Language, 42*, 203-217.

Rehak, A., Kaplan, J. A., Weylman, S. T., Kelly, B., Brownell, H. H., & Gardner, H. (1992). Story processing in right-hemisphere brain-damaged patients. *Brain and Language, 42*, 320-336.

Roman, M., Brownell, H. H., Potter, H., Seibold, M. S., & Gardner, H. (1987). Script knowledge in right hemishere-damaged and in normal elderly adults. *Brain and Language, 31*, 151-170.

Ross, E. D. (1981). The aprosodias: Functional-anatomical organization of the affective components of language in the right hemisphere. *Archives of Neurology, 38*, 561-569.

Ross, E. D. (1984). Right hemisphere's role in language, affective behavior and emotion. *Trends in Neurosciences, 7*, 342-346.

Ross, E. D. (1993). Nonverbal aspects of language. *Neurologic Clinics, 11*, 9-23.

Ross, E. D., Edmondson, J. A., Seibert, G. B., & Homan, R. W. (1988) Acoustic analysis of affective prosody during right-sided Wada test: A within-subjects verification of the right hemisphere's role in language. *Brain and Language, 33*, 128-145.

Ross, E. D., & Mesulam, M. M. (1979). Dormant language functions of the right hemisphere? Prosody and emotional gesturing. *Archives of Neurology, 26*, 144-148.

Ross, J. (1970). On declarative sentences. In R. A. Jacobs & P. S. Rosenbaum (Eds.), *Readings in English transformational grammar*. Waltham, MA: Ginn.

Sadock, J. M. (1974). *Toward a linguistic theory of speech acts*. New York: Academic Press.

Schlanger, B., Schlanger, P., & Gerstman, L. (1986). The perception of emotionally toned sentences by right hemisphere-damaged and aphasic patients. *Brain and Language, 3*, 396-403.

Schneiderman, E. I., Murasugi, K. G., & Saddy, J. D. (1992). Story arrangement ability in right brain-damaged patients. *Brain and Language, 43*, 107-120.

Schnitzer, M. (1989). Pragmatic-mode mediation of sentence comprehension among aphasic bilinguals and hispanophones. *Brain and Language, 36*, 76-91.

Shapiro, B. E., & Danly, M. (1985). The role of the right hemisphere in the controlof speech prosody in propositional and affective contexts. *Brain and Language, 25*, 19-36.

Stemmer, B. (1994). A pragmatic approach to neurolinguistics: Requests (re)considered. *Brain and Language, 46*, 565-591.

Stemmer, B., Giroux, F., & Joanette, Y. (1994). Production and evaluation of requests by right hemisphere brain-damaged individuals. *Brain and Language, 47*, 1-31.

Tompkins, C. A. (1990). Knowledge and strategies for processing lexical metaphor after right or left-hemisphere brain damage. *Journal of Speech and Hearing Research, 33*, 307-316.

Tompkins, C. A. (1991). Redundancy enhances emotional inferencing by right-and-left-hemisphere-damaged adults. *Journal of Speech and Hearing Research, 34*, 1142-1149.

Tompkins, C. A., Boada, R., & McGarry, K. (1992). The access and processing of familiar idioms by brain-damaged and normally aging adults. *Journal of Speech and Hearing Research, 35*, 626-637.

Tompkins, C. A., & Flowers, C. R. (1985). Perception of emotional intonation by brain-damaged adults: The influence of task processing levels. *Journal of Speech and Hearing Research, 28*, 527-538.

Tompkins, C. A., & Mateer, C. A. (1985). Right hemisphere appreciation of prosodic and linguistic indication of implicit attitude. *Brain and Language, 24*, 185-203.

Stickel, G. (Ed.), (1983). *Pragmatik in der Grammatik*. Berlin: Mouton de Gruyter.

van Dijk, T. A. (1981). *Studies in the pragmatics of discourse*. The Hague: Mouton.

Weintraub, S., Mesulam, M., & Kramer, L. (1981). Disturbance in prosody: A right hemisphere contribution to language. *Archives of Neurology, 38*, 742-744.

Weylman, S. T., Brownell, H. H., Roman, M., & Gardner, H. (1989). Appreciation of indirect requests by left and right brain-damaged patients: The effect of verbal context and conventionality of wording. *Brain and Language, 36*, 580-591.

Wilcox, M. J., Albyn Davis, G., & Leonard, L. B. (1978). Aphasics' comprehension of contextually conveyed meaning. *Brain and Language, 6*, 362-377.

Williams, S., & Canter, G. (1982). The influence of situational context on naming performance in aphasic syndromes. *Brain and Language, 17*, 92-106.

Winner, E., & Gardner, H. (1977). The comprehension of metaphor in brain damaged patients. *Brain, 100*, 717-729.

Wulfeck, B., Bates, E., Juarez, L., Opie, M., Federici, A., McWhinney, B., & Zurif, E. (1989). Pragmatics in aphasia: Cross-linguistic evidence. *Language and Speech, 32*, 315-336.

Pergamon

J. Neurolinguistics, Vol. 11, Nos 1–2, p. 11–20, 1998
© 1998 Published by Elsevier Science Ltd. All rights reserved
Printed in Great Britain
0911-6044/98 $19.00 + 0.00

PII: S0911-6044(98)00002-5

Pragmatic breakdown in patients with left and right brain damage: Clinical implications

*Ronald L. Bloom** and *Loraine K. Obler*[†]

*Department of Speech-Language-Hearing Sciences, Hofstra University;
[†]Program in Speech and Hearing Sciences, The Graduate School of the City University of New York

Abstract—Formal and functional approaches to pragmatics are based upon different assumptions about the nature of the language system. This paper examines these approaches to pragmatics and considers how each approach has been applied to understand the different language disorders that emerge from left and right brain-damage. The paper explores conflicting reports in the literature about pragmatic performance and hemispheric side of lesion. Finally, clinical methods based on both the formal and functional approaches to pragmatics are discussed.

Introduction

Two paradigms in linguistics provide different assumptions about the nature of pragmatics: a formal/structural approach and a functional approach. Formalists examine pragmatics as an autonomous system with the goal of uncovering the patterns and regularities that describe language behavior in context. Functionalists study pragmatics as the overarching framework that closely influences the internal organization of the linguistic system (Owens, 1991). Both formalist and functionalist perspectives have been useful in describing how pragmatic language behavior breaks down following unilateral damage to the brain. This paper examines the formalist and functionalist approaches to pragmatics and their application to understanding language disorders in brain damaged adults. Past studies employing diverse methodologies have led to conflicting descriptions of verbal pragmatic abilities in aphasic and right-brain-damaged patients. This paper examines the evidence suggesting that pragmatic performance breaks down differentially as a function of lesion side. We include a discussion of the clinical applications that emerge from both the formal and functional paradigms.

The formal and functional perspectives are based on different assumptions about pragmatics in describing the language system. Several theorists have contrasted these two perspectives. Leech (1983) for example, noted that formalists view language as a mental event whereas functionalists consider language as a societal phenomenon. Hymes (1974) suggested that for formalists, a description of linguistic structure is prerequisite to examining how language is used in context. For functionalists, Hymes (1974) noted the description of language use pinpoints the critical features and relationships to be examined. Thus, for functionalists, analysis of language use is primary in the description of linguistic structure. According to Schiffrin (1994) the functionalist position holds that there are functions external to the linguistic system that impinge upon the internal organization of the linguistic system. Schiffrin contrasts this with the formalist position that maintains that the social and cognitive functions of language do not influence its internal organization.

Linguists tend to use either the formal or functional approaches to describe language behavior. Similarly, speech-language pathologists tend to employ one approach or the other to analyze and treat language disorders in patients with brain-damage. The methods used to analyze pragmatics are many and vary according to the specific theoretical approach taken. With emphasis on the formal aspects of pragmatics, the analysis identifies the constituent elements that comprise discourse, describes the rules that govern the way constituents link together, and determines if organization of the constituents is well-formed or not. By contrast, with an emphasis on function, the communicative intent and actions accomplished by a speaker's language are analyzed with the goal of uncovering the effect of the discourse on the listener. Analysis of language structure or function alone creates numerous challenges for the researcher. Both of these approaches have significant implications for the clinical management of people with brain-damage.

A formal pragmatic perspective

Research from a formal pragmatic perspective emphasizes the structure of discourse and seeks to explain its internal organization. Anecdotal clinical reports often noted that aphasic patients communicate better than they talk (Holland, 1980) and that social communication was superior to performance on linguistic tasks. Many studies have examined what structural elements contribute to the observed discrepancy between social communicative function and linguistic performance in people with aphasia (e.g., Ulatowska, Allard, & Chapman, 1990; Ulatowska, Doyel, Stern, Hayes, & North, 1983; Ulatowska, Freeman-Stern, Doyel, Macaluso-Haynes, & North, 1983; Ulatowska, North, & Macaluso-Haynes, 1981). The findings of Ulatowska and her colleagues that global discourse structure is preserved in the face of morphosyntactic and phonologic deficits in mild and moderately impaired aphasic individuals were indeed striking. It is important to note that few of these studies reported a distinction in discourse skills between patients with anterior and posterior damage to the left hemisphere (Ulatowska et al., 1981; 1983). From a formal perspective, pragmatics appears to be a relative strength for the patient with aphasia.

One element common to research from a formal perspective is that it places the source of the pragmatic disorder within the speaker's language system. This view supports the idea that patients with aphasia are, underneath it all, competent communicators who cannot access the linguistic structures necessary to express their competence. For example, on experimental tasks the essential concepts of discourse are often recalled or stated by aphasic patients. This achievement in discourse production is made in spite of problems with sentence level formulation. This observation is consistent with the notion that aphasia is a deficit in language performance, not communicative competence. From the formalist perspective, pragmatic abilities appear to be a relative strength that helps to maintain the aphasic individuals' communicative competence in the face of their linguistic deficits. However, clinical observation of patients with mild-severe aphasia suggests that they may have linguistic deficits that undermine their communicative competence. Thus, pragmatic problems that arise in aphasia are thought to stem directly from linguistic deficits as well as compensations made to deal with these deficits (Newhoff & Appel, 1997).

Maximizing the aphasic individual's use of discourse macrostructure (i.e., the global organization of essential information in a discourse) is a key feature of the clinical approach suggested by Ulatowska and Chapman (1994). In this approach, the clinician controls the basic elements of story structure to parallel the level of processing demonstrated by the patient. Ulatowska and Chapman suggest that a description of an individual's macrostructure helps define competency in using the conceptual information and linguistic form of discourse. The work of Ulatowska and colleagues (1981, 1983,

1990, 1994) has consistently demonstrated a dissociation between a cognitive-based information system and a linguistic system that operates linguistic cohesion. For those patients whose primary difficulty is in manipulating discourse structure, systematic exposure to a range of tasks that vary in conceptual complexity from simple to more challenging is recommended. Thus, the simplest tasks would have all of the meaning explicitly stated and the more complex tasks would contain implicit information that required inferencing in order to extract the meaning. Ulatowska and Chapman (1994) further recommend exploring different modes of information presentation such as visual or verbal or both to identify what is most manageable for the patient. Treatment is then geared toward repairing the aphasic patient's impaired language processing system through measured exposure to certain cognitive-linguistic operations.

In light of the sparing of macrostructure in aphasia, the right hemisphere was a reasonable place to look for the neurological mechanism that placed the global elements of discourse into a coherent structure. This notion was supported by clinical reports suggesting patients with right-brain-damage were copious and tangential in their spontaneous language (Gardner, 1975; Perecman, 1983). Bloom, Borod, Obler and Gerstman (1992) examined the structure of discourse information in patients with unilateral left- and right-brain-damage. Besides replicating the findings of Ulatowska et al. (1981, 1983, 1990) for aphasic patients with left-brain-damage, findings revealed that patients with right-brain-damage produced discourse that contained less information, fewer concepts and more irrelevant remarks than normal control subjects.

Several studies from a formal perspective have reported that patients with right-brain-damage produce ambiguous and poorly structured discourse (Joanette, Goulet, Ska, & Nespoulous, 1989; Myers & Brookshire, 1994). On narrative tasks, right-brain-damaged patients have often been described as listing information, rather than interpreting events and relationships between characters in the story (Brownell, Potter, Bihrle, & Gardner, 1986; Hough, 1990). The formal pragmatic approach has been useful from a clinical perspective in that it has identified a host of abstract pragmatic structures that make useful targets for clinical intervention with right-brain-damaged patients. Observing the presence or absence of certain pragmatic structures such as topic shifts, topic initiation, revisions and conversational turns in a patient's discourse helps to identify treatment goals. The implication is that competent communicators use these pragmatic behaviors proficiently, and that increasing knowledge and use of these pragmatic behaviors in patients who do not is an important clinical objective.

Myers (1994, 1995) has outlined several treatment strategies designed to enhance discourse skills specifically for right-brain-damaged patients. One approach heightens the patient's awareness of story structure. Therapy is directed at improving the patient's impaired language system at the level of discourse. In this 4-step program, the patient is given a series of instructions and tasks designed to increase the number of elements included in a narrative and improve its organizational structure. Consistent with a formal pragmatic approach, Myers' clinical method (1994) rests upon the view that focused cognitive training on particular elements of story structure develops communicative competence.

Research from the formal pragmatic perspective seems to suggest that each side of the brain makes a separate contribution to discourse processing. Evidence from the formal pragmatic perspective indicates that the right side of the brain is required to produce discourse that is elaborate, well integrated and fully descriptive. For patients with left-brain-damage, the ability to utilize the essential elements of discourse is viewed as a communicative strength. In aphasia, the mental operations required for discourse processing are essentially spared but access to them is limited by linguistic deficits and pragmatic compensations. For patients with right-brain-damage, loss of knowledge of how to structure and organize discourse is regarded as a liability that may be hidden by proficiency with phonology and syntax. Because pragmatics is evaluated alongside linguistic form, this approach gives the impression that the language disorders that result from left-brain-damage are more severe than those that arise from damage to the right side of the brain.

A functional pragmatic perspective

The challenge for functionalists is that the abstract units of analysis are not particularly well defined and are crucially linked to the context in which they occur (Hymes, 1974). Moreover, to employ a functionalist approach, it would be critical to account for what the speaker intends to communicate, rather than to describe the pragmatic structures that are present within a speaker's discourse. Gricean pragmatics provides a first approximation for such an approach to discourse analysis. Grice (1975) proposed that communication is governed by four maxims or general laws of communication based on the principle of cooperation between the speaker and listener. Specifically Grice addressed the participant's Quantity (i.e., contribution should be as informative as required), Quality (i.e., contribution should not be false and should not lack evidence), Relation (i.e., contribution should be relevant) and Manner (i.e., contribution should be direct and unambiguous). Importantly, Grice noted that participants do not strictly follow these maxims in every communication situation. Rather, Grice's claim was that listeners interpret what they hear as if it conforms to these maxims. From this perspective, language is viewed from outside of the patient's impaired language system and inside the environment where communication occurs.

Grice's approach provided a theoretical framework for the Pragmatic Protocol developed by Prutting and Kirchner (1987). These authors note that pragmatic features are continuous throughout discourse and are derived from the listener's perception of a speaker's performance in a conversation. The Pragmatic Protocol has been used as a clinical tool to identify the appropriateness of 30 pragmatic abilities thought to be integral to communicative competence in adults with neurological impairments. The Pragmatic Protocol includes observation of three interacting communication components: verbal behaviors (e.g., topic selection, initiation, maintenance, message specificity and cohesion), paralinguistic behaviors (e.g., prosody, vocal quality and intelligibility) and nonverbal behaviors (e.g., facial expression, eye gaze and proximity of the speaker to the listener). Prutting and Kirchner specify that these aspects of pragmatics are to be observed in a natural, familiar conversational setting and scoring is to focus on the interaction between the speaker and listener rather than the production and comprehension linguistic structure.

Some interesting observations about site of lesion and pragmatic performance came out of the Prutting and Kirchner (1987) study. They reported that performance in fluent, non-fluent and mixed aphasic patients was generally pragmatically appropriate. Specifically 82% of the pragmatic behaviors of the left-brain-damaged group were judged to be pragmatically appropriate. Inappropriate pauses, a reduction in informativeness of messages and ambiguous word choice were particularly problematic for the left-brain-damaged patients. When Prutting and Kirchner examined the pragmatic abilities of right-brain-damaged individuals, 84% of their pragmatic behaviors were judged appropriate. Some difficulty with topic maintenance, informativeness, eye contact and prosody were noted in the right-brain-damaged groups. The finding that fluent and non-fluent aphasic patients were similar in the amount of pragmatic behaviors judged to be appropriate was striking in the face of the vast literature suggesting significant differences between these groups on purely linguistic measures. In addition, the finding that patients with right-brain-damage had pragmatic difficulties that were equivalent to those of patients with left-brain-damage also had major clinical implications.

The Prutting and Kirchner (1987) study indicated that patients show some strengths in pragmatic abilities regardless of the site of lesion (i.e., anterior vs. posterior) or side of damage (left hemisphere vs. right hemisphere). Additionally, all groups of brain-damaged subjects showed a measurable degree of pragmatic impairment. The difficulties demonstrated by the patients with right-brain-damage tended to cluster in the non-verbal domain whereas pragmatic difficulties of the left-brain-damaged patients seemed to be more verbal in nature. In contrast to the studies from the formal perspective, the Prutting and Kirchner study revealed considerable overlap in the individual pragmatic profiles of subjects in both groups. As no distinct profile of pragmatic impairment emerged in either group, these results support the idea that pragmatic deficits cannot be localized to a specific area of the brain.

The Prutting and Kirchner study is best described as taking a functional approach because the patient is examined in a typical conversational context and listener judgments are used to describe pragmatic appropriateness. Paralinguistic and non-verbal aspects of communication are considered so it is possible to examine adaptations that individuals might use to convey a message at the expense of their verbal abilities. However, this approach limits the investigation of verbal pragmatic abilities because it emphasizes compensations and does not systematically explore how pragmatics interacts with different environments or different types of contextual information.

Bloom, Borod, Obler and Gerstman (1993) took a functional approach to examine whether distinct profiles of verbal pragmatic behaviors emerge in patients with damage to either the left or right hemisphere. This approach differed from that of Prutting and Kirchner because it focused solely on the verbal pragmatic aspects of language without the benefit of contextual information or paralinguistic cues to support the listener in interpreting the speaker's intention. In this study, raters naive to the hypothesis of the investigation made judgements about the presence or absence of certain pragmatic features of discourse. Verbatim transcripts were used to avoid cues provided by facial expression and voice and limit the focus of the ratings to verbal pragmatic behavior. Bloom et al., (1993) noted that "conciseness" and "relevancy" tend to be compromised in right-brain-damaged patients whereas "lexical selection" and "quantity" tend to be impaired in left-brain-damaged aphasic individuals. Further, Bloom et al. noted a strong relationship between semantic content and verbal pragmatic performance in these patients. Specifically, emotional content in discourse facilitated appropriate verbal pragmatic behavior in the left-brain-damaged patients but suppressed pragmatic appropriateness in the right-brain-damaged patients. The authors concluded that patients with aphasia show deficits in the verbal aspects of pragmatics that vary with respect to semantic content and that interfere with the listeners' interpretation of the message. From the functional perspective it is evident that emotional content, although external to the linguistic system, influences pragmatics.

In our research, there was a measurable influence of content on both the global organization (Bloom et al., 1992) and pragmatic appropriateness (Bloom et al., 1993) of the discourse produced by the brain-damaged subjects. In contrast, our investigation of linguistic structure (or microstructure) alone (Bloom et al., 1995) conducted on the same group of subjects, found no measurable influence of content. Specifically, emotional, visuospatial or neutral content had no effect on discourse cohesion for either brain-damaged group (Bloom et al., 1995). In this study, a structural analysis was conducted where certain elements that contribute to discourse cohesion (e.g. connectives, anaphoric devices) were tallied and performance was compared to that of demographically matched aphasics, right-brain-damaged subjects and normal controls. While both brain-damaged groups demonstrated some problems with the use of cohesion, discourse content had no effect on this observation. Sensitivity to external influences such as emotionally loaded information may only become evident with a functional pragmatic analysis.

In contrast to the formal perspective, a strictly functional approach tends to place the source of the communication breakdown in the social context where it occurs. This is an appealing application because there is much to do clinically to modify the social context to make it more accessible to those individuals with communication disorders. There are perhaps two schools of thought among speech-language pathologists for conceptualizing functional pragmatic variables.

The first approach considers manipulation of the communicative context as the primary focus of aphasia treatment. Lubinski (1981) noted that certain environments (e.g., institutional settings, home) may present removable barriers to communication such as noise from loudspeakers or distracting lighting. Lyons (1992) emphasized working directly with a patient's significant communication partners. People in the patient's environment are taught to establish, accept and practice alternative modes of communication. Thus, the patient's pragmatic limitations are accepted and people in the environment are encouraged to alter their behavior to accommodate the patient. These clinical applications are important because they place primary emphasis on the social communication environment, not the individual's impaired language system. Several studies have identified the benefits of training the partners of aphasic adults (Lyons, 1992; Newhoff, Bugbee, & Ferreira 1981). However, when applied exclusively, these approaches to treatment neglect direct work on the patient's verbal pragmatic abilities.

In the second clinical application of functional pragmatics, the communicative context provides the background for the aphasic patient's active participation in it. The primary effort here is to empower the patient to increase the number and accuracy of the communicative intentions they produce. For example, Davis and Wilcox (1985) recommend that pragmatic treatment for aphasia be conducted within a therapeutic setting and utilize information from horizontal contexts (i.e., current people and events) and vertical contexts (i.e., settings events and people from the past) that are most meaningful to the patient. Further, they recommend that the speech-language pathologist gather information to maximize shared knowledge with the patient in order to increase the opportunity for successful communication. An important feature of this approach is that the patient and therapist participate equally as senders and receivers of new information. Information may be conveyed verbally, through the use of gesture or writing, or any combination of strategies. Only recently have the benefits of such an approach been investigated. Wilcox (1983) and Doyle, Oleyar and Goldstein (1989) implemented a behavioral method for improving speech acts in patients with aphasia. Preliminary case reports indicate that treatment of pragmatic difficulties may produce significant changes in a patient's communicative function (Doyle et. al., 1989.)

A preliminary problem with applying the verbal pragmatic approach to group studies with larger numbers of subjects is that at first glance, verbal pragmatic measurement seems highly subjective. For example, one rater's judgement that a discourse is relevant may vary greatly from another rater's impression of relevancy. Thus, problems in reliability and the psychometric construction of such a scale might limit its clinical application.

To address this problem, Bloom, Borod, Rorie, Pick, Andelman, Campbell, Obler, Welkowitz and Tweedy (1995) investigated pragmatic performance in discourse that varied with respect to condition (i.e., emotionality) and valence (i.e., pleasantness) in patients with right-hemisphere and left-hemisphere pathology. The purpose of this study was threefold. First, to reexamine the impact of negative and positive emotional content on pragmatic discourse performance. Second, to implement some methodological improvements to the pragmatic rating scale used in the Bloom et al. (1992) study and third, to carefully examine the psychometric aspects of the verbal pragmatic rating scale. Specifically, the internal consistency of the scale was determined and a factor analysis was conducted to evaluate the theoretical relationship among the different pragmatic features.

In this study narratives about recollected emotional and nonemotional experiences were elicited. Both positive and negative emotional narratives were obtained in order to examine hemispheric specialization as a function of valence. Each narrative was evaluated for appropriateness on the following six pragmatic features: Conciseness, Relevancy, Specificity, Topic Maintenance, Quantity, and Lexical Selection (Bloom et al., 1993; Grice, 1975). Three graduate students, naive to the study's hypotheses, rated each narrative via a 6-point Likert scale, where "1" = inappropriate and "6" = appropriate. Order of subjects, narratives, and pragmatic features were thoroughly randomized. Interrater reliability was 80.2% so ratings from the three judges were averaged for statistical analysis. Overall, LBDs and RBDs demonstrated deficits in verbal pragmatic performance relative to the NCs. There were also differences among the subjects' performance with respect to the individual pragmatic features, with Relevancy being the most appropriately produced feature and Lexical Selection being the least appropriately produced feature. There was also an interaction between Group and Condition, such that NCs and LBDs performed slightly better in the Emotional than in the Nonemotional condition, and RBDs performed worse in the Emotional than Nonemotional condition. This was especially the case for the pragmatic features of Quantity, Relevancy, Topic Maintenance, and Conciseness when the narratives were positive in valence. Results are consistent with findings in the literature that emotional context can facilitate performance in LBDs with aphasia.

When intraclass correlations were computed, findings revealed that the pragmatic rating scale has substantial internal consistency. This was the case across individual narratives and for the Emotional, Nonemotional, and mean Total rating scores. Further, there were consistently high item-total correlations for each of the pragmatic features, suggesting that these features are strongly related to one another. Consistent with the findings involving internal consistency, when a factor analysis was conducted, there was a substantial general factor before rotation that accounted for 83% of the variance which contained factor loadings ranging from .85 - .95 for all of the pragmatic features. Following rotation, the 5 factors indexed different aspects of pragmatic performance. Interestingly, the three factors that relate most to linguistic content loaded most highly on the first factor and were differentiated from factors indexing conceptual unity and conciseness. When evaluating pragmatic performance, raters apparently base their judgements on three major elements: presence of information, relatedness of concepts and amount of information. These three constructs are substantially represented in the Gricean model of pragmatics. Findings from this study strongly suggest that the measures of verbal pragmatic performance have substantial internal consistency and may be useful as a clinical measure of discourse production.

Future studies of pragmatic behavior from the functional perspective have the potential of exploring the way context influences the structure of discourse in brain-damaged patients. The notion of a facilitation effect is intriguing because it suggests that some aspect of context may be manipulated to produce a compensation or improvement in the language system. Several studies have suggested that in aphasic individuals, emotional content can facilitate pragmatic performance (Bloom, Borod, Obler, & Gerstman, 1993), as well as performance in auditory comprehension (Reuterskiold, 1991), reading, writing (Landis, Graves, & Goodglass, 1982) and oral-facial movement (Borod, Lorch, Koff, & Nicholas, 1987). Bloom et al. (1993) also reported a double dissociation whereby emotional content facilitated pragmatic performance in left-brain-damaged patients but suppressed it in right-brain-damaged patients. Acknowledging the role of emotional content on the communication of right-brain-damaged patients, Bloom (1994) advocated the use of role-playing to create challenging communicative interactions in treatment. A hierarchy of difficult situations, beginning with emotionally neutral ones and ending with highly emotional situations, could be used to increase a patient's communicative participation. Future studies should examine the extent to which pragmatic behaviors can be modified with clinical instruction and practice.

Functional approaches to pragmatics place emphasis on the patient's environment and how s/he interacts within it. Research from the functional perspective seems to suggest that pragmatic abilities may not easily be localized to either side of the brain. Patients with left- and right- brain-damage demonstrate both pragmatic abilities and deficiencies that vary according to the context where they are examined. From this vantage point patients with right-brain-damage have communication impairments (e.g., problems with detailed telephone messages, and recalling the gist of a favorite short story or movie) that may be as significant as those that arise from aphasia.

Conclusion

Models from theoretical linguistics continue to provide rich ground for application in Speech-Language Pathology. The formal and functional approaches to pragmatics are similar because they view discourse as the critical unit of language analysis. However, in the analysis these approaches emphasize different variables and different interactions among the variables. Both theoretical approaches provide challenging questions for researchers. An integrated neurological theory of language will require unifying the findings of both approaches.

There is conflicting evidence suggesting that pragmatic performance breaks down differentially as a function of lesion side. Based on formal studies of pragmatics in brain-damaged subjects, patients with aphasia demonstrate linguistic deficits that interfere with overall communicative competence. Many right-brain-damaged patients demonstrate a breakdown on discourse tasks but these impairments seem minor in comparison to the linguistic and pragmatic deficits associated with aphasia. These formal pragmatic studies suggest that the mental operations required to process language in context require the specialized contribution of both cerebral hemispheres. By contrast, functional studies suggest that pragmatic abilities may be insensitive to neurological localization. In these studies, subjects with brain-damage demonstrate pragmatic abilities that vary according to the specific communication context or language task. Functional pragmatic studies have begun to elucidate the impact of semantic and environmental variables on communication in patients with aphasia and right-brain-damage.

Clinical intervention will also require an integration of both approaches. Speech-language pathologists have adopted structural and functional approaches to describing the discourse of adults with brain-damage and are beginning to measure the clinical utility of these instruments. Recall that among the goals of research in Speech-Language Pathology are: to provide a strong rationale for clinical treatment, to measure the powerful impact of language disorder on a patient's social function, to objectively document communication behavior over time and, to infer preserved neurological structure from function in order to propose mechanisms of recovery. Clearly both the formal and functional paradigms play a central role in reaching these goals.

With emphasis on the formal aspects of pragmatics, treatment promotes the patient's successful use of the rules that characterize well-formed discourse structure. In the functional pragmatic approach, the patient's environment, communication partners and pragmatic communicative strategies are the primary focus of intervention. Patients with left-brain-damage and right-brain-damage are good candidates for a combination of approaches designed to alter limitations that are faced when communication abilities are compromised. In clinical practice, a combination of both approaches can create linguistic and pragmatic targets for intervention and provide the opportunity to practice these skills in a variety of environments.

Studies on the outcome of formal and functional approaches to language improvement are critical. Facilitation effects show promise for treatment as they suggest new avenues for language rehabilitation. Case reports and small-group longitudinal studies may be the best vehicle for examining the influence of context on information structure in brain-damaged patients. At the same time analytic discourse techniques continue to provide interesting ways to measure the relationship of language form and use.

References

Bloom, R., Borod, J., Rorie, K., Pick, L., Andelman, F., Campbell, A., Obler, L., Welkowitz, J., & Tweedy, J. (1995). Measurement of pragmatic performance following unilateral stroke. Poster presented at ASHA Conference. Seattle, WA.

Bloom, R. (1994). Hemispheric responsibility and discourse production: Contrasting patients with unilateral left and right brain damage. In R. Bloom, L. Obler, S. DeSanti, and J. Ehrlich (Eds.), *Discourse analysis and applications: Studies in adult clinical populations* (pp. 81-94). Hillsdale, NJ: Lawrence Erlbaum Associates.

Bloom, R., Borod, J., Obler, L., & Gerstman, L. (1993). Suppression and facilitation of pragmatic performance: Effects of emotional content on discourse following right and left brain damage. *Journal of Speech and Hearing Research, 36*, 1227-1235.

Bloom, R., Borod, J., Obler, L., Haywood, C., & Pick, L. (1995). Left and right hemispheric contributions to discourse coherence and cohesion. *International Journal of Neuroscience, 88*, 125-140.

Borod, J. Koff, E., Lorch, M., & Nicholas, M. (1987). The effect of emotional context on bucco-facial apraxia. *Journal of Clinical and Experimental Neuropsychology, 9*, 147-153.

Brownell, H., Potter, H., Bihrle, A., & Gardner, H. (1986). Inference deficits in right brain-damaged patients. *Brain and Language, 27*, 310-321.

Davis, G., & Wilcox, M. (1985). *Adult aphasia rehabilitation: Applied pragmatics*. San Diego, CA: College Hill Press.

Doyle, P., Oleyar, K., & Goldstein, H. (1989). Facilitating functional conversational skills in aphasia: An experimental analysis of a generalization training procedure. In T. Prescott (Ed.), *Clinical Aphasiology, 19*, 229-242. Austin, TX: Pro-Ed.

Gardner, H. (1975). *The shattered mind: The person after brain damage*. New York: Alfred A. Knopf.

Grice, H. (1975). Logic and conversation. In P. Cole and J. Morgan (Eds.), *Studies in syntax and semantics: Speech acts*. New York: Academic Press.

Holland, A. (1980). *Communicative abilities in daily living: A test of functional communication for aphasic adults*. Baltimore, Maryland: University Park Press.

Hough, M. (1990). Narrative comprehension in adults with right and left hemisphere brain-damage: Theme organization. *Brain and Language, 38*, 253-277.

Hymes, D. (1974). Linguistic theory and functions in speech. In *Foundations in Sociolinguistics: an Ethnographic Approach*. Philadelphia: University of Pennsylvania Press.

Joanette, Y., Goulet, P., Ska, B., & Nespoulous, J. (1989). Informative content of narrative discourse in right brain-damaged right-handers. *Brain and Language, 29*, 81-105.

Landis, T., Graves, R., & Goodglass, H. (1982). Aphasic reading and writing: Possible evidence for right hemisphere participation. *Cortex, 18*, 105-112.

Leech, G. (1983). *Principles of Pragmatics*. London: Longman.

Lubinski, R. (1981). Environmental language intervention. In R. Chapey (Ed.), *Language Intervention for Adult Aphasia* (pp. 223-245). Baltimore: Williams & Wilkins.

Lyons, J. (1992). Communication use and participation in life for adults with aphasia in natural settings: The scope of the problem. *American Journal of Speech-Language Pathology, 1*, 7-14.

Myers, P. (1994). Communication disorders associated with right-hemisphere brain damage. In R. Chapey (Ed.), *Language Intervention Strategies in Adult Aphasia* (pp. 201-225). Baltimore, Maryland: Williams & Wilkins.

Myers, P. (1995). Right hemisphere syndrome. In L. LaPointe (Ed.), *Aphasia and Related Neurogenic Language Disorders*. New York: Thieme.

Myers, P., & Brookshire, R. (1994). The effects of visual and inferential complexity on the picture descriptions of non-brain-damaged and right-hemisphere-damaged adults. *Clinical Aphasiology, 22,* 25-34.

Newhoff, M., & Apel, K. (1997). Impairments in pragmatics. In L. LaPointe (Ed.), *Aphasia and related neurogenic language disorders* (pp. 250-264). New York: Thieme.

Newhoff, M., Bugbee, J., & Ferreira, A. (1981). A change of PACE: Spouses as treatment targets. In R. Brookshire (Ed.), *Clinical Aphasiology Conference Proceedings*. Minneapolis, MN: BRK.

Owens, R. E. (1991). *Language disorders: A functional approach to assessment and intervention*. New York: Macmillan.

Perecman, E. (1983). *Cognitive functions in the right hemisphere*. New York: Academic Press.

Prutting, C., & Kirchner, D. (1987). A clinical appraisal of the pragmatic aspects of language. *Journal of Speech and Hearing Disorders, 52,* 105-109.

Reuterskiold, C. (1991). The effects of emotionality on auditory comprehension in aphasia. *Cortex, 27,* 595-604.

Schiffrin, D. (1994). *Approaches to discourse*. Cambridge, Massachusetts: Blackwell.

Ulatowska, H., Allard, L., & Chapman, S. (1990). Narrative and procedural discourse in aphasia. In Y. Joanette and H. Brownell (Eds.), *Discourse ability and brain damage* (pp. 171-190). New York: Springer-Verlag.

Ulatowska, H., Doyel, A., Stern, R., Hayes, S., & North, A. (1983). Production of procedural discourse in aphasia. *Brain and Language, 18,* 315-341.

Ulatowska, H., Freedman-Stern, Doyel, A., Macaluso-Haynes, S., & North, A. (1983). Production of narrative discourse in aphasia. *Brain and Language, 19,* 317-334.

Ulatowska, H., North, A., & Macaluso-Haynes, S. (1981). Production of narrative and procedural discourse in aphasia. *Brain and Language, 13,* 345-371.

Ulatowska, H., & Chapman, S. (1994). Discourse macrostructure in aphasia. In R. Bloom, L. Obler, S. DeSanti, and J. Ehrlich (Eds.), *Discourse analysis and applications: Studies in adult clinical populations*. Hillsdale, NJ: Lawrence A. Erlbaum.

Wilcox, M. (1983). Aphasia: Pragmatic considerations. In K. Butler (Ed.), *Topics in language disorders (vol. 3)*. Gaithersburg, MD: Aspen Systems Corporation.

Pergamon

J. Neurolinguistics, Vol. 11, Nos 1–2, p. 21–32, 1998
0911-6044/98 $19.00 + 0.00

PII: S0911-6044(98)00003-7

Conversational abilities in patients with right hemisphere damage

Yves Chantraine, Yves Joanette and Bernadette Ska

Centre de recherche de l'Institut universitaire de Gériatrie de Montréal;
École d'orthophonie et d'audiologie, Faculté de Médecine, Université de Montréal.

Abstract—Research on the conversation of subjects with right hemisphere damage has hitherto used tasks too far removed from natural communication. Referential communication is proposed as a means to investigate RHD verbal communication impairments. The RHD patient is asked to instruct a research associate, situated on the other side of a screen, on how to display a series of a priori unnamed visual stimuli in the order in which they appear on his/her page. This exercise is repeated several times in succession. Some RHD patients do not have any problems but others are observed to have referential and/or more qualitative difficulties (similar to RHD conversational impairments). This heterogeneity is discussed within the framework provided by Joanette, Goulet and Daoust (1991).

Introduction

> Conversation is the fundamental site of language use. For many people, even for whole societies, it is the only site [...]. How, then, do speaking and understanding work in conversation? For psychologists this ought to be a central question, but surprisingly, it has not been. (Clark & Wilkes-Gibbs, 1986, p. 1).

In the field of neurolinguistics and neuropsychology of language, the central question of how speaking and understanding work in conversation also requires input concerning the contribution of the right hemisphere. One way of addressing this question is to look for possible conversational impairments in individuals who have suffered Right Hemisphere Damage (RHD). In fact, work that studies both right hemisphere impairments and conversational abilities is very rare in current on-line scientific indexes, except for a few papers in journals such as *Clinical Aphasia* or books in which discourse is discussed (e.g. Code, 1987; Joanette, Goulet & Hannequin, 1990; Tompkins, 1994). No such papers are noted for the 1967-1983 period on PsycInfo, while there are only 10 for 1984-1997; Medline lists 5 papers on this topic between 1987 and 1997. Although RHD discourse abilities have been studied more often in recent years, only narrative and procedural discourses have been addressed. In these types of discourse, pragmatic factors, which are suspected of being an important dimension of RHD communicative problems, play a much smaller role than in conversation. For the purposes of this paper, pragmatics will be defined as "the study of relations between the language and the context in which it is used" (Joanette, Goulet & Hannequin, 1990, p. 160). In the experiments we will discuss, referential communication, a psycholinguistic paradigm, was used to stress the importance of general and immediate context and reveal how it is managed by RHD patients communicating with an interlocutor. Although it is not in itself connected speech, referential communication takes into account the three major points that Levelt (1989, p. 30) believes define conversation: "Canonical setting for speech in all human societies, a conversation is defined by as one in which speakers *interact*, in a shared spatio-temporal *environment*, for some *purpose*[1]."

This paper will therefore summarize some of the recent work conducted by our group using a referential task among RHD individuals. We shall first briefly review the general features of verbal communication impairments in RHD patients.

Right hemisphere damage and communicative impairment

In RHD patients with communicative disorders, the phonological and syntactic components are usually preserved while the lexico-semantic and verbal communication components[2] may be affected. Frequently, family members or clinicians report that in RHD patients' connected speech, part of the conversational context does not seem to be taken into account (Joanette, Goulet & Hannequin, 1990).

Generally, these RHD patients seem not to respect the conversational cooperation "contract", uttering irrelevant comments, unrelated to the topic of the conversation, or answering in a way that appears unrelated to the question (Joanette, Goulet, & Hannequin, 1990). Moreover, they seem to have difficulty respecting turn-taking rules (interrupting or speaking at the same time as the current speaker) and they do not seem to pay much attention to their conversational partners. In the course of natural conversation or in experimental conditions, digressions are observed (tangential discourse), often due to the subject's focusing on a specific detail (Rehak, Kaplan & Gardner, 1992). Some interpretative rigidity (difficulty changing a first interpretation) has also been mentioned (Brownell, Potter & Bihrle, 1986). When patients must interpret a situation by choosing from among several possibilities, not all of which are equally plausible, they tend to choose a possible but less plausible solution (Gardner, Brownell, Wapner, & Michelow, 1983). Finally, in narrative situations, RHD subjects with communicative impairments provide less information, omit the complication and produce less coherent narratives than controls (Joanette, Goulet, Ska & Nespoulous, 1986).

The lack of collaboration with the interlocutor, irrelevant utterances, implausible interpretations and lack of respect for turn-taking rules seem to suggest that some RHD patients have difficulties taking the communicative context into account. However, most of the published research has used experimental tasks (e.g. choice among several vignettes) or situations (e.g. narration to an experimenter who already knows the story) which are very unlike natural communication, thereby biasing the picture. To provide a more naturalistic evaluation of the above mentioned problems, a situation where patients produce discourse more freely is necessary. One such situation is the referential communication task.

The referential communication task

A referential communication task is a naturalistic situation, which simulates the lexico-semantic clarifications observable in natural conversation, but is constrained (by the

---

[1] Emphasis added.

[2] The language impairments of RHD patients are far from being uniform. In a study of word naming, sentence completion and narration, Joanette, Goulet and Daoust (1991) noted that some studies looking at similar aspects of verbal communication reported that RHD subjects had difficulties, whereas other studies observed a deficit in only half of their RHD subjects. In Joanette et al.'s own subjects, different patterns were observed: some were comparable to controls, some had difficulties with all four tasks and some presented double dissociations between tasks, showing that verbal communication deficits, when present, were of different kinds.

stimuli) and oriented (same ranking as goal of the task). Two subjects, separated by an opaque screen that prevents non-verbal communication, receive the same complex visual stimuli, but in a different order. Their goal is to arrange the stimuli in the same order. One of the subjects is assigned a "director" role, which means that the order of the stimuli s/he receives must be replicated by the other speaker, the "matcher". For technical reasons, the director must describe the visual stimuli from left to right. At the end of each trial, the interlocutors must remain silent. The experimenter signals the results of the previous trial to both interlocutors (usually everything is correct), gives the new order to the director and the next trial begins. With the repetition of the trials (usually 4 to 8 consecutive trials, with the same speakers and the same stimuli in different orders), a typical evolution of the referential negotiation process is usually observed: the first trial gives rise to a fairly complex exchange, with several descriptions and categorizations of the stimuli, including clarification requests by the matcher and answers by the director. This collaborative work to establish referential communication is built on in the next trials. Thus, from one trial to the next, the interlocutors come to mutually agree on common references. As a consequence, the number of words used by the director (before the matcher intervenes or in all the utterances concerning the picture) and the matcher, the number of content units used by the director, the number of speaking turns and the amount of time usually decreases while the number of definite references and referential labels increases[3]. An example of a reference to an unnamed picture during six trials is given in Figure 1.

Stimulus 20-Trial 1

D   Well, huh, the fourth, it looks like somebody in a praying position, but this time looking to the right, huh, instead of the left, his head is connected to his body, bent, it is like an *angel* with wings on his back

M   yes, that's it and we can see something like his feet//

D   //like his feet, yes, that's it

M   drawn on the right

D   on the right, yes

M   it's OK, yes, I have it

Stimulus 20-Trial 2

D   Well, the second one is the one I described to you as an angel with wings on his back, like an angel,

M   yes that's it with the two feet there, OK

D   yes with the two feet

M   it's all right

Stimulus 20-Trial 3

D   My last one is *the* angel

M   that's it

Stimulus 20-Trial 4

D   Well, the next one is the angel

M   yes

Stimulus 20-Trial 5

D   Angel

M   yes

Stimulus 20-Trial 6

D   The angel

M   yes

Figure 1.   Excerpt from a referential communication task between 2 students speaking of a picture during 6 successive trials (translated from French)

The experiments we will discuss here are based on the referential communication methodology, with one major exception. The matcher subject in these cases is an associate of the experimenter. It was, of course, not possible for patients to play the roles of both

---

[3] Definite references comprise the use by the director of definite articles and pronouns ("the angel", "my graduate student", etc.), among other grammatical structures, to indicate that s/he expects the coming reference to be known to the matcher. When definite references are short, they are called referential labels.

director and matcher and, since we were interested in the patients' conversational discourse, the patients and the corresponding controls, were always given the director role. To restrict the potential variance caused by the matchers, it was decided to keep their number to a minimum and to limit their speech as much as possible by forcing them to ask uninformative questions. They were not allowed to help the RHD subjects on a "content" level and were trained to remain as neutral as possible. This means that, for each difficult stimulus (first line) and for one easy stimulus, one unimportant or trivial question was prepared and systematically asked by the matcher (e.g. "is there a square in this picture?" when all the pictures have a square, or "what is the top of the picture like?" when the discriminating element is at the bottom of the picture). This prepared question planning was followed by all the matchers in all cases. When necessary (e.g. ambiguous utterance by the director), the matcher could ask extra questions. The goal was to make the patients—and the controls—believe they were participating in a dialogue while from our perspective it looked more like an enriched monologue.

In our case, the interlocutors had to speak about Tangram stimuli[4] (see Figure 2, where the first line is made up of pictures considered difficult to process while the second contains pictures considered to be easy). The pictures were presented on a single sheet of paper for the director and on cards for the matcher. All the subjects had the same order for each trial. The task was audio-taped.

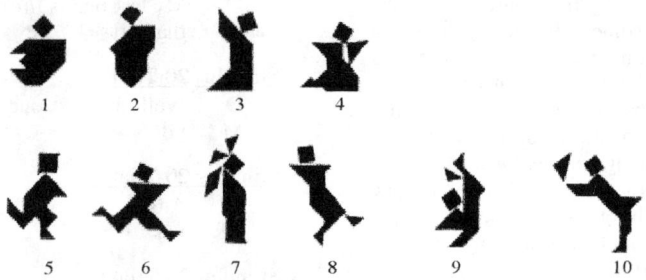

Figure 2.   The Tangram stimuli. The first line is made up of difficult stimuli while the first four stimuli of the second line are considered easy (see Hupet, Seron & Chantraine (1991) for explanations). The last two stimuli of the second line (one difficult and one easy) were placed there to avoid utterances like "and the last one, well, it is the last one". They were not analyzed.

On such a task, an (in)adequate taking into account of the context can be identified based on several factors:

- The screen makes gestural references useless.
- Apart from the non-transparency of the denomination of the stimuli, their array constitutes a context, as some stimuli can be confused with others. Ambiguous utterances must be avoided because they lead to clarification requests or errors by the matcher.

---

[4] These stimuli, taken from an ancient Chinese game, are a priori innominate and the speakers have to find a way to refer to them. The speakers can also be given stimuli with existing names such as lettuces, locomotives or cutlery (e.g. Chantraine, 1993); in that case, the distinguishing criteria must be very specific (locomotives and spoons were used as warm-up tests in the current experiments). Finally, spatial perspectives can also be used in maze references (e.g. Garrod and Anderson, 1987).

- At the beginning of the task, the low level of common references and points of view should induce the director not to cling to her/his own perspective but rather to try to take the matcher's point of view into account and avoid idiosyncratic references.
- On the other hand, with successive trials, the increase in common references and points of view should be manifested lexico-semantically with the use of referential expressions for which an agreement has been reached.
- On the structural side, taking the context into account is also observed through respect of speaking turns since changes of turn cannot take place just anywhere (Sacks, Schegloff & Jefferson, 1978).
- The utterance of self-corrections by the director (Levelt, 1983) and her/his management of the dialogue through pauses or questions directed at the matcher is also analyzed.
- Finally, on the formal side, the definite marking of old information ("the, this," etc.) compared to the use of indefinite references for new information at the beginning ("a, a sort of," etc.), generally indicates that the director believes the referent is known to the matcher (Kleiber, 1981). This change from indefinite to definite reference is usually observed in referential communication tasks involving non-brain-damaged individuals.

The cognitively complex and effortful character of the referential task (It calls upon attention, perception, memory, linguistic and pragmatic factors, among others) makes it difficult to evaluate the verbal communication impairments of RHD subjects. The goal of our research was to confirm or disconfirm the presence of impairments observed with other methodologies, especially when taking conversational context into account.

To sum up, the goal is to observe how RHD patients take into account, or fail to take into account, "existing" knowledge that is shared with their non-neurologically-impaired matchers. The existence of impairments reflected in the non-respect of conversational cooperation by some RHD subjects is postulated. However, not all subjects will necessarily show these impairments. Based on what is known from the literature as well as the kinds of verbal communication disorders RHD patients tend to present, the following predictions can be made.

   a. When only partially accounting for the whole array of references, RHD subjects will utter more ambiguous references that potentially designate more than one stimulus (e.g. "the next one is running..." = stimuli 5 or 6).

   b. RHD patients will utter more irrelevant references (centered on a small detail or linked to their personal life as in "the next [stimulus 5] is a dinosaur...").

   c. RHD subjects will make digressions that slow down the progress of the dialogue (e.g. "the next [stimulus 7], well, it reminds me of my last holidays in Montana, with the children when we visited a reservation and...").

   d. RHD patients self-correct less than controls.

   e. As a consequence of these four predictions, the matchers will have to intervene (e.g. make clarification requests) more with the RHDs than with controls.

   f. RHD subjects will have a tendency not to take these interventions into account, which will slow down the progress of the dialogue.

   g. RHD patients will tend to interrupt their matchers in non-transition-relevant places.

   h. RHD subjects will not tend to take their matcher's comprehension into account.

   i. RHD patients will tend to ineffectively mark the information they present (e.g. indefinite pronouns for old information or/and definite pronouns for new information).

j.   Finally, RHD subjects will tend not to re-use referential expressions for which an
     agreement seemed to have been made with the matcher (e.g. stimulus 5 described as
     knight (tr01), dinosaur (tr02) or dog (tr04)).

Some results of two studies carried out by our group will now be presented to show the
advantages of the conversational referential communication task in assessing RHD
patients.

Referential communication and RHD patients

A first study was conducted by Arès (1997) with two RHD subjects and two controls
(Table 1). The design was a multiple single case (Shallice, 1988). The general
methodology of a referential communication test with 4 trials was followed, except that
the matcher was an associate of the researcher. However, a change of matcher was
instituted at the end of the second trial, when the first matcher was supposedly summoned
to the telephone and replaced by a new one. For this second matcher, no knowledge of the
Tangrams was shared with the RHD director. Therefore, the director was expected to re-
initiate a referential negotiation process in this third trial, possibly facilitated by the two
previous trials (e.g. slightly fewer words or seconds), but without any impact on the
referential labels which were to be proposed again. Three specific predictions were made
concerning the RHD subjects[5] and the change of matcher.

Subject	Sex	Age[1]	School[2]	Laterality[3]	Post CVA[4]	Hospital
RHD1	F	60:4	15	100	6	yes
Con1a	F	55:5	15	100	---	yes
Con1b	F	67:11	14	100	---	yes
RHD2	M	66:3	13	83	21	no
Con2a	M	60:4	18	90	---	no
Con2b	M	65:8	15	100	---	no

[1]Age: years and months; [2]School: years of schooling; [3]Laterality: handedness; [4]Post CVA: months

Table 1. Characteristics of subjects of experiment 1

First, [1a] the RHDs were expected to perform the task in the same way as the controls
(decrease in time, words, and speaking turns and increase in definite references and
referential labels between the first and second trials). However, [1b] their performance
would still be rated higher for the first three factors and lower for the last two than the
controls'.
   Second, it was predicted that [2] the RHD subjects would not take into account the
shared knowledge established with the matcher in the same way as the controls. They
would still use new content elements to refer to the pictures, especially in the second trial,
and this behavior would prevent a normal increase in referential labels.

---

[5] As a general requirement, all subjects had to be right-handed and native speakers of French. For
experimental subjects, the lesion had to be the first one. No patients with psychiatric or neurological deficits
were accepted. Moreover, all the subjects were able to perform the referential communication task, and this
implied the preliminary use of some tests of the PENO ("Protocole d'Evaluation Neuropsychologique Optimal"
proposed by Joanette and collaborators in 1995) to exclude inappropriate subjects.

Third, it was hypothesized that [3] the RHDs would not take into account the change of matcher after the second trial. Therefore, the evolution of the 5 referential variables was expected to occur as if no change had happened, with a decrease in the first three variables and an increase in the last two in the third trial.

Actually, in the case of RHD1, a decrease in time and speaking turns is observed in trial 2 while an increase is observed for words. For the third trial, these three variables increase and remain at that level, and there is even an increase for time in the fourth trial. All the values are higher than for the controls. As far as new content is concerned, no decrease is observed in trial 2 and the decrease in trials 3 and 4 is lower than for the controls. As for definite references and referential labels, a single definite reference is made in the fourth trial.

For RHD2, a decrease in time, speaking turns and words in trial 2 was observed, but all the values were greater than the controls'. In the third trial, a slight decrease or pause was observed, and in the fourth there was a decrease, but all values, as with RHD1, were greater than for the controls. With regard to new content, this subject was comparable to one of his controls while the other control seemed to have difficulty in keeping his score low. This second control subject and RHD2 do not produce either definite references and referential labels.

By comparing their results, we can see that the values for the three initial decrease for RHD2 (prediction 1a confirmed), but not for RHD1[6], and that these values are greater in both RHD subjects than in the controls (prediction 1b confirmed for both subjects).

When it comes to not adding new content, RHD2 is comparable to one of his controls but not the other, who still proposes new elements after the first trial; this means that an adaptation to the knowledge shared with the matcher is possible for RHD patients but also that considerable variation exists, not only among RHD subjects, but also among control subjects. The second prediction is thus not confirmed for RHD2, who succeeds in retaining most of his old content elements after the first trial. For the same variable, however, RHD1, and one RHD2 control, fulfill the prediction, being unable to decrease the number of new elements. However, due to the complete absence of definite references and referential labels in both RHD subjects, one cannot speak of real adaptation (in this case, prediction 2 is confirmed for both RHD subjects).

The third prediction is not confirmed since all subjects reacted to the matcher change (adaptation) but at the same time no adaptation (slowing down of the first three variables, increase in the two others and no evolution in content) was observed in RHD1. Adaptation to the change of matcher is neither confirmed nor disconfirmed and further research is needed.

As mentioned, Arès's (1997) work was a preliminary study. It seems clear that the number of patients (n = 2) should be increased, especially given the dissonant results between RHD subjects but also to allow clearer differences from control subjects appear. If the patients do not become too fatigued, changing to a six-trial task should be better, especially to allow the expected break before the third trial (which would now be before the fourth trial) to show up (appearance of a new matcher and expected changes in referential utterances).

Another study was then undertaken, where all four trials were conducted with the same matcher. Table 2 presents the patients' and their controls' characteristics. Some of the quantitative (referential) disorders observed in the fourth trial and some of the qualitative problems (see predictions above) are presented here.

---

[6] This is explained by the fact that RHD1's second matcher asked more questions, and then both used more speaking turns and words.

Subject	Sex	Age[1]	Sch.[2]	Lat.[3]	Post CVA[4]	Hosp.[5]	Lesion
RHD1	M	60:10	7	100	2	yes	sylvian-parietal; cortical and sub-cortical with spared caudate nucleus
RHD2	M	68:03	7	83	3	yes	aneurysm clipping right internal carotid + frontal
RHD3	M	69:04	7	100	2	yes	abnormal motricity
Con23	M	67:04	7	100	---	yes	
RHD4	M	74:08	10	100	3	yes	parietal on hypertension + haematoma drain + frontal features
Con4	M	75:08	15	100	---	---	
RHD5	M	83:06	4	83	1	yes	sylvian cortical + cingulate gyrus hypodensity
RHD6	M	51:10	12	100	24	no	parietal after therapeutic embolism for artero-venous malformation + rolandic haematoma drain
Con6A	M	40:05	15	83	---	no	
Con6B	M	46:04	12	66	---	no	
Con6C	M	47:07	11	75	---	no	
RHD7	F	80:00	9	100	2	yes	temporo-parietal; cortical and sub-cortical with lacunae head of caudate nucleus
Con7A	F	76:02	14	100	---	yes	
Con7B	F	75:01	12	100	---	yes	
Con7C	F	80:01	11	100	---	yes	
RHD8	F	51:03	11	92	12	no	frontal
Con8A	F	52:08	12	75	---	no	
Con8B	F	52:00	11	83	---	no	
Con8C	F	49:08	11	100	---	no	
RHD9	M	55:00	9	100	11	no	embolism in temporo-parietal lobe
RHD10	F	50:00	15	75	9	no	frontal with sylvian embolism + aorta malformation
Con10A	F	49:05	15	100	---	no	
Con10B	F	47:01	15	100	---	no	
Con10C	F	54:03	14	100	---	no	

Table 2: Characteristics of subjects of experiment 2.
[1] Age: years and months; [2] School: years of schooling; [3] Laterality: handedness; [4] Post CVA: months; [5] Hospitalization

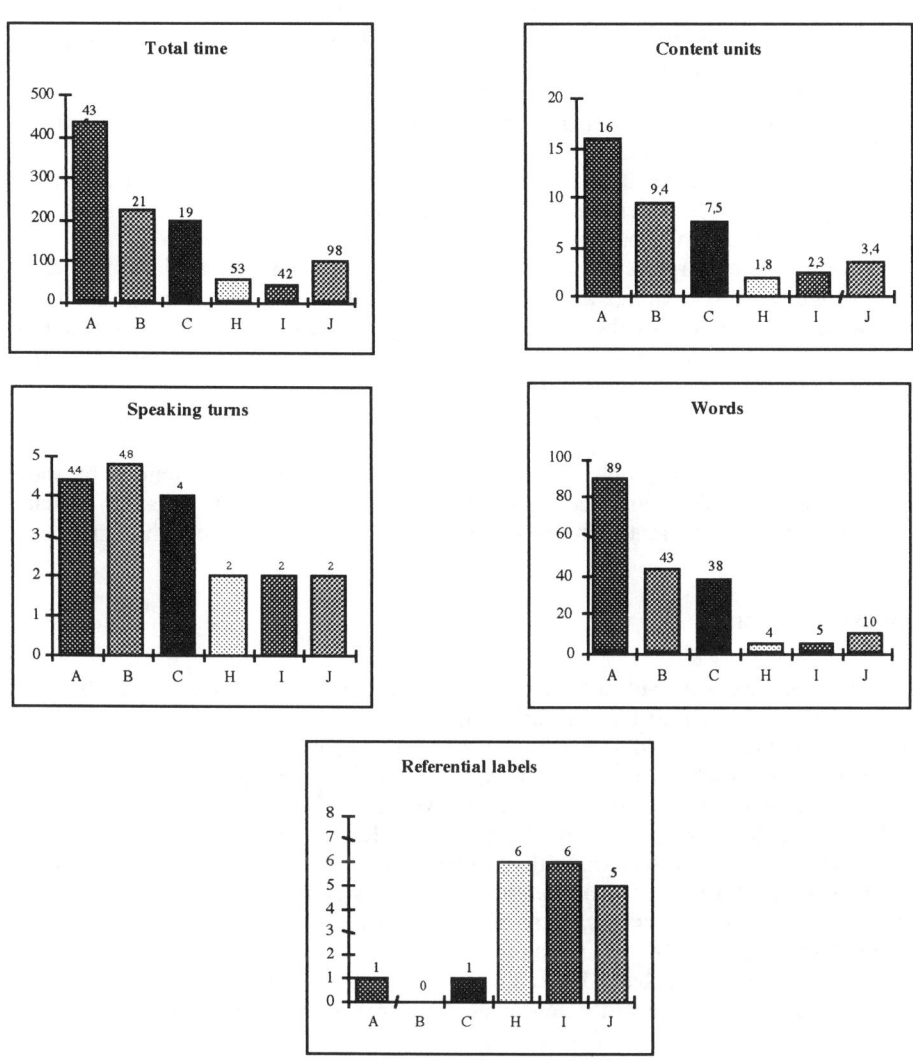

Figure 3. Mean contrasts between subjects with (A, B and C), and without (H, I and J) referential problems during the 4th trial.

As far as qualitative problems are concerned, three classes can be distinguished on the basis of our 10 predictions ((a) to (j) above). Three predictions were not validated. First, (d) the directors did make self-corrections and (h) check for the matcher's comprehension and these predictions were not confirmed; nor was (g) interruptions in non-transition-relevant places. Second, two predictions were confirmed both for some RHD subjects *and* for some controls; it happened that (f) some subjects did not respond to some interventions by the matchers and continued to over-explain themselves, as if the matchers' neutrality was too strong to allow immediate acceptance. As far as (i) the definite marking of shared information is concerned, it seems that some RHDs, and some controls, did not use definite pronouns to refer to mutually known stimuli. The rest of the predictions applied only to the RHD subjects; the most interesting are presented below.

- Two subjects initially had problems integrating the instructions.
- Another described all the angles and the directions of the lines for the difficult figures during the 4 trials.
- Another subject included (b) many idiosyncratic references in his initial turns and (j) changed most of his references for the second trial.
- During the four trials, one subject chose (a) the questionable method of taking the square as the centerpiece of the 10 references, even if more appropriate methods of distinguishing the stimuli appeared in his own references, sometimes even before the first intervention by the matcher. It also appeared that this subject was unable to conceptualize some of the easy pictures as a whole. This last finding also held true for another subject.
- Finally, one patient had trouble with all five criteria in question.

The other patients had problems with the referential criteria (descriptions of some difficult pictures, trouble seeing some pictures in the first trials, slight memory deficits) but not as severe as the patients described above; thus they actually form a third group.

General discussion

The results of more control subjects (for some RHD patients, no suitable controls have yet been found) are still needed before we can come to a firm conclusion. However, the RHD patients for whom we have at least one control subject are characterized by their heterogeneity. The studies' results are comparable to those of by Joanette, Goulet and Daoust (1991). Some subjects do not seem to have any difficulties with the task. On the other hand, some have problems with referential variables but not qualitative ones, whereas others present an opposite pattern, with qualitative but not referential variables. Finally, some subjects exhibit major impairments in both types of variables. This implies that, even for impaired RHD subjects, patterns of deficit can differ.

In the first study, the two RHD patients are different from their control subjects but RHD1 seems more affected than RHD2, although the post-CVA periods are not the same. RHD2 performs well with regard to new content, whereas one of his controls cannot help adding new information after the first trial.

In the second study, three of the 10 patients do not seem to manifest any difficulties with conversation, and they sometimes look "better" than their controls. The other seven do use inaccurate referential discourse, with some referential but not qualitative problems, some the reverse, and some both. Three of these seven subjects, although they have some problems, do not seem to belong to the second problematic group. Focusing on the control subjects, a similar pattern is observed, at least in terms of referential difficulties (e.g. one subject took more than 35 minutes to perform the 4 trials, while another, generally very good, rarely used referential labels).

It is therefore clear that some RHD patients do not have conversation problems, at least of the referential kind, while others do. These patients' impairments may be task-linked (e.g. excessive amount of time or speaking turns even after several trials or absence of referential labels) and/or of a more qualitative nature. In the latter category, idiosyncratic and ambiguous references, partial view of the stimulus and poor re-use of "common" references seem to be frequent.

A pragmatic impairment (not taking common ground with the matcher into account) could therefore be diagnosed for this problematic group. These subjects do not realize that using, and re-using, non-idiosyncratic definite references is the best way to share knowledge with their partner. On the contrary, they change their references across trials, use angles and line directions, focus on useless parts of the pictures or refuse to acknowledge the existence of common knowledge once it has been established.

On the other hand, poor cooperation, digressions, lack of self-correction, unauthorized interruptions or failure to answer clarification requests are rare or nonexistent, possibly due to the rigidity of the task.

Conclusion

After a lesion in the right hemisphere, some subjects have difficulties with verbal communication, although their core grammar (phonology, syntax, part of semantics) remains intact. These RHD patients must go beyond the literal meaning of an utterance and make contextual inferences in order to discover the full meaning of what is said to them. The various kinds of non-literal speech, where contextual inference is imperative (discourse, humor, indirect speech acts or metaphors for instance) allow one to observe the difficulties some RHD patients have.

In discourse, the most often referred to RHD problems are ambiguous remarks, irrelevant comments, digressions, inferential problems, disregard for the interlocutor's statements, difficulty in choosing plausible solutions and difficulty in revising first impressions[7]. In research, however, only experiments related to sentence and discourse structures, rather unrelated to realistic speech contexts, were done on these patients. For several reasons, including the impossibility of comparing subjects in conversational tasks without common referents, this kind of study was rarely carried out with conversations, even though spontaneous speech is so important.

From a more ecological point of view, the introduction of the referential communication task responds to these objections since the presence of common references which permit comparisons between subjects (the stimuli and the goal are the same for all subjects) is a given but the speech of the RHD subjects is still completely free. What has this test shown about the RHD patients and their controls?

Without going into detail, Arès's (1997) main prediction is not confirmed (the subjects do accommodate the change of matcher) but variation is clearly present. Not only are the two RHD subjects different, but one of them performs better than his control in some respects.

The second test reveals heterogeneity with regards to referential and qualitative variables. In the first case, 3 subjects with no referential problems at all are opposed to 7 others who still manifest referential problems in the 4th trial (a mean of more than 7 minutes, 4.4 speaking turns with 16 content units and 89 words to utter only 1 referential label, in the case of the worst subject). In the second case, RHD-specific impairments for such a task were observed both with subjects with relatively few problems and with subjects displaying far greater problems (these latter subjects were often also members of the referentially impaired group).

Therefore, the desirability of studying discourse-related impairments in the RHD patients, or at least some of them, with a conversational task such as the referential communication task appears clear.

---

[7] For further information, see Joanette, Goulet and Hannequin's (1990) chapter on pragmatics.

Clinically, however, if one wishes to study the conversational problems of RHD patients, only the problematic subjects should be evaluated. However, to the best of our knowledge, there is no test available to identify these patients. For the time being, clinicians must live with this heterogeneity of conversation disorders and adapt to each patient by trying to separate the affected and protected aspects of each one's conversational discourse.

Acknowledgments—We would like to thank Geneviève Arès for her contribution to part of this paper based on her M.A. Thesis and Caroline Gingras and Caroline Tatta for their participation. The help of several Montreal hospitals is also much appreciated: we would like to thank the Centre hospitalier Côte-des-Neiges (Drs. Barzauskas, Blouin and J. Roy), the Institut de Réadaptation de Montréal (J. Blumberger and D. Forté), the Centre de réadaptation Lucie Bruneau (R. Longpré) and, last but not least, the Villa Medica Hospital (P. Beaudoin). Yves Chantraine is a Posdoctoral fellow of the Medical research council of Canada and is grateful for the Council's support.

References

Arès, G. (1997). Etude de l'adaptation de sujets cérébrolésés droits au savoir partagé, lors de l'établissement de références communes. Unpublished Masters Thesis, Université de Montréal.

Brownell, H. H., Potter, H. H., & Bihrle, A. M. (1986). Inference deficits in right brain-damaged patients. *Brain and Language, 27,* 310-321.

Chantraine, Y. (1993). Autonomie et collaboration dans l'établissement dialogique de la référence. Doctoral Thesis in Psychology. Université de Louvain, Belgium.

Clark, H. H., & Wilkes-Gibbs, D. (1986). Referring as a collaborative process, *Cognition, 22,* 1-39.

Code, C. (1987). *Language aphasia and the right hemisphere.* Chichester: J. Wiley.

Gardner, H., Brownell, H. H., Wapner, W., & Michelow, D. (1983). Missing the point: The role of the right hemisphere in the processing of complex linguistic material. In E. Perecman (Ed.) *Cognitive processing in the right hemisphere* (pp. 169-191). New York: Academic Press.

Garrod, S., & Anderson, A. (1987). Saying what you mean in dialogue: A study in conceptual and semantic co-ordination. *Cognition, 27,* 181-218.

Hupet, M., Seron, X., & Chantraine, Y. (1991). The effects of codability and discriminability of the referents on the collaborative referring procedure. *British Journal of Psychology, 82,* 449-462.

Joanette, Y., Goulet, P., & Daoust, H. (1991). Incidence et profils des troubles de la communication verbale chez les cérébrolésés droits. *Revue de Neuropsychologie, 1,* 3-27.

Joanette, Y., Goulet, P., & Hannequin, D. (1990). *Right hemisphere and verbal communication.* New York: Springer.

Joanette, Y., Goulet, P., Ska, B., & Nespoulous, J.-L. (1986). Informative content of narrative discourse in right-brain-damaged right handers. *Brain and Language, 29,* 81-105.

Joanette, Y., Ska, B., Poissant, A., Belleville, S., Lecours, A.-R., & Peretz, I. (1995). Evaluation neuropsychologique dans la démence de type Alzheimer: Un compromis optimal. *L'Année Gérontologique, 2,* 69-83.

Kaplan, J. A., Brownell, H. H., Jacobs, J. R., & Gardner, H. (1990). The effects of right hemisphere damage on the pragmatic interpretation of conversational remarks. *Brain and Language, 38,* 315-333.

Kleiber, G. (1981). *Problèmes de référence: Descriptions définies et noms propres.* Paris: Klincksieck.

Levelt, W. J. M. (1983). Monitoring and self-repair in speech. *Cognition, 14,* 41-104.

Levelt, W. J. M. (1989). *Speaking: From intention to articulation.* Cambridge, MA: The MIT Press.

Oldfield, O. D. (1971). The assessment and analysis of handedness: The Edinburgh inventory. *Neuropsychologia, 9,* 97-113.

Rehak, A., Kaplan, J. A., & Gardner, H. (1992). Sensitivity to conversational deviance in right-hemisphere-damaged patients. *Brain and Language, 42,* 203-217.

Sacks, H., Schegloff, E., & Jefferson, G. (1974). A simplest systematic for the organisation of turn taking for conversation. In *Language, 50,* 696-735.

Shallice, T. (1988). *From neuropsychology to mental structure.* (Chap. 10). Cambridge: Cambridge University Press.

Tompkins, C.A. (1994). *Right hemisphere communication disorders: Theory and management.* San Diego: Singular Press.

Pergamon

J. Neurolinguistics, Vol. 11, Nos 1–2, p. 33–53, 1998
© 1998 Published by Elsevier Science Ltd. All rights reserved
Printed in Great Britain
0911-6044/98 $19.00 + 0.00

PII: S0911-6044(98)00004-9

Conversing in dementia:
A conversation analytic approach

Lisa Perkins, Anne Whitworth and Ruth Lesser

Department of Speech, University of Newcastle upon Tyne

Abstract—Pragmatic impairment is a major source of disruption to communication between people with dementia and their caregivers. Speech and language therapists have an important role to play in providing education and advice to caregivers that will facilitate more effective communication. This aspect of therapy has become central to the management of communication difficulties in dementia and comprehensive lists of general strategies can be found in the literature. Currently, however, there are no procedures available to guide clinicians in individually targeting advice that take into account both the range of communication problems that can arise and the unique interaction that occurs between two individuals. Furthermore, suggested communication strategies have been clinician-driven rather than led by the patient or caregiver, thereby failing to incorporate the knowledge and skill that the caregiver has already developed. Finally, there are limited empirical data about the effects of modifying linguistic variables in communication with people with dementia. A review of existing research in this area is provided and the potential contribution of conversation analysis to the assessment and management of pragmatic disorders in dementia targeted at an individual level is explored. Using a conversation analytic framework, a methodology will be described that (a) identifies the interactional difficulties from the perspective of the individual patient and caregiver, (b) obtains information on the strategies currently being used at home and (c) determines the degree of their success. Data are presented and the implications for individually targeted education and advice to caregivers are discussed.

Introduction

Investigations of pragmatic ability in dementia have received increasing attention in the literature. Studies of this nature have been fuelled by the possible insights that pragmatic impairment may offer, first, to understanding both the role of pragmatic behaviour in normal, non-disordered communication and in the profile of progressive loss of communication in dementia, and, second, to identifying clinical applications this area may have for people with dementia. A semantic deficit is generally considered to be a central feature of communication impairment in dementia (Obler, 1983; Bayles & Kaszniak, 1987; Chertkow & Bub, 1990), together with a breakdown in pragmatic ability. Both seem to be more vulnerable to disruption than syntax or phonology (Ehrlich, 1995). With the bulk of the work in this area focusing on dementia of the Alzheimer type (DAT), research into pragmatic behaviour has primarily been carried out through the analysis of the ability of the person with dementia to produce different forms of discourse, including picture description, story telling, procedural discourse and clinical interviews. In doing so, attention has largely focused on the pragmatic abilities of the person with dementia in isolation from the social contexts in which everyday communication takes place, with little examination of the behaviour of the interlocutor who may be both intentionally and inadvertently influencing how language is used in the communication process.

This paper addresses the importance of examining pragmatic ability within the social context of the person with dementia, including an examination of the role of the conversational partner in interaction. In doing so, the discussion focuses on the analysis of conversational data between people with dementia and their caregivers. The merits of Conversation Analysis (CA) for characterising the pragmatic abilities and impairments of people with dementia are highlighted. Based on CA, a methodology is presented that both assists in determining the conversational profile of the person with dementia and drives individually targeted intervention that aims to facilitate more effective communication with key conversational partners.

Current approaches to investigating pragmatics and dementia

The majority of studies addressing the breakdown of pragmatic abilities in dementia have attempted to characterise the nature of discourse production. These have provided descriptions of language output beyond the level of the sentence that aim to capture the deterioration in communicative abilities. As different types of discourse require very different skills, a wide range of discourse features have been investigated. It is not the intention of this discussion to provide a full critique of these studies (see Bloom et al, 1995, and Hamilton, 1994, for detailed reviews of this literature). It is pertinent, however, to highlight that studies focusing on picture description, story telling and procedural discourse have used such indices as cohesive devices (e.g., Ripich & Terrell, 1988; De Santi et al, 1995), coherence (e.g., Ripich & Terrell, 1988), use of anaphora (e.g., Hier et al, 1985), length of communication unit (e.g., Shekim & LaPointe, 1984, cited in Ehrlich, 1995), rate of speech (e.g., Hier et al, 1985), and numbers of information units (e.g., Hier et al, 1985; Beeson et al, 1987, cited in Ehrlich, 1995) or propositional forms (e.g., Ulatowska et al, 1988; Ripich & Terrell, 1988) to quantify what is happening when language is used in discourse. These studies have provided descriptions of the language production of people with dementia in relation to particular surface manifestations. Ehrlich (1995), for example, in a review of the discourse production studies undertaken of dementia, suggested that the discourse of people with DAT

> "may be marked by fewer substantives, more circumlocutions and digressions from the topic. This profile of 'empty' speech in discourse is also characteristically egocentric and concrete with ideational perseverations, and either excessive speech or little or no speech in later stages" (p. 151).

Other studies have focused on conversational discourse and addressed such features as the management of topic (e.g., Hutchinson & Jensen, 1980; Mentis et al, 1995), turn-taking (e.g., Hutchinson & Jensen, 1980; Sabat et al, 1984; Ripich et al 1991; Sabat, 1991; Garcia & Joanette, 1994; Causino Lamar et al, 1995) and/or repair (e.g., Hamilton, 1994) (see Hamilton, 1994, for a review of this literature). Findings from some of these studies will be explored later in this paper.

When attempting, however, both to characterise the nature of discourse in dementia and to gather clinically useful information that may be directly relevant to managing the communication deficits, a number of issues arise that are central to achieving these aims. The first of these relates to the methodology and context in which discourse is sampled. Almost without exclusion, the above studies have been carried out in an experimental paradigm, where the sample of analysed discourse has been collected in an artificial situation (see Hamilton, 1994, for an exception to this). Procedures that require people to describe a picture, tell a story or engage in a "topic-directed interview" (Ripich & Terrell, 1988) with a person who is not a key conversational partner (e.g., an examiner) result in a discourse sample that is unlikely to reflect what happens between people with dementia and their caregivers on a daily basis. The often ambiguous link between the assessment procedure and the use of language in the person's social context further make interpretation both highly subjective and speculative. This point is illustrated in Hamilton's discussion of Bayles' interpretation of a person with dementia's response to describing a button during assessment (Hamilton, 1994: 18). While the task was designed to elicit creative speech, Hamilton reported that failure may reflect more about a person's lack of understanding of the demands of the testing situation than provide informative or diagnostic information on discourse abilities. Sampling methodologies that allow a transparent view of how language is used in a person's usual social context are crucial.

Attention must also be given to the impact of the communication environment on language use. This relates both to the broader social and physical environment as well as to the conversational partner. Ripich *et al* (1991) reported that

> "although the literature reveals little information regarding discourse of SDAT [Alzheimer's] patients, even less is reported about the discourse of their partners. Knowledge of partners' discourse features is critical since communication is reciprocal with each participant shaping the interaction. (p. 332)"

Recent work has recognised the importance of understanding discourse abilities not as solely attributable to the deficits of the person with dementia but also in terms of the interactions in which they are displayed, since the communication partner can greatly influence the ability of the person with dementia to produce discourse (Sabat, 1991; Ramanthan-Abbott, 1994). Addressing the interactive role of the interlocutor, Ramanthan-Abbott (1994) analysed the discourse of conversational partners, particularly in relation to the prompting that they gave during conversation with a person with DAT. The discourse sample contrasted narrative sequences elicited during conversation with the spouse to those elicited by the researcher. Ramanthan-Abbott concluded that, even though narrative ability was compromised in dementia, the impact of this during interaction with the unfamiliar interlocutor was less pronounced than during conversation with the spouse, highlighting the role of the conversational partner in the nature of the interaction.

Given the complexity of the communication and cognitive deficits experienced by the person with dementia, it is also essential to address any explicit link between the cognitive deficits and the pragmatic impairment (Mentis *et al*, 1995). Ehrlich (1995), for example, argued for the need to explore the loci of narrative deficits by manipulating the experimental conditions, stating that

> "The unearthing of the interplay between cognitive and linguistic processes which underlie narrative production represents a long-term goal in neurolinguistic investigations of the language of DAT adults. (p.158)"

In dementia, communication is usually compromised by a breakdown in both language processing and other cognitive processes, e.g., memory, that underpin effective communicative behaviour. Breakdown of semantic processing[1] is regarded as central to the

---

[1]Obler (1983) discusses this impairment in relation to *semantic processing* in a linguistic sense while Bayles and Kaszniak (1987) interpret the language deficits seen in dementia within a model of memory and identify a breakdown in *semantic memory* as the key disruption (see Ehrlich, 1995, for a review of this distinction).

communication deficit seen in dementia (e.g., Fischer *et al*, 1988; Snowden *et al*, 1989; Chertkow & Bub, 1990; Perkins *et al*, 1996). While syntactic and phonological processes have been reported to be relatively preserved for both comprehension and production in dementia of the Alzheimer type early in the disease, these often present later in the disease process (Whitaker, 1976; Schwartz *et al*, 1979; Appell *et al*, 1982; Kempler *et al*, 1987). A substantial literature also documents other cognitive deficits, e.g., memory, attention, psychotic symptoms, in different forms of dementia (see Hart & Semple, 1990, for a review). There has been comparatively little work, however, exploring the impact of these on communication and how they may interplay with pragmatic abilities and impairment. Compromise to each of the above areas will interact with how language is used by individuals across different situations and create a unique pattern of language use for each person involved. The individual nature of the interplay amongst all factors is a further aspect to consider with respect to the reporting of group profiles. While particular constellations of deficits have been associated with different forms of dementia (e.g., Hodges *et al*, 1991; Randolf *et al*, 1993; Snowden, 1994), there is a vast amount of individual variability within diagnostic categories that is potentially obscured by presentation of group data (Perkins *et al*, 1996).

A final consideration here is that of the deficit-focused approach that is used almost exclusively in the literature. Most previous studies have examined the deficits present in pragmatic behaviour rather than the retention of abilities and the possibilities that these may have to offer in managing communication breakdown. Given the observation that some people still have available certain conversational abilities even in the late stages of dementia (Bayles *et al*, 1982; Bayles & Kaszniak, 1987; Hamilton, 1994; Causino Lamar *et al*, 1995), these abilities can be drawn upon to maintain communication with caregivers. Albert (1980) referred to these as "pockets of strength" (p.145).

It is in the context of these observations that this paper presents a methodology that aims to enhance our insight into the interactions that take place between people with dementia and their caregivers. From a sampling perspective, we propose that the context of conversation offers the most representative sample of discourse that takes place between people in their everyday social context, and that sampling conversation between familiar conversational partners is the most ecologically valid and least artificial. From an understanding of conversational management, supported by a framework for analysis, we aim to show how an enhanced insight may be used to, first, inform the pragmatic profile of people with dementia and, second, drive clinical intervention. The importance of the environment, with respect to both sampling and examining conversational data with familiar conversational partners and taking into consideration wider environmental influences, is now discussed, prior to presentation of an explicit methodology for use in clinical practice.

Conversation analysis as a method to examine discourse ability in dementia

Conversation analysis (CA) complements existing approaches to the investigation of discourse in dementia by overcoming some of the limitations that have been highlighted. Its object of study is naturally occurring conversation without an attempt to manipulate the conditions of data collection. It can therefore handle the most common type of discourse, namely the person with dementia interacting in his or her social setting. This gives it high validity with valuable implications for speech and language therapy intervention, as it provides analysis at the level at which intervention is ultimately targeted.

CA is a data-driven approach, emphasising the description of observable behaviour and seeking evidence of communicative success or failure from the sequential context, i.e., the mutual responses of the conversational partners. Evidence of problems arising in conversation, strategies employed to deal with them and the outcome of the strategies is available in the data. This approach permits the analyst to move away from prescriptive judgements of appropriacy based on subjective judgements of normalcy. The person with cognitive impairment does not have 'normal' interactional resources but deviation from what is normal does not necessarily equate with failure or communicative ineffectiveness (Perkins, 1995a, 1995b).

CA focuses on conversation as a collaborative achievement (Schegloff, 1982), recognising the joint responsibility of both the person with dementia and his or her conversational partners for the success or failure of discourse. While working with the caregiver to facilitate interaction has been recognised as central to the management of this population (e.g., Bayles & Kaszniak, 1987; Mace & Rabins, 1987; Dodd *et al*, 1990; Enderby, 1990; Orange, 1991), several of the experimental paradigms used to investigate discourse deficits, such as narrative or picture description, do not include a discourse partner. Research which has focused upon conversational discourse has sometimes included analysis of the behaviour of the conversational partner, but this is frequently treated as an independent variable. CA's emphasis on conversation as a collaborative achievement demonstrates that interaction is more than the addition of the contributions of the two halves, each contribution to conversation is built upon and responds to the partner's previous contribution. This has been highlighted for discourse in dementia by Ramanthan-Abbott's (1994) finding, discussed earlier, that the narrative ability of a person with DAT differed greatly with two different conversational partners. This study demonstrated that the person with dementia's ability to produce extended and meaningful speech is not simply dictated by underlying cognitive deficits but is, in part, interactionally produced.

The focus on conversation as collaboratively achieved also allows attention to be paid to the social role of language. With their emphasis on identification of deficits, studies of discourse have tended to focus upon the communication of information. A large amount of interaction deals with wider social functions such as building and maintaining relationships and self identity and, as reviewed later, deterioration in these functions is often the most distressing aspect of change in discourse ability for both people with dementia and their caregivers. The data-driven approach of CA allows examination of how interaction is maintained and allows identification of interactional strengths as well as weaknesses.

In order to explore the insights that CA provides into the discourse of people with dementia, three key aspects of conversational management, turn taking, repair and topic management, are examined here. The discussion draws upon data collected for a research study investigating the language and conversational abilities of people with dementia with Lewy Bodies and people with DAT (Perkins et al., 1996). The data consists of recordings of people with dementia and their key conversational partners in their home setting. Recordings were made using a radio microphone without the researcher being present in order to capture everyday interaction. The discussion of these areas also draws on other studies reported in the literature. Much of this earlier research, while focusing on conversational management procedures, has not, however, used the data-driven approach of CA, instead using rating scales. This research also predominantly uses conversational data collected in experimental settings with an experimenter for conversational partner and, in some studies, predetermined conversational topics. While it therefore provides useful information about the discourse deficits that can occur for people with dementia in the areas of turn taking, repair and topic management, it does not provide insight into the collaborative construction of discourse and the role of both conversational partners in handling the manifestations of the cognitive deficits of dementia. A notable exception to these limitations is Hamilton (1994).

Turn taking

Turn taking is fundamental to conversation. Overwhelmingly one interlocutor speaks at a time with frequent split second transition from one speaker to the next. Sacks *et al* (1974) proposed that the mechanism which accounts for this split second timing is a set of rules which operate on a turn by turn basis as a sharing device for the right to take the conversational floor (see Levinson, 1983 for an accessible review of this work). In relation to dementia, Hamilton (1994) suggested that the mechanical task of turn taking may be relatively preserved even in the later stages of DAT. Given the split second timing which turn taking requires, however, cognitive deficits may compromise the ability of the person with dementia to secure the conversational floor or to hold onto it. Particularly in multi-party conversations, where there is more competition for the conversational floor, the person with dementia may have difficulty producing a turn quickly enough to secure a right to the floor. The impact of this would be to impede his or her ability to initiate in the conversation, thus resulting in a passive role where contributions may only be possible when the impaired speaker is explicitly given the floor by the current speaker asking a question. Causino Lamar *et al* (1995), in their study of the conversational abilities of ten patients in the end stages of DAT, reported that the people in their study seldom initiated a conversation or were responsible for prolonging it. Initiation will be considered further in relation to topic management.

Even when a person with dementia is explicitly given the conversational floor, he or she may be impaired in producing a response. Attentional deficits may result in an individual not processing what has been said and therefore failing to respond. This phenomenon has lead to the common advice given to caregivers to establish joint attention through touch, eye gaze and use of the person's name before they start talking to the person with dementia. While failure to respond may be interpreted as failure to process the utterance, other reasons may also underlie such an interactional outcome. Other cognitive deficits, for example linguistic processing deficits (e.g., impaired word finding, impaired linguistic comprehension) or generally slowed cognitive processing (as reported in Parkinson's disease) may preclude the person with dementia from being able to produce a turn quickly. Under the pressure of the normal turn taking rules which result in silences after questions being treated as attributable and accountable (Sacks *et al*, 1974), the conversational partner may not allow enough time to respond before taking a further turn. This is seen in the following extract taken from a conversation between an 84 year old woman diagnosed with dementia with Lewy bodies and her nephew. P refers to the person with dementia, R to the relative (the symbols <...> indicate overlapping speech; the numbers in parentheses record the duration of a pause in seconds with (.) indicating a micropause):

1   R   what did you get did you not get your nails cut did you get them cut at the club?
        (1.0)
2   R   your nails <that's what's doing that Win> is your nails
3   P   <that that my nails I know>

In this extract, P fails to respond when selected as the next speaker by a question in turn 1 (T1). After a one second silence, R takes a further turn. The strategy he uses is not to reinitiate the question but continue with the topic. The outcome of this is that P responds in T3 in overlap of R's T2. The content of this turn indicates that she has indeed processed R's two previous turns. Her failure to respond, however, does not emerge only as a consequence of her cognitive deficits. A different interactional outcome may have been seen if R had allowed P more time to respond, as is seen in another part of their conversation:

1   R   did you see the cat yesterday then? (3.0)
2   P   the cat no

In this extract R allows time for P to respond which she does after three seconds.

Delayed turn taking latencies of people with DAT have been reported in the literature (e.g., Sabat, 1991; Causino Lamar *et al*, 1995). Obler (1981) proposed that the phenomenon of silence or muteness characteristic of some DAT patients, besides being associated with decreased initiative typical of the later stages of the disease, may be further explained as an artefact of long response lags and the normal conversational partner not allowing enough time for a response. Sabat's (1991) findings concur with this suggestion that failure to respond may be interactionally produced. He proposed that the adjustment of his own turn taking behaviour to allow time for a patient with DAT to find words and organise her thoughts was central to influencing the relative success of their interaction.

Cognitive impairments may also give rise to people with dementia leaving long pauses within their conversational turns. Again the interactional outcome of these will be dependent on the tolerance of the conversational partners. In the following extract taken from a conversation between a woman with a diagnosis of dementia with Lewy bodies and her niece, recorded at the niece's home, R is tolerant of a six second silence which seems to arise as a consequence of either a word finding or memory search:

1   P   oh Marks and Spencers she's er they were ma- oh excuse me there's erm (6.0) that was at it's like Bath Lane
2   R   yes that's right
3   P   and that's at right at the top was Marks and Spencers

As a consequence of the tolerance, P successfully completes her turn and so makes a contribution to the conversation without the need for collaborative repair work. In the next extract between a woman with DAT and her daughter, toleration of silence within a turn is much less:

1   P   there there's (1.0)
2   R   you what?
3   P   hh oh
4   R   what were you going to say?
5   P   I was going to say did you see the [bisəz] the er lady who was (2.0) come to our meeting
6   R   Mrs. Johnson I saw I recognised her she was with two other ladies

After a one second silence after the start of T1, R initiates repair in T2. The outcome of this is not immediately successful as, in T4, R again initiates repair on the incomplete T1. Finally, in T5, P manages to repair her incomplete turn. A second mid-turn unfilled pause occurs in T5. On this occasion, R tolerates it. The outcome of this strategy is successful as the person with dementia successfully completes her turn.

Minimal turns such as "mm", "aha", "yeah", "right" pervade conversation. The exploitation of such turns by people with aphasia has been described in the literature (Lesser & Milroy, 1993; Perkins, 1995a). Lesser and Milroy comment that, in view of the interactional function of minimal turns, their limited linguistic substance and lack of semantic content, it is not surprising that speakers with aphasia make extensive use of them. In their usage, people with aphasia can participate in conversation by placing the onus of conversation on the conversational partner. Such a strategy avoids the need for them to produce more linguistically complex full turns, a task which is compromised by the impact of aphasia. In a longitudinal study of Elsie, a woman with DAT, Hamilton (1994) describes a final passive stage in her interactional ability in which all responses are confined to minimal responses. This is illustrated in the following extract of a conversation between Elsie and the researcher, Heidi (Hamilton, 1994: 159)[2].

1	Heidi	Here I'll show you a picture. Hmm? Do you wanna see a picture?
2	Elsie	Mhm.
3	Heidi	Yeah? I'll see if you know who this is. Just a second [leaves to get photograph and returns] Look at this. Who's that?
4	Elsie	Mmmmm. [high to low pitch contour]
5	Heidi	Isn't that . . Is that a nice man?
6	Elsie	Mhm.
7	Heidi	Who is that?
8	Elsie	Mhm.
9	Heidi	Is that your husband? [leaves to get another photograph and returns] Look at this one. Do you know this person?
10	Elsie	Mhm. Mhm.
11	Heidi	[chuckles] That's you!
12	Elsie	Mm Hm.

Hamilton stresses that despite Elsie's limited communicative repertoire, she is still able to achieve a range of interactional functions including requesting repetition of her conversational partner's utterance, taking conversational turns appropriately and indicating that she recognises personally important topics.

Repair

Repair is an important interactional resource to conversation. It provides a mechanism to deal with any trouble source which emerges in the interaction, so permitting the discourse to progress. Examples of trouble sources include the need to change the message, false starts, dysfluencies, mishearings and misunderstandings. Conversation analysts use the term *trouble source* in preference to error since there is no one-to-one relationship between the two. Speakers may revise their utterances when there is no hearable error and conversely may ignore an error or ambiguity if it does not impede the ability to produce a sequentially relevant next turn. This is an important point for dementia, since it suggests that not all errors arising from cognitive impairments will necessarily be trouble sources which require repair work. Given the variety of potential trouble sources which may impede the progression of conversation, repair is a particularly important device for the

---

[2] The transcription conventions used by Hamilton (1994) have been retained for this illustration.

communication disordered population (Milroy & Perkins, 1992).

Schegloff *et al* (1977), in their analysis of repair in normal interaction, make two important distinctions: first, *self-initiated* versus *other-initiated* repair which refers to repair by a speaker respectively with or without prompting; second, *self-repair* carried out by the speaker versus *other-repair* carried out by another participant. Repairs are organised according to the participants' opportunities to carry them out and in normal interaction, repair work is overwhelmingly carried out within the turn in which the trouble source appears (self-initiated self-repair), with less preferred forms of repair (other-initiated and other-repair) usually being resolved within two further turns. In dementia, given the range of cognitive deficits (including compromised linguistic, mnesic or attentional processing) which could give rise to interactional trouble sources, repair is an important interactional resource. The ability to repair rapidly and effectively, however, may in itself be compromised by the same cognitive deficits.

Hamilton (1994) proposed that the ability to self-initiate repair demonstrates the ability to take on the role of others by indicating the speaker's awareness that something that he or she has said needs to be adjusted to help the listener achieve mutual understanding. Illes (1989) reported evidence of self-initiation of repair for people in the early and middle stages of DAT, indicating awareness of their own verbal difficulties and the needs of their interlocutor. By the middle stages, self correction attempts were increasingly made. At this stage, however, a significant increase in aborted phrases was found, which may reflect that cognitive impairment compromises the ability to successfully carry out self-repair work. The following two extracts illustrate self-initiation of repair. In the first, the person with DAT is able to successfully carry out the repair herself:

```
1  P  just b- hot er boiled potatoes and peas you know cooked peas
2  R  mhm
```

In T1, P successfully carries out a self-initiated self-repair with a cut off and then replacement of 'hot' with 'boiled' followed by expansion of 'peas' to 'cooked peas'.

In the following extract, the person with DAT shows awareness of the need to repair with a filled pause and 'oh'. She is not able to resolve the word finding difficulty, however, and a protracted collaborative repair sequence ensues with her daughter:

```
1  R  well what are you having for your lunch?
2  P  oh I'm having erm oh
3  R  oh dear
4  P  oh oh hhhhhhhh (1.0)
5  R  you're having what?
6  P  I'm having my (3.0) oh (4.0) oh (3.0)
7  R  a fish pie?
8  P  eh?
9  R  are you having the fish pie?
10 P  no I'm having a (2.0) big one
12 R  a big one of what?
13 R  oh I don't know
14 P  you don't know
15 R  something meaty?
16 P  something what?
17 R  something with meat in?
18 P  yes there's there's got the er chicken in it
```

19 R   chicken and pasta?
20 P   something like that

The ability to initiate repair on the conversational partner's turns is an important skill to compensate for cognitive deficits, including memory failures and failures in comprehension which may compromise the person with dementia's ability to produce a next appropriate response. The following extract demonstrates the ability of a woman with DAT to request clarification of a referent from her daughter, which may reflect a memory failure for a previously shared referent:

1 R   little Johnny's dying for you to get better he wants to take you for a run in the country
2 P   which is Johnny?
3 R   Johnny Blair little Johnny Blair
4 P   oh aye what does my mother say when she saw him?

The ability to initiate repair on her conversational partner's turn allows the person with dementia to establish a shared referent and the topic can then be developed. This contrasts with the following extract where a man with DAT, in conversation with his son, fails to initiate repair on a memory failure for a previously shared referent:

1 R   and Yvette is working her last week
2 P   where?
3 R   this is her last week at work Yvette (.) you know who Yvette is?
4 P   no no not
5 R   she's your granddaughter

In T1, R tells the person with cognitive impairment about his granddaughter. P does not initiate repair on this turn although it becomes clear that he does not know who 'Yvette' is. R realises this and the strategy he employs in T3 is to redo T1 and then check understanding. The outcome of this is that the failure in understanding who is being talked about becomes clear in T4 and R then orients him to the identity of Yvette.

In Hamilton's (1994) study, discussed earlier, there was a change from using six types of clarification requests in the earlier parts of the study to the use of only non-specific requests in the later stages, suggesting a progressive decline in the ability to initiate repair. As this ability declines, it is likely, as is seen in the extract above, that the unimpaired conversational partner will take more responsibility for ensuring that repair work is initiated when it is necessary. An alternative interpretation of the above extract is not that the person with dementia cannot initiate repair but that he chooses not to as this will reveal his cognitive deficits. Analysis has shown that conversationalists orient to repair work as socially sensitive. If the problem necessitating repair can be traced back to some personal insufficiency, it becomes an event which threatens face (Goffman, 1955; Couper-Kuhlen, 1992). In order to save face, the person with dementia may avoid repair when this will reveal a memory or comprehension failure. Given this interpretation, his avoidance could be seen as evidence of awareness of social expectations and his own failure to meet these. The conversational partner's check of his knowledge of the referent can therefore be seen as a face-threatening act by its exposure of the memory failure.

Jefferson (1987) demonstrated that a characteristic of other-repair is an accounting of lapses in conduct which have given rise to repair work. Hamilton (1994) reported that, in her longitudinal study, repairs due to lapses of memory were found in the early conversations between herself and Elsie, but that these disappeared from her conversation as her condition deteriorated. Hamilton suggested that one possible account for this is that, with increasing cognitive deterioration, Elsie was no longer aware of the gap between her abilities and social expectations.

Given the sensitivity of protracted repair in exposing failures in competence, interlocutors can deal with a potential trouble source by passing over it altogether (Heritage & Atkinson, 1984). This option is open to both the person with dementia and the conversational partner. While the occasional glossing over of a potential trouble source may not have great interactional consequences, Perkins (submitted) presents data on interaction between a person with aphasia and her relative that shows the impact it does have under some circumstances. For the person with aphasia, the analysis highlights that her conversational partner's strategy of not initiating repair severely limits her ability to actively contribute to the interaction. As she does not successfully make a contribution to conversation, she is not able to influence the development of the topic and as a consequence is forced into a passive role. In the event of dementia, the conversational partner's reluctance to pursue clarification of a person with dementia's turn may be related to expectations of his or her ability to be able to respond to the clarification and resolve the trouble source. The following extract shows retained ability to respond to other-initiated repair of a woman with DAT talking to her daughter:

1   P   she came from the pit houses didn't she Colliery Road?
2   R   who?
3   P   Margaret
4   R   not that I'm aware of I didn't know she came from there Mam I didn't
        know where she came from.

In T2, R initiates repair on P's T1, requesting clarification of the pronoun. P is able to self-repair in T3, the success of this being marked by R being able to respond to her question in T4.

Topic management

Topic can be loosely defined as "what is talked about through some series of turns at talk" (Lesser & Milroy, 1993, p. 204). Topical coherence can be seen to be something that is constructed across turns by the collaboration of participants. As discussed by Garcia and Joanette (1995), conversational topic management draws upon many linguistic and psychological functions unavailable to the neurologically impaired population. As will be discussed below, a number of aspects of topic management may be problematic in interaction with people with dementia. Mentis *et al* (1995) suggest that these can be accounted for by cohesion problems arising from breakdown in the syntactic and semantic devices used to establish cohesive ties. Alternatively they may arise from the production of confabulatory and contradictory units and failure to specify the relevance of ideas arising from deficits in judgement, memory or attention.

A number of studies have reported reduction in topic initiation by people with DAT in conversations with an unfamiliar interviewer (Mentis *et al*, 1995; Causino Lamar *et al*, 1995; Garcia & Joanette, 1995). In contrast, Hutchinson and Jensen (1980) found, in their study of the discourse skills of five women with DAT, that new topics were initiated more frequently than the normal elderly control group and that initiation was done in the absence of appropriate closing of the previous topic.

Impaired ability to orient the conversational partner to new topics has also been reported in the literature (Mentis *et al*, 1995). Garcia and Joanette (1995) reported less topic shading in conversations with subjects with DAT in comparison to normal elderly control subjects. They proposed that topic shading requires the speaker to hold the previous topic in memory and relate the new topic to the old one. The memory deficits of people with DAT may therefore compromise the ability to carry out topic shading. Hamilton (1994) proposed that difficulty in taking the role of the other to determine what his or her conversational partner may view as an appropriate change in topic may underlie topic shifts perceived as inappropriate. The impact on interaction of failure to orient the conversational partner to new topics will depend upon the way that the conversational partner chooses to treat the new topics. In the following extract taken from a conversation between a woman with DAT and her daughter-in-law, R initiates repair work to establish the relevance of a new topic:

1   R   I was telling Jean about having been out to the Co-op last week
2   P   no cats?
3   R   heh?
4   P   there was no cats?
5   R   which cat Grandma?
6   P   well the cat you were looking for the big cat
7   R   that was a long time ago with the cat Grandma wasn't it you remember it was when you lived down Stephen Street with the cat wasn't it so I don't think we'll see the cat any more
8   P   hm it's gone
9   R   mhm (4.0)
10 R   you know Bob said it was the black cat that you got from the fire wasn't it?
11 P   aha in the cross house fire
12 R   yes mhm
13 P   it walked in great big cat great big huge cat
14 R   was it but that was about seventy years ago wasn't it?
15 P   it's a long time ago

P initiates a new topic at T2 which has no topical link with the previous topic of T1. Evidence for failure to orient her conversational partner to the new topic is provided by R's general repair initiator in T2 and a more specific request for clarification in T5. The strategy used by R is therefore to embark on repair work to establish the relevance of the topic. The outcome of this is that the topic is developed with both conversationalists contributing. Indeed the topic of cats continues to be discussed for a further ten turns.

Compromised topic maintenance has been reported in a number of studies investigating the discourse abilities of people with DAT in conversation with a researcher (Mentis *et al*, 1995; Garcia & Joanette, 1995). As noted in both of these studies, reduced ability to maintain a topic increases the responsibility of the conversational partner to keep the conversation going. Garcia and Joanette reported that, in their study, the experimental conversational partner was found to change topics because of failure to maintain topics more frequently with the people with DAT than with normal elderly control subjects. The role of the conversational partner in maintaining the conversation can be seen in the following extract from a conversation between a woman with DAT and her daughter-in-law:

1   R   and Jean said she was coming down tomorrow but they usually go away at half

```
         term I didn't ask her about that
2   P    mm (1.0)
3   R    did she say she was going away?
4   P    no
5   R    oh (1.0)
6   R    because it would be the start of half term
7   P    mm
8   R    and then the lights go back the hour goes back on Saturday night
9   P    mhm
10  R    so it'll be darker in the morning (1.0)
11  R    will you stay in bed any longer?
12  P    I don't think so Janice
```

P uses minimal turns repeatedly (turns 2, 4, 7 and 9) and does not make more major contributions to the conversation at points where it lapses (after turns 5 and 10). R attempts to elicit a greater contribution to the topics by asking P questions (turns 3 and 11). While P does respond to these, her responses are minimal and do not allow the topic to be developed further.

The following extract (part of which has already been presented and discussed above) from a conversation between a man with DAT and his son suggests that the P may exploit the use of questioning to get the conversational partner to maintain the topic:

```
1   R    and Yvette's working her last week
2   P    where?
3   R    this is her last week at work Yvette (.) you know who Yvette is?
4   P    no no not
5   R    she's your granddaughter
6   P    where is she?
7   R    that's Yvette up there look (pointing to photograph)
8   P    where?
9   R    there and she's over there (1.5)
10  R    that's Yvette
11  P    aye (.) <where's> she live then?
12  R    <she's>
13  R    Durham
14  P    oh aye (1.0)
15  P    oh <yeah>
16  R    <anyway> its her last week at work before (.) the baby's born she goes this week
         and then she's not going anymore (.) until after the baby's born
17  P    where's she at?
18  R    she lives in Durham
19  P    Durham aye
20  R    but she works from Peterson (.) Grammar (.) and she's a (.) now what do they call
         it (1.0) she's a special needs advisor
```

In this extract, P repeatedly uses the question 'where' as seen in turns 2, 6, 8, 11 and 17. This is a pervasive pattern throughout his conversation and appears to be a strategy to continue the conversation and pass the conversational floor back to R. On the first use of 'where' in T2, R does not respond to the question but instead redoes his previous turn and checks understanding of who is being talked about. The interpretation of the use of 'where' as an avoidance of some difficulty is shown as correct, as in T4 P reveals that he does not know who Yvette is. R orients him in T5 and he responds again with a 'where' question in T6. R uses this to do further orientation work, using photographs in the room to establish Yvette's identity. P then asks a more specific 'where' question in T13 which the R answers. Four turns later, he asks a further 'where' question and R provides him with the same answer again.

Hamilton (1994) reported in the longitudinal study of Elsie already referred to earlier, that Elsie's discourse abilities with personally important topics appeared to be more flexible than more banal ones. For one of the participants with dementia with Lewy bodies studied by Perkins *et al* (1996), when in conversation with her nephew, there was a clear pattern of greater topic maintenance for those topics she initiated in contrast to those initiated by her conversational partner, to which she predominantly responded with minimal turns. An examination of the topics that she did initiate demonstrated that she repeatedly initiated the same topic with a question about the researcher who had just visited. The following extracts all occurred within a ten minute sample of conversation:

(i)
1 P   aye it seems like it where does she live then where does she live?
2 R   that's Dr. I think that's I think they call her I think that's Dr. Perkins that you know that's the lady I was telling you about that rung up you know

(ii)
1 P   where does where does she live then?
2 R   I don't know where she lives er Win

(iii)
1 P   where does she live then?
2 R   I don't know Win I I haven't a clue I don't know

(iv)
1 P   I wonder where she lives then
2 R   I don't know

R responds each time without challenging the repetition of the topic. P appears to have ideational perseveration (Bayles *et al*, 1985) which has been explained by Shindler *et al* (1984) as arising from the person's inability to monitor his or her own speech output or from inability to change mental sets.

Cognitive deficits may manifest themselves in the topic management of people with dementia through the initiation of topics based on hallucinations or delusions, an example of which is seen in the extract below from the conversation of a woman with dementia with Lewy bodies and her son:

1 R   Jane's said you've lost your teeth
2 P   yeah got pinched
3 R   oh get away who's going to pinch <teeth?>
4 P   <don't> be soft I'm not that daft they took your two ten pounds note away (1.0)
5 R   I didn't have two ten pound notes
6 P   two ten pound you've got I've got one from Harwich Lodge and I've got one from Leicester
7 R   oh I didn't know that do you want me to write back and say thanks?

8 P   no I couldn't
9 R   do you want me to <write back> and say thanks?
10 P   yes please yes
11 R   right that's Leicester and Harwich
12 P   I thought you would know
13 R   well no I didn't know
14 P   well they did I got one from Leicester and one from
15 R   no I didn't know
16 P   er Harwich
17 R   I didn't know
18 P   two ten pounds and I folded them up and put them away but they were found

P has a delusion around being robbed in the nursing home that she lives in. This manifests itself several times in a ten minute conversational sample and can be seen in turns 1 and 4. The strategy employed by R is to challenge the delusional belief as seen in turns 3 and 5. The outcome of this is that P does not accept the challenge but instead provides information in support (turns 4 and 6). A second strategy employed by R, as seen in T7, is to pick up on factual information in the turn expressing the delusion and make this the focus of the topic. The outcome of this is that the following turns orient to this focus but eventually P returns to her belief that she has been robbed in T18.

The preceding discussion illustrates the usefulness of CA, in its examination of turn-taking, repair and management of topic, as a method to investigate the interactional consequences of dementia. The analysis has highlighted the importance of paying attention to the collaborative construction of discourse, with the conversational partner having a central role in the way that the linguistic and cognitive deficits are managed. The influence of the conversational partner, however, extends beyond the immediate conversation to the wider environment. This encompasses overall perceptions and attitudes towards the person with dementia and the way in which these interact to create the social and physical environment in which communication takes place. These factors will be discussed briefly before addressing the application of CA to clinical assessment and management.

Environmental influences on pragmatic breakdown in dementia

The broader environment plays an integral role in the way in which language is used and in the way in which communication may break down. Lubinski (1991) has, for example, proposed that in addition to individual variables affecting changes in the communication abilities of the person with dementia (e.g., patterns of physiological change, pre-morbid communication abilities, communicative needs), learned behaviours may also arise as a consequence of the perceptions or beliefs of others. In particular, a commonly held perception of incompetence that is associated with dementia will influence the social and communication opportunities available to the person with dementia. Lubinski proposed that *learned helplessness* and *a cycle of incompetence* can be triggered by the diagnosis of dementia. Learned helplessness occurs when people believe that events and outcomes are independent of their actions and that any further action is futile. The stereotyping of helplessness in dementia by caregivers may lead to overgeneralisation of incompetency, thus restricting the opportunity to demonstrate intact skills and promoting dependency. Often influenced by their own reactions and difficulties in coping, caregivers do not expect the person to perform with competence and may give feedback concerning both actual and potential failure (e.g., not sharing new information). Similarly, the structure of the environment (e.g., seating positions) which may be highly controlled sends cues to the person that he or she is no longer expected to behave or interact competently. Together these social and environmental cues help set up minimal expectancies in people with dementia from both themselves and significant others in the environment. Skills cannot be maintained when there are reduced opportunities to perform them. This reduction in demonstration of skills reinforces the stereotype of helplessness and so the cycle of incompetence is maintained.

The consequences of such cycles are not felt by the person with dementia alone. Muir (1996) reported evidence suggesting that "the loss of meaningful interactive and conversational skills is more distressing to caregivers than the developing of behaviours upon which many professionals focus their attention, e.g., 'aggression', wandering and incontinence" (p. 222). Increased caregiver stress as a direct result of communication problems experienced with people with dementia has been extensively reported (Rabins, 1982; Poulshock & Deimling, 1984; Kinney & Stephens, 1989; Rau, 1991; Stephens et al, 1991). In Orange's (1991) survey of family members of people with DAT, almost half of the respondents noted a change in their relationship with the person with dementia as a direct result of their relative's communication problems, reporting feelings of frustration, loneliness, guilt, embarrassment and social isolation.

The importance, therefore, of working with both the person with dementia and his or her conversational partner in managing communication breakdown is paramount. An approach has been developed by the current authors which uses a partnership approach to both characterise the interaction that is taking place between the person with dementia and his or her caregiver and directly drive intervention. The assessment tool used in the approach, the *Conversation Analysis Profile for People with Cognitive Impairment (CAPPCI)* (Perkins, Whitworth & Lesser, 1997), was developed as a resource for speech and language pathologists/therapists for use with people with generalised cognitive impairment, as in dementia, and their key conversational partners or family caregivers. This is discussed below as a methodology for sampling and examining interaction within its relevant social context.

Conversation analysis profile for people with cognitive impairment (CAPPCI)

Driven by the methodology of CA, the *CAPPCI* looks directly at the interaction between the conversational partners, as opposed to the deficits only of the person with dementia, and works closely with the caregiver to identify effective strategies to facilitate or manage breakdown in communication. It is an individualised approach that recognises the range of communication problems that can arise, the spontaneous development of strategies by caregivers and the need for careful evaluation of those strategies. The specific objectives of the *CAPPCI* are:

(a) to determine the caregiver's perception of the current conversational abilities of the person with dementia,
(b) to determine the strategies being employed in interaction and their success,
(c) to assess change from premorbid styles and opportunities of interaction, and
(d) to capture the relationship between the caregiver's perceptions and what actually occurs in a sample of conversation.

The *CAPPCI* consists of an interview with the caregiver and a method of analysing a sample of conversation between the conversational partners. These combine to provide an overall profile of what is occurring in the interaction between the person with dementia and his or her caregiver, from which intervention can proceed. The interview section seeks to elicit information from the caregiver on the person with dementia's conversational management procedures of initiation, turn taking, topic management and repair. It further examines the impact of speech, linguistic and other cognitive impairments on the interaction. For each question, the conversational partner is asked to rate the frequency with which a behaviour occurs. If the frequency meets the criterion of potentially differing from what would be expected in normal conversational management, further information is elicited on the strategies used to manage the behaviour, the outcome of the strategies, and how much the caregiver considers the behaviour to be a problem. The interview also elicits information on premorbid and current interactional styles and opportunities, enabling a comparison to be made between the two time periods. As discussed earlier, with the onset of communication impairment, opportunities for interaction are often reduced (Lubinski, 1991), an important factor to consider in the analysis of the handicap which results from dementia. Knowledge of pre-morbid styles is also important to understanding the change that has taken place in communication and ensuring the suitability of treatment goals (Green, 1984).

Complementing the interview is a sample of conversation recorded between the person with dementia and his or her caregiver at home. Around ten minutes of the conversation are transcribed and a conversation analysis is undertaken to look for evidence of the behaviour, the subsequent conversational management and the strategies reported by the caregiver during the interview. This therefore allows an evaluation of the accuracy of the caregiver's perceptions as well as providing an opportunity to observe strategies not reported by the caregiver. Each of the areas of conversational management discussed earlier is scrutinised along with any observable impact of speech, linguistic and other cognitive impairments on the interaction.

Examination of the dual sources of data allows exploration of a number of key issues which can directly inform the intervention process and which embody the principle of partnership between the person with dementia and the caregiver.

(a) Identification of current conversational management

Both intact ability and impairment of conversational management between the person with dementia and the caregiver are captured, first, by the caregiver's rating of the frequency of a behaviour and, second, in the analysis of the conversational sample. Comparison of the findings of the CA to the caregiver's frequency rating can indicate whether or not the caregiver has a realistic picture of the conversational abilities of the person with dementia, providing important information for establishing the starting point of intervention. It is from this information that the therapist may draw conclusions about whether the caregiver may be over-estimating or under-estimating the person's abilities. If over-estimation of abilities gives rise to breakdown in conversation, stemming from poor understanding of the cognitive changes that have taken place in the person with dementia, education may be necessary. If under-estimation of abilities is occurring, this may be promoting a learned helplessness cycle, and discussion to promote a more accurate and positive perception of the person's interactional abilities may be helpful.

(b) Caregiver's perception of problem

The interview also establishes a rating of problem severity for all impaired behaviours. The relationship between this rating and that for frequency requires careful interpretation. The problem severity rating will be influenced by the strategies upon which the caregiver draws. For example, where a caregiver has developed successful strategies, a particular aspect of impaired conversational management may not be perceived as a problem. The problem severity rating may also indicate the level of acceptance that the caregiver has reached. A high problem severity rating may be indicative of a lack of acceptance which the therapist may wish to address in management. A low problem severity rating may reflect that the caregiver has accepted the cognitive changes. In the piloting of the *CAPPCI*, a common report was that a behaviour that had been a major problem in the past was no longer a problem because the caregiver had learned to accept it. Although this may be considered a positive feature from a management perspective, it is important to consider whether, as a consequence, the caregiver is underestimating the person's abilities which may, once again, be feeding into a learned helplessness cycle.

(c) Caregiver's strategies

The caregiver's strategies and their outcome will have a large impact on the maintenance of successful interaction. Qualitative information about strategies is elicited in the interview and also identified from the analysis of the sample of conversation. The conversation analysis permits validation of the caregiver's report and allows exploration of how the person with dementia responds to the strategies employed. The conversation analysis may also identify strategies that the conversational partner uses but has not reported. In participating in the interview, the caregiver's awareness is heightened of positive steps that he or she is already undertaking to deal with difficulties. The therapist, in management, can positively reinforce productive strategies. In certain instances, no further intervention may be warranted beyond this. In areas where strategy use is not found to be successful, the therapist and caregiver can work together to identify possible alternative approaches.

(d) Comparison of premorbid and current interactional styles and opportunities

The *CAPPCI* also provides information on the change in interactional styles and opportunities of the person with dementia. It provides the therapist, first, with a wider picture within which to interpret the findings of the analysis of current conversational abilities. For example, lack of initiation in conversation will differ in significance between someone who premorbidly had a quiet and passive interactional style and someone who had a talkative, dominant style. Second, it allows assessment of the degree of change with which the caregiver has to deal. Analysis of change in interactional opportunities (people, situations and topics) has management implications in considering possible modification of the interactional environment.

In summary, the *CAPPCI* is designed to provide accurate information on the specific conversational strengths and weaknesses of the person with dementia, the caregiver's knowledge and perception of these and the strategies already being employed by the conversationalists. By sampling the discourse of the person with dementia within his or her own social context and with his or her usual conversational partners, coupled with rationally motivated examination of the caregiver's perceptions of how conversation takes place, the *CAPPCI* provides a basis for the therapist to develop targeted intervention which is client-led and incorporates both the knowledge and skills already developed by the caregiver and the communication strengths of the person with dementia.

Conclusion

This paper has advocated the use of a conversation analytic approach in examining the everyday discourse abilities of people with dementia in jointly constructing communication with their conversational partners. We have stressed the importance of the communication environment, with particular regard to the conversational partner and the impact he or she has on the course of interaction. Through discussion of an explicit methodology, we have attempted to highlight how insights gained through the application of CA may both inform the pragmatic profile of people with dementia and, second, drive clinical intervention with this population.

References

Albert, M. L. (1980). Language in normal and dementing elderly. In L. Obler and M. Albert (Eds.) *Language and communication in the elderly* (pp. 145-150). Lexington, MA: D. C. Heath.

Appell, J., Kertesz, A., & Fisman, M. (1982). A study of language functioning in Alzheimer patients. *Brain and Language, 17*, 73-91.

Bayles, K., Tomoeda, C., & Caffrey, J. (1982). Language and dementia producing diseases. *Communicative Disorders, 7*, 131-146.

Bayles, K., & Kaszniak, A. (1987). *Communication and cognition in normal aging and dementia*. Boston: College-Hill Press.

Beeson, P. M., Bayles, K. A., Tomoeda, C. K., & Slauson, T. J. (1987). Oral discourse in demented, aphasic, and elderly individuals: Content analysis. Paper presented at annual meeting of ASLHA, New Orleans, cited in J. S. Ehrlich (1995). Studies of discourse production in adults with Alzheimer's disease. In R. L. Bloom, L. K. Obler, S. De Santi, and J. S. Ehrlich (Eds.), *Discourse analysis and applications: Studies in adult clinical populations* (pp. 149-160). Hillsdale: LEA.

Bloom, R. L., Obler, L. K., De Santi, S., & Ehrlich, J. S. (Eds.) (1995). *Discourse analysis and applications: Studies in adult clinical populations*. Hillsdale: LEA.

Button, G., & Casey, N. (1984). Generating topic: The use of topic initial elicitors. In J. M. Atkinson and J. Heritage (Eds.) *Structures of social action: Studies in social action* (pp. 167-190). Cambridge: CUP.

Causino Lamar, M. A., Obler, L. K., Knoefel, J. E., & Albert, M. L. (1995). Communication patterns in end-stage Alzheimer's disease: Pragmatic analyses. In R. L. Bloom, L. K. Obler, S. De Santi, and J. S. Ehrlich (Eds.), *Discourse analysis and applications: Studies in adult clinical populations* (pp. 217-236). Hillsdale: LEA.

Chertkow, H., & Bub, D. (1990). Semantic memory loss in dementia of Alzheimer's type: What do various measures measure? *Brain, 113*, 397-417.

Couper-Kuhlen, E. (1992). Contextualising discourse: The prosody of interactive repair. In P. Auer and A. Di Luzio (Eds.), *The contextualisation of language*. Amsterdam: Benjamins.

De Santi, S., Koenig, L., Obler, L. K., & Goldberger, J. (1995). Cohesive devices and conversational discourse in Alzheimer's disease. In R. L. Bloom, L. K. Obler, S. De Santi, and J. S. Ehrlich (Eds.), *Discourse analysis and applications: Studies in adult clinical populations* (pp. 201-216). Hillsdale: LEA.

Dodd, B., Worrall, L., & Hickson, L. (1990). *Communication: A guide for residential care staff*. Canberra: Australian Government Publishing Service.

Ehrlich, J. S. (1995). Studies of discourse production in adults with Alzheimer's disease. In R. L. Bloom, L. K. Obler, S. De Santi, and J. S. Ehrlich (Eds.), *Discourse analysis and applications: Studies in adult clinical populations* (pp. 149-160). Hillsdale: LEA.

Enderby, P. (1990). Promoting communication in patients with dementia. In G. Stokes and F. Goudie (Eds.), *Working with dementia* (pp. 128-133). Bicester, Oxon: Winslow Press.

Fischer, P., Gatterer, G., Marterer, G., & Danielczyk, W. (1988). Non specificity of semantic impairment in dementia of the Alzheimer's type. *Archives of Neurology, 45*, 1341-1343.

Garcia, L. J., & Joanette, Y. (1995). Conversational topic-shifting analysis in dementia. In R. L. Bloom, L. K. Obler, S. De Santi, and J. S. Ehrlich (Eds.), *Discourse analysis and applications: Studies in adult clinical populations* (pp. 185-200). Hillsdale: LEA.

Green, G. (1984). Communication in aphasia therapy: Some of the procedures and issues involved. *British Journal of Disorders of Communication, 19,* 35-46.

Goffman, E. (1955). On face work. *Psychiatry, 18,* 213-231.

Hamilton, H. E. (1994). *Conversations with an Alzheimer's patient.* Cambridge: CUP.

Hart, S., & Semple, J. M. (1990). *Neuropsychology and the dementias.* London: Taylor and Francis.

Heritage, J., & Atkinson, J. M. (1984). Introduction in J. M. Atkinson and J. Heritage (Eds.), *Structure of social action: Studies in conversation analysis* (pp. 1-16). Cambridge: CUP.

Hier, D., Hagenlocker, K., & Shindler, S. (1985). Language disintegration in dementia: Effects of etiology and severity. *Brain and Language, 25,* 117-133.

Hodges, J. R., Salmon, D. P., & Butters, N. (1991). The nature of the naming deficit in Alzheimer's and Huntington's disease. *Brain, 114,* 1547-1559.

Hutchinson, J., & Jensen, M. (1980). A pragmatic evaluation of discourse communication in normal and senile elderly in a nursing home. In L. Obler and M. Albert (Eds.), *Language and communication in the elderly* (pp. 59-74). Lexington MA: D. C. Health.

Illes, J. (1989). Neurolinguistic features of spontaneous language production dissociate three forms of neurogenic disease: Alzheimer's, Huntington's, and Parkinson's. *Brain and Language, 37,* 628-642.

Jefferson, G. (1987). On exposed and embedded correction in conversation. In G. Button and J. R. E. Lee (Eds.), *Talk and social organisation* (pp. 86-100). Clevedon: Multilingual Matters.

Kinney, J. M., & Stephens, M. A. (1989). Care giving hassles scale: Assessing the daily hassles of caring for a family member with dementia. *Gerontologist, 29,* 328-332.

Kempler, D., Curtiss, S., & Jackson, C. (1987). Syntactic preservation in Alzheimer's disease. *Journal of Speech and Hearing Research, 30,* 343-350.

Lesser, R., & Milroy, L. (1993). *Linguistics and aphasia: Psycholinguistic and pragmatic aspects of intervention.* London: Longman.

Lubinski, R. (1991). Learned helplessness: Application to communication of the elderly. In R. Lubinski (Ed.), *Dementia and communication* (pp. 142-151). Philadelphia: B. C. Decker, Inc.

Mace, N., & Rabins, P. (1987). *The thirty-six hour day.* Baltimore: John Hopkins University Press.

Mentis, M., Briggs-Whittaker, J., & Gramigna, G. D. (1995). Discourse topic management in senile dementia of the Alzheimer's type. *Journal of Speech and Hearing Research, 38,* 1054-1066.

Milroy, L., & Perkins, L. (1992). Repair in aphasic discourse: Towards a collaborative model. *Clinical Linguistics and Phonetics, 6,* 27-40.

Muir, N. (1996). Management approaches involving caregivers. In K. Bryan and J. Maxim (Eds.), *Communication disability and the psychiatry of old age* (pp. 221-242). London: Whurr.

Obler, L. (1981). Review of *Le Langage des déments* by Luce Irigaray. *Brain and Language, 12,* 375-386.

Obler, L. (1983). Language and brain dysfunction in dementia. In S. Segalowitz (Ed.), *Language functions and brain organization.* (pp. 267-282). New York: Academic Press.

Orange, J. B. (1991). Perspectives of family members regarding communication changes. In R. Lubinski (Ed.), *Dementia and communication* (pp. 168-187). Philadelphia: B. C. Decker, Inc.

Perkins, L. (1995a). Applying conversation analysis to aphasia: Clinical implications and analytic issues. *European Journal of Disorders of Communication, 30,* 372-383.

Perkins, L. (1995b). An exploration of the impact of psycholinguistic impairments on conversational ability in aphasia. *International Journal of Psycholinguistics, 11,* 167-188.

Perkins, L. (submitted). Negotiating repair in aphasia: Interactional issues.

Perkins, L., Lesser, R., & McKeith, I. (1996). Language as a possible diagnostic medium for dementia with Lewy bodies (DLB.). Poster presentation, The Lancet Conference 1996: The Challenge of the Dementias. Edinburgh.

Perkins, L., Whitworth, A., & Lesser, R. (1997). *Conversation analysis profile for people with cognitive impairments (CAPPCI).* London: Whurr.

Poulshock, S. W., & Deimling, G. T. (1984). Families caring for elders in residence: Issues in the management of burden. *Journal of Gerontology, 39,* 230-239.

Ramanthan-Abbott, V. (1994). Interactional differences in Alzheimer's discourse: An examination of AD speech across two audiences. *Language in Society, 23,* 31-58.

Rabins, P. V. (1982). Management of irreversible dementia. *Psychomatics, 22,* 591-597.

Randolf, C., Braun, A. R., & Goldberg, T. E. (1993). Semantic fluency in Alzheimer's, Parkinson's and Huntington's disease: Dissociation of storage and retrieval deficits. *Neuropsychology, 7,* 82-88.

Rau, M. T. (1991). Impact on families. In R. Lubinski (Ed.), *Dementia and Communication* (pp. 152-167). Philadelphia: B. C. Decker, Inc.

Ripich, D., & Terrell, B. (1988). Patterns of discourse cohesion and coherence in Alzheimer' disease. *Journal of Speech and Hearing Disorders, 53,* 8-15.

Ripich, D. N., Vertes, D., Whitehouse, P., Fulton, S., & Ekelman, B. (1991). Turn-taking and speech act patterns in the discourse of senile dementia of the Alzheimer's type patients. *Brain and Language, 40,* 330-343.

Sabat, S. (1991). Turn-taking, turn-giving and Alzheimer's disease: A case study in conversation, *Georgetown Journal of Language and Linguistics, 2,* 161-175.

Sabat, S., Wiggs, C., & Pinizzotto, A. (1984). Alzheimer's disease: Clinical vs. observational studies of cognitive ability. *Journal of Clinical Experimental Gerontology, 6,* 337-359.

Sacks, H. (1992). *Lectures on conversation.* Oxford: Blackwell.

Sacks, H., Schegloff, E., & Jefferson, G. (1974). A simplest systematics for the organisation of turntaking in conversation. *Language, 50,* 696-735.

Schegloff, E. A. (1979). The relevance of repair to syntax-for-conversation. In T. Givon (Ed.), *Syntax and semantics 12: Discourse and syntax* (pp. 261-286). New York: Academic Press.

Schegloff, E., Jefferson, G., & Sacks, H. (1977). The preference for self-correction in the organisation of repair in conversation. *Language, 53,* 361-382.

Schegloff, E. A. (1982). Discourse as an interactional achievement: Some uses of "uh huh" and other things that come between sentences. In D. Tannen (Ed.), *Georgetown Roundtable on Language and Linguistics 93* (pp. 71-93). Georgetown: University Press.

Schwartz, M., Marin, O., & Saffran, E. (1979). Dissociations of language function: A case study. *Brain and Language, 7,* 277-306.

Shindler, A. G., Caplan, L. R., & Hier, D. B. (1984). Intrusions and perseverations. *Brain and Language, 23,* 148-158.

Shekim, L. O., & LaPointe, L. L. (1984). Production of discourse in individuals with Alzheimer's disease. Paper presented at 12th annual meeting of the International Neuropsychological Society, Houston, cited in J. S. Ehrlich (1995). Studies of discourse production in adults with Alzheimer's disease. In R. L. Bloom, L. K. Obler, S. De Santi, & J. S. Ehrlich (Eds.), *Discourse analysis and applications: Studies in adult clinical populations* (pp. 149-160). Hillsdale: LEA.

Snowden, J. S., Goulding, P. J., & Neary, D. (1989). Semantic dementia: A form of circumscribed cerebral atrophy. *Behavioural Neurology, 2,* 258-271.

Snowden, J. S. (1994). Contribution to the differential diagnosis of dementias: I Neuropscyhology. *Reviews in Clinical Gerontology, 4,* 227-234.

Stephens, M. A., Kinney, J. M., & Ogrocki, P. K. (1991). Stressors and well-being among caregivers to older adults with dementia: The in-home versus nursing home experience. *Gerontologist, 31,* 217-223.

Ulatowska, H. K., Allard, L., Donnell, A., Bristow, J., Haynes, S., Flower, A., & North, A. J. (1988). Discourse performance in subjects with dementia of the Alzheimer's type. In H. A. Whitaker (Ed), *Neuropsychological studies of non-focal brain damage* (pp. 108-131). New York: Springer-Verlag.

Whitaker, H. (1976). A case of isolation of the language function. In H. Whitaker and H. A. Whitaker (Eds.), *Studies in neurolinguistics* (vol. 2, pp. 1-58). New York: Academic Press.

Pergamon

J. Neurolinguistics, Vol. 11, Nos 1–2, p. 55–78, 1998
© 1998 Published by Elsevier Science Ltd. All rights reserved
Printed in Great Britain
0911-6044/98 $19.00 + 0.00

PII: S0911-6044(98)00005-0

Discourse in fluent aphasia and Alzheimer's disease: Linguistic and pragmatic considerations

Sandra Bond Chapman, Amy Peterson Highley, Jennifer L. Thompson

University of Texas at Dallas; Callier Center for Communication Disorders, Dallas, Texas

Abstract—This paper compares discourse performance across three groups, i.e., patients with mild to high-moderate aphasia (APH), mild to early-moderate stage Alzheimer's disease (AD), and normal control subjects (NC) across tasks that elicited discourse texts of varying linguistic and pragmatic difficulty using fables, single-frame pictures, and proverbs. We investigated discourse performance in terms of linguistic formulation and three pragmatic aspects including inferencing, interpreting communicative intentions, and ratio of language to information. The results revealed that the APH group received significantly lower scores than both the AD and NC groups on linguistic formulation. The patients with AD exhibited significant difficulties on the pragmatic domain of drawing inferences as compared to the APH and NC groups. For the majority of tasks, there were no significant group differences on the communicative intentions measure. However, there were significant differences on the domain of language-information balance for APH and AD groups on most tasks as compared to the NC group. The disparity in linguistic formulation and ability to draw inferences between AD and APH groups suggests that discourse differences at mild levels of impairment for these two neurological diseases are qualitatively different. Explanations for these difficulties as well as theoretical and clinical implications are delineated.

Introduction

Historically, the most widely adopted methodologies to characterize communication ability in various adult neurogenic populations have evolved from the study of aphasia (Chapman & Ulatowska, 1994). Since the most prominent symptoms of aphasia are exhibited in the formal linguistic aspects of the language system (i.e., the phonological, semantic, and syntactic components), these domains have commanded the most attention in assessment and treatment of communication deficits in aphasia. This management bias has been mirrored across other adult neurogenic populations, including dementia. A dissatisfaction has grown over the inadequacy of language measures to predict functional communication capabilities in aphasia and dementia. Patients with aphasia are commonly noted to communicate better than they talk and patients with AD are observed to talk better than they communicate (Holland, 1982).

Since performance on measures of specific linguistic function has proven to be a poor index of communicative competence, clinicians and researchers alike have extended their focus beyond purely linguistic abilities to assess the contribution of paralinguistic phenomena, namely pragmatic ability (Carlomagno, 1994; Levinson, 1983; Smith & Leinonen, 1992). One question commonly raised is whether linguistic and pragmatic ability can operate independently to support or hinder communication in neurogenic populations. Perhaps many patients with aphasia are able to compensate for their linguistic deficits, maintaining functional communication through a reliance on relatively preserved pragmatic abilities (Carlomagno, 1994; Davis, 1986). However, patients with dementia appear less able to enlist their preserved linguistic function to maintain communicative competence because of impaired pragmatic abilities (Chapman and Ulatowska, 1991; Ulatowska & Chapman, 1995). The precise nature of the pragmatic disruption in dementia is poorly understood and controversial because pragmatics is not a single phenomena (Garcia & Joanette, 1994). Some evidence suggests that patients with dementia are able to maintain social interactions even when their cognitive and linguistic resources become severely depleted by relying on formulaic social routines (Chapman & Ulatowska, 1997; Hamilton, 1994; De Santi et al., 1994).

The role of pragmatics has become foregrounded as a key domain in further elucidating communicative competence to complement the already widely-accepted practice of evaluating specific linguistic function. Therefore, appropriate methods are sought to examine both parameters. Recent evidence indicates that discourse measures offer one of the most promising ways to examine the contribution of both linguistic and pragmatic factors on communication (Chapman & Ulatowska, 1992; Ulatowska & Chapman, 1994). Discourse provides a highly instructive methodology for elucidating the complex associations across linguistic, pragmatic, and cognitive processes as well as the potential dissociations either across or within clinical populations (Chapman & Ulatowska, 1994; Ulatowska & Chapman, 1994).

Purpose of study

This paper addresses the contribution of pragmatic and linguistic abilities to discourse production using a contrastive approach between two adult neurogenic populations, i.e., patients with fluent aphasia and patients with Alzheimer's disease (AD). To accomplish this goal, the paper is organized into four major sections. First, a rationale is given for comparing the two populations. Second, a multi-level conceptual framework is proposed to compare discourse function in fluent aphasia and AD. Third, the results of a recent study that directly compared aspects of discourse function in aphasic and demented populations are summarized. This information is the major focus of this paper and provides empirical evidence to guide the selection of behaviors which are the most relevant for defining communicative competence, namely linguistic and pragmatic abilities in these two distinct clinical groups. Fourth, clinical and theoretical implications are delineated to highlight the most salient findings. Clinically, the comparison of fluent aphasia and AD can guide the development of appropriate discourse measures to distinguish individual patterns and the planning of effective treatment. Theoretically, this information will further our knowledge of the neurobiological basis of language and how component aspects of complex behaviors such as discourse function can be altered by different types of brain damage in an intriguing array of associations and dissociations.

Rationale for contrasting fluent aphasia and Alzheimer's disease

Over the last decade, a number of studies report evidence suggesting that aspects of verbal communication can be affected in similar ways in patients with fluent aphasia and individuals with AD (Bayles et al., 1989; Obler, 1983; Obler & Albert, 1984). The shared disturbances include comprehension deficits, word-finding deficits, verbosity (disproportionate amount of language when compared to the amount of information conveyed), empty content, and incoherent responses. These disturbances are most apparent on discourse measures (Hier, Hagenlocker, & Shindler, 1985; Nicholas et al., 1985). The evidence of like-symptomatology is somewhat surprising given the prominence of deficits in the formal language domains in fluent aphasia that overshadow their cognitive deficits. This pattern is contrary to the profile of prominent cognitive problems in AD deemed as secondary to their language difficulties.

Language disturbances associated with AD received only cursory attention until recently. Current descriptions of language have confirmed that language disturbances are pervasive even in the earliest stages of AD (Bayles et al., 1989; Henderson, 1996; Hodges et al., 1996; Kontiola et al., 1990). The impairments are localized predominately in the semantic domain marked by semantic paraphasic errors (Bayles, 1982) and by a reduced vocabulary which is particularly evident on confrontation naming tasks (Hier et al., 1985; Martin & Fedio, 1983). With regard to syntax, the results in AD are equivocal. A number of researchers note a surface disruption to sentential structure giving rise to behaviors reminiscent of the paragrammatic errors characteristic of fluent aphasic patients. It is postulated that these paragrammatic errors reflect formulation deficits rather than representing an underlying impairment in use of grammatical structures. As such, the incorrect grammatic strings occur inadvertently, resulting from revisions in information (Chapman & Ulatowska, 1991). With regard to syntactical complexity, consistent findings show that syntax is simplified in AD (Hier et al., 1985; Ulatowska et al., 1988). This simplification in syntax may reflect changes in memory function that disrupt the ability to use complex grammatical structures (Chapman & Ulatowska, 1997).

Evidence of similarities in verbal output can be traced to the earliest descriptions of fluent aphasia by Wernicke and the first description of AD by Alois Alzheimer (Matthews et al., 1994). Since these earliest characterizations of language in fluent aphasia and AD, investigators have amassed fairly consistent data that the two populations manifest similar disruptions in language behavior on a variety of verbal measures. The similarities between patients with fluent aphasia and patients with AD are paradoxical given the disparity between the communicative competence of the two populations. Blanken and colleagues (1987) suggested that the similar abnormalities of impoverished vocabulary, word finding difficulties, verbal paraphasias, circumlocutions, empty language, and impaired coherence may be due to the limited number of ways that language can break down.

One perplexing issue is how a focal lesion in fluent aphasia can give rise to similar disturbances in verbal output as that manifested in AD, a disease with a diffuse cerebral pathology (Matthews, Obler, & Albert, 1994). The evidence that patients with fluent aphasia and patients with AD have separate etiologies, divergent progressions, and disparity between cognitive and linguistic ability motivates examination of the extent of the shared disturbances and clarification of disparities in other domains of communication. Investigations comparing discourse function in these two distinct groups should shed light on whether the discourse similarities are superficial, reflecting different underlying mechanisms in each patient population. Caramazza (1984) claimed that language studies in brain-damaged populations have provided a means of identifying component processes of complex behavior that are otherwise not apparent in normal behavior (labeled *fractionation*). Moreover, he claimed that localized brain damage affects certain functions and spares others to reveal potential associations and dissociations (labeled *transparency*). Clearly, more studies are needed that directly compare the two populations along different levels of communicative function using the same experimental paradigm for clinical and theoretical reasons.

Conceptual framework for discourse production

Current research indicates that similarities and divergences in communicative function across adult clinical populations, such as aphasia and dementia, are most likely to be revealed through examination of the different aspects of discourse representation (Blanken et al., 1987; Cardebat, Demonet, & Doyon, 1993; Chapman & Ulatowska, 1994; Kempler & Zelinski, 1994). Contrastive studies of discourse production in various brain-damaged populations could provide a window to view the linguistic planning and production processes (Blanken et al., 1987). The parameters typically depicted in discourse production models include linguistic, pragmatic, and conceptual functions. These parameters are signified in the following definition of discourse: a unit of language used to convey a message where the expression of the message is governed by (a) *linguistic facility* in expressing the information, (b) *pragmatic aspects* that relate to paralinguistic phenomena, and (c) *cognitive function* required to effectively manipulate information to communicate a message.

Our conceptual framework for characterizing discourse production incorporates selected aspects of linguistic, pragmatic, and cognitive domains. It is important to note that the framework does not incorporate all possible components of discourse processing. Rather, we selected specific parameters based on theoretical and clinical evidence relevant to the discourse profiles of patients with fluent aphasia and patients with AD (Bayles et al., 1989; Chapman & Ulatowska, 1991; Chapman et al., 1985; Obler & Albert, 1984; Ulatowska & Chapman, 1994). The specific parameters are defined in Table 1 along with examples of possible disturbances that may occur in each of these levels.

Level	Definition	Examples of disruptions
Linguistic level	Facility with lexical and grammatical systems to formulate verbal response	Hesitations, circumlocutions, semantic paraphasic errors, paragrammatic errors, neologisms
Pragmatic level	Paralinguistic phenomena (a) Communicative Intentions: ability to interpret intentions of speaker (b) Drawing Inferences: ability to draw inferences across the linguistic content, the situational context, and real world knowledge (c) Language/Information Balance: appropriate amount of language relative to the amount of information conveyed	(a) a response that does not meet task demands (b) a literal response when a generalized one is required (c) excessive language with limited content
Cognitive level	Cognitive processes (such as memory, attention, problem-solving) related to manipulating discourse information for communicative purposes	Unable to hold global semantic meaning in memory to guide text production

Table 1. Conceptual framework for levels of discourse processing

From a theoretical perspective, this framework was derived from our own work and models proposed by Blanken et al. (1987) and Cardebat et al. (1993). These two research teams utilized a different multi-level framework to characterize discourse function, but nonetheless adopted an approach that included linguistic and pragmatic parameters. Blanken and colleagues' (1987) discourse production model consisted of three major levels labeled and defined as follows: (a) pragmatic-conceptual apparatus (the formation of the conceptual structure or meaning of the intended speech act), (b) formulation apparatus (the lexicalization and grammaticalization, and pre-articulatory processing components), and (c) articulation apparatus (planning and execution of articulatory movements (p. 251)). Their model proposed that discourse first emerges at a prelinguistic phase when the global semantic meaning or the idea of the intended speech act (pragmatic conceptual aspect) is conceived. Then, the global semantic meaning is coded linguistically and expressed through speech. Cardebat's team also used a multi-level analysis of discourse production with a linguistic, pragmatic (defined as the strategy used to respond to tasks such as description versus narrative), and global semantic meaning (defined as the central event/complicating action).

We modified these researchers' discourse production models predominately by expanding the pragmatic level. According to our framework (Table 1), we included the pragmatic dimensions of *communicative intentions* (ability to interpret the intended speech act), *pragmatic ability to draw inferences* (ability to go beyond the explicit meaning of individual words and sentences to derive a more global semantic representation through drawing inferences between the text and real world knowledge), and *language/information distribution* (ability to use an appropriate distribution of language relative to the amount of information conveyed).

From a clinical perspective, we selected parameters based on the literature defining population-specific profiles. Evidence has shown that careful evaluation of the pragmatic and linguistic levels of discourse may discriminate patients with fluent aphasia from patients with AD (Blanken et al., 1987; Ehrlich, Obler, & Clark, 1997), despite superficial similarities in discourse production. For example, the ability to determine the conceptual structure/meaning and to fulfill the intended speech act at a pragmatic-conceptual level was found to be disturbed independent of the intact linguistic formulation system in patients with AD (Blanken et al., 1987). For patients with fluent aphasia, the converse pattern was found. In the same vein, Ehrlich et al. (1997) found discourse to be impaired at the level of ideational planning in AD patients. We interpret this finding as support for the possibility that AD speakers are impaired at the level of formulating the communication intent and the general idea of what they plan to say rather than at the linguistic formulation stage. Research by our team has identified difficulties in the pragmatic domain of drawing inferences in AD in constructing condensed, generalized versions of the central idea of a narrative (Chapman & Ulatowska, 1997; Chapman et al., 1995). The ability to draw inferences was found to be relatively intact in aphasia (Ulatowska & Chapman, 1994).

In the study described below, we focused primarily on the linguistic and pragmatic levels. Our goals were to identify associations and dissociations in discourse processing between patients with fluent aphasia and patients with AD. Evaluation of the linguistic, pragmatic, and conceptual aspects of discourse will advance our empirical knowledge of parallels and divergences between the discourse in aphasia and dementia. We propose that similar disruptions to discourse may be due to different underlying mechanism in the two populations.

Discourse study

In this section, we describe a study that compared discourse abilities across three groups, i.e., patients with mild fluent aphasia, patients with early AD, and normal control subjects. The primary goal was to determine whether discourse production across these groups could be distinguished along the dimensions defined in Table 1. Specifically, performance was compared on the linguistic formulation level and on three pragmatic aspects of discourse: (a) communicative intentions, (b) drawing inferences, and (c) language/information balance. We also considered the role of cognitive abilities such as memory and attention and their effect on discourse.

Our hypotheses were that the two neurologically impaired groups would show various convergent and divergent patterns across the domains of discourse production. In particular, we predicted that patients with fluent aphasia would show greater deficits at the linguistic level whereas the patients with AD would have more difficulty at the pragmatic level. We also postulated that longer responses and increased conceptual complexity of the task would contribute to between-group patterns of association and dissociation. In contrast to the overall group performances, we predicted that some individuals within a group would not conform to the expected pattern. For example, individual patients with aphasia may overlap performance of AD patients and vice versa. This prediction is based on previous evidence that individual performance on cognitive and linguistic measures may be more directly associated with discourse performance than group classification (Chapman et al., 1997).

Method

Subjects

Thirty subjects participated in this study, all of whom were selected from a larger investigation of discourse processing in various elderly populations. The participants in the current project included 10 adults with fluent aphasia, 10 adults with AD, and 10 normal control adults (NC). The groups were selected based on a mild to high-moderate level of severity. Additionally, the subjects were required to exhibit some level of verbal abstraction ability by attaining a minimum scaled score of 6 on the similarities subtest of the *Wechsler Adult Intelligence Scale-Revised (WAIS-R)* (Wechsler, 1981). The three groups did not differ across the variables of age, gender, socioeconomic, educational, and occupational levels (Table 2).

Group:	Mild fluent aphasia	Early stage AD	Normal control
Number:	n = 10	n = 10	n = 10
Gender (M/F):	6/4	5/5	5/5
Age:			
Mean	65	65	65
Range	47-74	49-74	47-75
Education:			
Mean	15	15	16
Range	12-19	7-19	13-20
MMSE (mean):	26	22	30
Ravens (mean):	19	14	21

Table 2. Demographics for subjects

Our patients with fluent aphasia met the following criteria: facility with articulation, long strings of words in phrases, and a variety of grammatical markers. These verbal behaviors are claimed to be characteristic of fluent aphasia (Goodglass & Kaplan, 1983). All ten patients with aphasia exhibited brain injury localized predominately to the left posterior temporo-parietal cortices. Eight of the subjects experienced a single cerebrovascular accident, one suffered a focal traumatic brain injury when hit with a blunt object, and one had a focal brain tumor that had been removed more than 4 years prior to evaluation. Although comprehension was mildly impaired for all subjects, these patients were able to understand the instructions required for the tasks as implicated by a relatively high performance on the comprehension of sentences and paragraphs subtest of the *Boston Diagnostic Aphasia Examination* (Goodglass & Kaplan, 1983) achieving nine points or better out of a possible twelve. Severity ratings for the aphasic groups ranged from three to four on the *BDAE* severity rating scale, indicating mild to high-moderate impairment.

The diagnosis of probable AD was made based on neurologic, cognitive, and behavioral testing using National Institute Neurological and Communicative Disorders—ADRDA criteria (McKhann et al., 1984). All subjects were judged to be in the mild to early moderate stages of the disease as assessed by a performance of 19 or higher on the *Mini-Mental State Examination* (*MMSE*) (Folstein, Folstein, & McHugh, 1975) and by a rating of one or lower on the *Clinical Dementia Rating Scale* (Hughes, Berg, & Danziger, 1982) indicating relatively maintained activities of daily living.

Experimental tasks

Stimuli

The experimental paradigm consisted of a battery of discourse tasks that have been shown to place different demands on linguistic formulation and pragmatic processes including communicative intentions, drawing inferences, and language/information balance (Adams et al., 1990; Chapman et al., 1995; Chapman et al., 1997; Ulatowska & Chapman, 1991; Ulatowska & Chapman, 1994; Ulatowska, Chapman, Highley et al., in press). The stimuli included (a) three fables, (b) four proverbs, and (c) a single frame, contextually rich picture by the American painter Norman Rockwell. These stimuli have been identified as salient to older populations (Ulatowska & Chapman, 1991; Chapman et al., 1995). The fables and proverbs are provided in Appendix A along with a description of the single frame picture.

Tasks

Fables:
As shown in Table 3, multiple discourse tasks for the same stimulus were utilized to evaluate the subject's linguistic abilities and performance on selected pragmatic aspects of the discourse task (Chapman et al., 1997; Ulatowska & Chapman, 1994; Ulatowska, Chapman, & Johnson., in press). Each of the tasks place different requirements on communicating the appropriate message in terms of linguistic formulation and pragmatic constraints. Using the fable stimuli, the subjects were asked to construct texts in the form of retells, gists, and lessons.

Single frame pictures:
Similar tasks as used for the fables were designed to elicit responses using the single frame print. Specifically the subjects were asked to generate a dynamic story based on the static events visually depicted in the single frame print followed by tasks eliciting a possible gist and theme or life value depicted in the picture.

Proverbs:
The four proverbs, two familiar and two unfamiliar, utilized in the present study were selected from the California Proverb Test (Delis, Kramer, & Kaplan, 1984) based on prior evidence that these four were the most discriminating of the 10 original proverbs in distinguishing across aphasic, demented, and normal control populations (Chapman et al., 1997). These four proverbs were administered in two formats, a spontaneous and multiple choice condition. In the spontaneous condition, the subjects were asked to verbally explain the intended meaning of the proverb. In the multiple choice condition, the subjects were asked to select the best possible meaning when presented with four choices (i.e., correct abstract, correct concrete, abstract foil, and semantic foil as shown in Appendix A).

Presentation

The fables and proverbs were presented simultaneously in auditory and written form. The single frame picture was shown to the subject, who was allowed to view it for as long as necessary. In most instances, the picture was then turned face down and the subject was asked to create a story about possible events for the characters in the picture. We have found that the tendency to produce a picture description is minimized when the subject is not viewing the picture while generating their response. However, subjects were allowed to view the picture while producing a story if their short-term memory precluded remembering the information from the picture long enough to try to tell a story.

FABLES	(Stimulus: Fable written in simplified language presented simultaneously in visual and auditory form)
Retell	Subject is asked to retell the fable as close to the original as they remember it.
Gist	Subject is asked to give the main idea of the fable.
Lesson	Subject is asked to give a lesson which could be learned from the fable.
SINGLE FRAME PICTURE	(Stimulus: A single frame, contextually rich picture)
Story Generation	Subject is asked to observe the details of a picture, formulate a story based on those details, and relate the story to the examiner with the picture face down.
Gist	Subject is asked to give the main idea of their story.
Thematic Life Value	Subject is asked to generate a human value that is depicted in the picture.
PROVERBS	(Stimulus: Four proverbs, two familiar and two unfamiliar, presented simultaneously in visual and auditory form)
Spontaneous Interpretation	Subject is asked to provide a spontaneous interpretation of the proverb.
Multiple Choice	Subject is asked choose the best possible meaning of the proverb from among four choices (correct abstract, correct concrete, semantic foil, abstract foil).

Table 3. Discourse tasks

Rationale for tasks

Over the last 15 years of research experience in discourse processing across adult populations, we have found the experimental tasks described above to be well-suited for evaluating (a) linguistic formulation ability (Chapman & Ulatowska, 1991; Ulatowska & Chapman, 1994), (b) communicative intentions (Ulatowska, Chapman, Johnson, & Branch, submitted), (c) drawing inferences (Adams, 1991; Adams, et al., 1997; Chapman et al., 1995; Chapman et al., 1997; Ulatowska & Chapman, 1994; Ulatowska, Chapman, Highley et al., in press), and (d) language/information balance (Chapman & Ulatowska, 1991; Ulatowska & Chapman, 1995).

Linguistic formulation of discourse responses

The experimental discourse tasks (using fables, proverbs, and a single frame picture) provided a way to address the effects of both text length and conceptual complexity of the discourse text on linguistic formulation. For example, the story retells involved production of longer units of connected language than did the gist and lesson formulations. The conceptual complexity of the gist and lesson statements was greater than the story retells since considerable inferencing and transforming of information is required to produce a story gist and lesson. In contrast, minimal transformation of information was required on the retell task (Ulatowska & Chapman, 1994). Using familiar and unfamiliar proverbs provides a way to examine the effects of conceptual complexity on linguistic formulation since explicating the meaning for unfamiliar proverbs involves some degree of problem solving whereas explaining the meaning for familiar proverbs may be comparable to defining familiar words (Ulatowska et al., 1995). The complementation of the spontaneous task with a multiple choice task allowed comparison of performance when the linguistic formulation was minimized given choices.

Pragmatic aspects of discourse processing

As shown in Table 1, we evaluated three aspects of pragmatic ability including communicative intentions, drawing inferences, and language/information balance. The different types of information requested across the discourse tasks provided a means to evaluate ability to derive the *communicative intentions*. For example, we examined whether the subjects recognized the intent of the probe and responded appropriately in the form of a retell, story generation, gist, lesson, and proverb explanation.

Communicative difficulties can arise from difficulties in drawing inferences between textual knowledge and real world knowledge. A window to the subject's pragmatic ability to *draw inferences* is particularly transparent on tasks requiring transformation of information at a higher level of generalization than in the original stimulus. Tasks that require paraphrasing of information are represented by tasks such as gist construction, fable lesson formulation, proverb interpretation, and story generation from a static picture. Each task requires the individual to go beyond the linguistic meaning to achieve the necessary interpretation. For example, the ability to construct a fable gist, fable lesson, and proverb explanation requires that the individual understand the surface meaning and derive the central or nonliteral meaning through drawing inferences between the textual information and deep truths about life (Chapman et al., 1997; Ulatowska & Chapman, 1995;

Ulatowska, Chapman, Highley et al., in press). In addition, the ability to generate a dynamic story from a static picture is achieved through drawing inferences between the depicted information and internalized knowledge structures referred to as frames. Frames are established through real life experience and define a set of possible actions and roles for the characters in the picture (Chapman et al., 1995). Thus, not only are most individuals able to attend to the explicitly depicted participants and props in pictures, recognize the salient cues, and arrive at a global frame of interpretation almost instantaneously, but they are also able to conjecture about preceding or following events even when not apparent (Goffman, 1974).

The third pragmatic variable of language/information balance was also rated. The discourse tasks varied in the amount of information necessary to convey the intended speech act with the story retells/generations requiring longer responses than gist, lesson/value, and proverb responses. As a result, the tasks lend themselves to examining the acceptability of the response in terms of *language/information balance*. To a large degree, the feature of language/information balance is congruent with Grice's Co-operative principle of manner (1975) that the speaker should try to be as brief and as informative as possible avoiding obscurity and ambiguity.

Discourse analysis

Response coding

The general methods for analyzing the discourse responses for the fables, single frame picture, and proverbs are delineated in Table 4 in terms of ratings and definitions. Criteria for ratings are presented along the dimensions of linguistic formulation ability, communicative intentions, and drawing inferences. The rating for language/information balance was judged as either normal or impaired, i.e., increased amount of language used to convey the information. Responses for the thirty subjects were randomly mixed into one large group and the raters were blinded as to the group classification.

Rating	Linguistic	Communicative intentions	Drawing inferences
4	Minimal to no formulation difficulties	Recognizes the demands of the task	Interpretive response that is achieved through making inferences between the central meaning and an individual's value system
3	Mild formulation difficulties manifested by hesitations and minor revisions	Partially recognizes the demands of the task	Generalized response that integrates information from stimuli with world knowledge
2	Moderate formulation difficulties manifested in hesitations, semantic paraphasic errors, circumlocutions, and ambiguous pronouns	Marked difficulty recognizing the demands of the task. Needs clarification of task	Central meaning is conveyed through literal explanation
1	Severe formulation difficulties resulting in inability to convey message characterized by above deficits	Fails to recognize the demands of the task. Makes tangential comments.	Partial or tangential explanation which does not convey the central meaning
0	No response	No response	No response or incorrect response

Table 4. Scoring criteria for fables, single frame picture, and proverbs

Results

To examine the similarities and differences in discourse production of patients with aphasia and patients with Alzheimer's disease, we analyzed the responses across the majority of discourse tasks (excluding the language/information balance and proverb multiple choice variables) using a repeated-measures analysis of variance. A p value of less than .05 was considered significant. As outlined in Table 3, the tasks included (a) the fable retell tasks, (b) the fable gist tasks, (c) the fable lesson tasks, (d) the picture story generation task, (e) the picture story gist task, (f) picture story thematic life-value task, (g) the explanations for proverbs, and (h) the selection of the abstract meaning for the proverbs.

Test	Task	Linguistic	Drawing inferences	Communicative intentions	Language/information balance*
Fable	Retell	APH<AD=NC	AD<APH=NC	APH=AD=NC	APH=1, AD=0, NC=0
	Gist	APH<AD=NC	AD<NC APH=NC, AD	APH=AD=NC	APH=2, AD=0, NC=0
	Lesson	APH<AD=NC	AD<NC APH=NC, AD	APH=AD=NC	APH=1, AD=2, NC=0
Single frame pictures	Generation	APH<AD=NC	AD<APH=NC	APH=NC, AD AD<NC	APH=5, AD=8, NC=2
	Gist	APH<AD=NC	APH=AD=NC	APH=AD=NC	APH=2, AD=1, NC=1
	Value	APH=AD=NC	APH=AD=NC	APH=AD=NC	APH=4, AD=5, NC=0
Proverbs	Explanation	APH<AD=NC	AD=APH<NC	APH=AD=NC	APH=17, AD=8, NC=4
	Multiple choice		AD<APH=NC		

*the number of responses in each group for each task that failed language/information balance by providing too much or too little information

Table 5. Group comparisons on specific discourse tasks

Linguistic formulation level

As shown in Table 5, the patients with aphasia received significantly lower ratings than both the Alzheimer group and the normal control group. The patients with Alzheimer's disease did not differ significantly from the normal control group on the linguistic formulation rating for any task. This suggests that the individuals with aphasia exhibited marked difficulty in lexicalizing their ideas as manifested by increased hesitations, revisions, circumlocutions, paraphasic errors, and ambiguous pronouns as compared with normal control subjects. This formulation difficulty disrupted the coherence of the various texts. Responses for a patient with fluent aphasia are illustrated in Table 6 in sample responses on the fable retell, gist, and lesson tasks and are contrasted with the minimal linguistic problems exhibited in the samples for both a normal control and a patient with Alzheimer's disease.

	Normal Control	Fluent Aphasia	Alzheimer's Dementia
Retell	Well, you have this raven, and you know that ravens are black. He saw a lot of pigeons in a pigeon coop, and he said he wanted to be there with them, but his color made him different, and uh painted his feathers white, and tried to come in there and dilly-dally with the pigeons, but when he started crowing instead of cooing, they knew he was a fake, so they chased him out. So hey, he had to go back home where the other ravens were, but he had this white paint on him. The ravens looked at him and said, "Hey man, what are you? You're white, get outta here." So he was ousted. So he had no place to go. He was persona non grata.	A raven saw a, saw the paig- (6 seconds) (E: pigeons) of pigeon, pigeons in the ka-coop. The-the raven looked and they had of food. The raven where he got th-th-th uh paint to paint his feathers white so he can go into the cope-coop and have some food then. Well he got in there, into the coop and as soon as he wanted foo, no he-he wanted to, not talk, well he wanted to crow. The pigeon, "No-ho, that's not a cousin." So they threw him out and he had to run away. And then he went back to the raven-raven family and-and they looked at him, his color isn't right. It was white so he had to tea- they threw him out.	Uh, you know, the raven uh, like any-any bird has got to eat, and the way they get their food is uh, apparently by guile. By uh, uh, confusing the uh, other...whatever the hell it was. I forgot. Um, it was a raven and what else? (E: The raven and the pigeon, okay.) Pigeon, okay. Uh, the raven uh-uh, was gonna do whatever he's needed to do to accomplish the goal uh, which was to get some food uh, when this uh, rabbit had plenty of food and he didn't, so he uh-uh, used his head and figured the program out and uh...achieved the end and that he was after, the more food.
Gist	Oh, the gist of the story is that a raven wanted to be other than what he was and he had suffered the consequences.	Don't change, or like that? Don't change the color...don't change, don't change your, don't change, don't change yourself, what you are to get some money, ***to get uh food.	Uh, well, just the main idea is that uh, any animal or person or anything else will do whatever they have to do to get the food they need to survive. Survival of the fittest, I guess.
Lesson	Oh, go back and read Shakespeare and some of the things, unto thine own self be true and be natural, don't try to be something that you aren't. Be yourself, thyself.	Don't change, don't change, don't change things to, don't-don't-don't take, don't take, don't change, don't change your life just to get what you want to.	Uh... I guess work hard and persevere and you won't have to uh, steal your food.
Comments	Absence of formulation difficulties. Subject clearly isolates central meaning by integrating textual information with real world knowledge. Subject responds appropriately to task demands, recognizing requirements for condensing the text in gist and lesson form. Pragmatic informativity is normal. Although in the retell the subject gives more information than contained in the original, the balance between amount of language and information is normal since elaboration enriches story and reflects normal stylistic differences.	Moderate formulation difficulties manifested in repetitions, revisions, hesitations, and semantic paraphasias. Subject is able to make inferences, but it is less clear as to how these inferences relate to the text. Responses are appropriate to the task demands for each task. While the pragmatic informativity is normal for the retell, there is excessive language in relation to amount of information conveyed for the gist and lesson tasks due to repetitions and revisions.	Minimal formulation difficulties. Some evidence of anomia on the retell task probably due to memory deficits in that he has forgotten the characters and the rabbit character represents an intrusion from the previously presented fable. Also perhaps due to memory deficits, his responses do not convey the central meaning. He appears to be relying more on world knowledge than the context of the story to give a generalized response for the task demands. He clearly recognizes the task requirements for the gist and lesson tasks. Language information balance is normal.

Table 6. Sample responses for "Raven & Pigeons" fable

Pragmatic level

Communicative intentions

For the majority of pragmatic measures, there were no significant group differences on the *communicative intentions* measure. That is, the three groups were able to recognize the intent of the probes and respond appropriately to the task with minimal, if any, difficulty. The only exception of this pattern was the deficit observed in the group with Alzheimer's disease on the story generation task using the single frame picture. A significant number of AD patients exhibited difficulty going beyond the static picture to generate a dynamic narrative, instead producing a picture description. This tendency toward descriptive discourse is exemplified in the response by a patient with Alzheimer's disease presented in Table 7.

	Single frame picture: Story generation	Comments
AD	Well, uh there is a man there and there is uh looks like a child and um there's a dog and I think there's an older man there too. I think so. Um I think that's an older man in the picture. Um it looks it's an autumn day out there and it's kind of overcast, kinda like it is today. Um, and the man and the boy are talking and uh I think that the boy, the little guy, I think that they are going fishing or they're thinking about going fishing.	The patient produced a description of the explicit information contained in the picture (i.e., there is a man there... looks like a child and um there's a dog...) with few inferences (i.e., they are going fishing.) Moreover, the inference that "the boy and man are fishing" was an atypical interpretation that did not integrate the salient props. Additionally, this patient with AD failed to define the relationship between the two characters. This patient also exhibited minimal linguistic difficulties such as a few hesitation and some repetition of thoughts.
APH	Let's see, it must have been during the 1930's uh a young man going to college, university, state university. And they were sitting on a a kind of a, on a small, on a ru-ru-rubber, rubber, rudder, sitting on a rubber of a car. They used to have that. And uh this uh, there was a, this young man he has a friend, a friend which was a dog, probably sheep-shep-shepard. And dad was smoking and he was thinking well, "you've got to say good-bye." And uh I think he was a farmer. And so the man is thinking, "I'm going to leave and going to college or in the university." And dad is gonna wis-miss and so will the dog miss this young man.	The patient had moderate word finding problems as evidenced through hesitation, repetitions and reformulations. Despite these linguistic difficulties, he produced a typical frame of interpretation. He expressed the central meaning, (i.e., young man going to college) conveyed the relationships of the characters (father and son). He also made inferences with his real world knowledge to infer the time period of this picture (i.e., 1930's) as well as inferring the emotions of the characters.
NC	They had their son, and they're waiting for a train that's gonna take the boy away to college. Anyway he has a lunch that his mother has finished for him. His collie dog senses that the boy is gonna be going away. And the boy is really duded up cause he has him a pair of argyle socks, and regular black shoes and a matching sort of tie and white shirt. And he's looking with anticipation on his face, and you see the father's face, "well there goes one of my hands so I'm gonna do all the chores that he's been doing" but they're waiting for the train.	The normal subject produced a narrative with a typical frame of interpretation without any linguistic deficits. The normal subject expressed the central meaning of the story (the boy is going off to college), conveyed the relationship between the characters (father and son), and gave background information (waiting for a train). His response revealed proficiency I making inferences between the depicted information and real-life knowledge. When he interjected descriptive information (i.e., he has him a pair of argyle socks and regular black shoes and matching sort of tie and white shirt), it was used to support his inferences, e.g., that the boy was dressed up for the big occasion.

Table 7. Single frame picture story generation

Drawing inferences

The most relevant measure for distinguishing the patients with Alzheimer's disease from the aphasic or normal control groups was the *pragmatic inferencing* variable. That is, the patients with AD exhibited significant difficulties on this domain due to problems drawing the necessary inferences between the text and real world knowledge. Consequently, the texts of the AD patients were less coherent than the normal control or aphasic groups. This pattern was particularly apparent on the fable tasks (see examples in Table 6) and story generation for the single frame picture (see examples in Table 7). On the single frame picture, half the subjects with AD failed to make the necessary inferences between the information depicted and real world knowledge to establish the typical frame of interpretation (i.e., going to college).

With regard to proverb explanation, both the aphasic and Alzheimer's groups performed significantly below the normal control group on the pragmatic aspect of drawing inferences. On this task, both patient groups exhibited significant difficulty formulating correct nonliteral explanations for proverbs. It is interesting to note that even though the two patient populations overlapped on the drawing inferences dimension, the patients with aphasia continued to show greater linguistic difficulties than the AD group or normal control group (Table 5). The multiple choice proverb interpretation task was less difficult than the spontaneous explanation task for the patients with fluent aphasia. That is, most patients with fluent aphasia were able to choose the correct abstract interpretation, suggesting that they can appreciate the abstract meaning of proverbs when linguistic formulation demands are minimized. Conversely, patients with AD continue to exhibit difficulty given choices, often choosing the abstract foil (in the form of another proverb).

The following sample responses from a patient with fluent aphasia and a patient with Alzheimer's disease for the unfamiliar proverb *"One swallow doesn't make a summer"* demonstrates the similar difficulties in drawing the necessary inferences.

Patient with fluent aphasia:
All right, (laughs) so what does that mean, um, well it takes all summer. Uh, what, uh, well, a swallow doesn't make a summer. Right. Um, it takes time, uh let's see time, uh, autumn could also be just as well, so the swallow, what about it, if he goes in the summer time, well, that's great, but um, how about in the autumn, he goes south. I'm not doing well with this at all. I mean it doesn't um, I don't know, maybe the birds, uh, one swallow, that's a bird isn't it? (E: Can you think of a reason why one swallow wouldn't make a summer?) Right. Nope, I'm sorry.

Note the similarities between the above explanation and the following by a patient with Alzheimer's disease:

Um...one swallow doesn't make a summer. Mmmm, I don't know what to say to that. Uh (5 sec) Oh, we're not talking about the swallow as a bird, are we? One swallow doesn't make a summer. Umm...one swallow doesn't make a summer- one swallow- one bird doesn't make a...one, more than one way to skin a cat, is it? Um...well, just because someone is doing this job, doing it that way doesn't mean you can't do it a different way and uh, and reach the same conclusion.

Neither explanation incorporates the necessary inferences to convey the correct, generalized meaning of the proverb. The patient with aphasia seems unable to move beyond the context of "birds" to the more global meaning of the proverb. That is, he exhibits difficulty in reformulating the proverb outside the original language of the proverb by accessing and selecting his own words. Notice that the patient with AD attempts to use an unrelated proverb, "There's more than one way to skin a cat," to help explain the proverb "One swallow doesn't make a summer," indicating that he realizes that a proverb should be interpreted abstractly, but fails to make the correct inferences that integrate the textual information with real world knowledge.

Language/information balance

On this pragmatic domain, both patient groups tended to construct texts in which the distribution between language and information was judged to be impaired. The patients with aphasia and dementia produced less information than normal controls, but did not use less language. The number of instances in which subjects were rated to use more language to convey the information than deemed appropriate for the task is indicated in Table 5. This tendency to use excessive language (verbosity) was particularly interesting for the proverb task in that the aphasic group appeared to have more difficulties on the proverb explanation task than the AD group.

Clinical validation of disparate profiles

A secondary goal of the study was to determine whether or not the profiles identified by the results of this study could be used to clinically distinguish the two groups. Clinicians were blinded to the diagnosis and attempted to classify the subjects' discourse responses according to whether the response seemed consistent with that of a normal control, a patient with early AD, or a patient with fluent aphasia. At the time of attempted classification, the clinicians were provided with information related to the expected profile for each diagnostic group. Specifically, the raters were instructed to classify the patients as aphasic if they exhibited linguistic formulation difficulties, but still seemed to be able to draw inferences between the textual and real world knowledge across most tasks. In contrast, classification of AD was recommended for individuals who showed deficits in drawing inferences between the textual and world knowledge with only minimal linguistic formulations problems. If no visible symptoms of linguistic formulation difficulties outside a normal range or deficits in drawing inferences were present, the individual should be classified as normal.

The quantitative results revealed that there was no difficulty differentiating the normal controls from the two patient populations. All normal subjects were identified as normal on all tasks. For the aphasic group, three patients were misclassified as producing responses similar to patients with AD for fables; one was misclassified on the single picture task, and five patients were misclassified on the proverb explanation task. Two of the aphasic patients were misclassified on both the fable and proverb tasks. For the patients with AD, two were judged to have aphasia on the basis of their fables responses, two were misclassified as aphasic on the single picture task and none were misclassified on the proverb task. Table 8 presents atypical responses for patients who were misclassified.

	Fluent Aphasia	Alzheimer's Disease
FABLES ("Raven & Pigeons")	(RETELL) *This was a story of the raven that...painted his feathers white. And got over, to get over the pigeons him and made him get away. But he left, let me see, some of them was still there but the pigeons got them all out of the way.* (E: *Some of who were still there?*) *The pigeons- the ravens, the ravens was there. But they finally got them all out. They all had uh yellow-yellow-no white, the was painted white.* (E: *And how did the pigeons know...that this was a raven and not a pigeon?*) *Because they s-l don't know, it's, they sounded different* (laughs). (LESSON) *Uh don't, don't, don't let the ravens get in the chicken coop.*	(RETELL) *Well, you had now- it was a bird, it was uh, raven uh, who was uh, is a raven and uh I believe he noticed that uh uh the crows uh was a getting something to eat. Uh, so he uh, got uh painted or they painted him, figured he could get more, a better deal on that, that's right—I believe that the uh crows that's what it was, kinda knocked him out of that game.* (E: *And what happened?*) *So he went back to his uh* (4 seconds), *he went back to his own business.* (7 seconds) *Took care of him it, got his own food.* (LESSON) *There's no uh uh, it's no one, you can't get it for nothing, you can't get anything for nothing.*
Comments	Responses contain mild formulation difficulties manifested by pauses, revisions, and repetition of words. Responses do not convey the central meaning of the fable. However, subject attempts to fulfill task demands by formulating responses in terms of a retell and lesson, respectively.	Mild to moderate formulation difficulties manifested in semantic paraphasias, revisions, hesitations, and ambiguous pronouns. Retell response is vague and must be interpreted by listener, though most key elements are included. Lesson response is a generalized statement which seems to reflect the individual's value system. Subject fulfills task of lesson.
PROVERBS	(ONE SWALLOW DOESN'T MAKE A SUMMER) *I don't know if that's talking about a bird or* (laughs) *if-if I got hot. Uh, that- that means that uh, they was a lot more birds. Or there's more of anything. Let's see, a swallow, one swallow does not make a summer. Uh-uh* (5 sec). *Sw-uh...uh...b-uh.* (E: *One swallow doesn't make a summer.*) (10 sec) *See* (6 sec) *swallow....*(E: *Can you think of an example of when you might use a saying like that?*) *Uh...there's not a uh, let's see, there's not a uh-uh drop of rain, there's not any r-uh-rain-uh...let's see. Swallow* (5 sec) *a...I don't know ***.*	(TOO MANY COOKS SPOIL THE BROTH) *Too many cooks spoil the broth...too many cooks spoil the broth...* (E: *Got a guess on it?*) *Well, there's bound to be something there. I know it happens* (laughs). *I know that's happened. Uh...*(E: *What does it mean in general?*) *Yeah, well uh...you can't get a whole lot of things done with a whole lot of people. You've got to come up with some kind of a thing that's "dooble" and that some people may not be involved with it. Uh...you don't want to have uh-you just can't ex-take 'em all.* (E: *How come?*) *Well, they got too many peole get involved in too-gonna uh, the cooks too as well-too many peole get involved in something ***that way responsibility had to be uh and uh...you can't have twelve people getting involved with something and somebody's got to do it.*
Comments	Only mild language formulation difficulties, manifested in incomplete thoughts, pauses, and fillers. Unable to make the inferences necessary to convey the global meaning of the proverb.	Despite formulation difficulties, is able to convey the general meaning of the proverb by making inferences between the text and his real world knowledge.

Table 8. Discourse responses from patients who were misclassified

Discussion

This paper presented a framework for examining salient components of discourse function relevant to defining communicative competence and applied the framework to clinical characterization in patients with fluent aphasia and patients with early stage AD. Our premise was that communicative competence could best be illuminated by characterizing discourse abilities at multiple levels of representation, including both linguistic and pragmatic aspects.

The results of the study revealed that discourse was equally impaired across a range of tasks for both the mild aphasia group and the early stage Alzheimer's disease group when compared to normal control subjects. At a global level of description, the discourse performance in both groups could be characterized as sharing word finding deficits, reduced content, incoherent responses, and verbosity. The most important finding, however, was the disparate pattern of breakdown in linguistic and pragmatic domains of discourse production between patients with mild fluent aphasia and patients with early stage AD as compared to normal control subjects. The patients with aphasia were significantly impaired in the linguistic domain, whereas the patients with AD showed greater disturbances in the pragmatic domain of drawing inferences.

Our data confirm previous evidence that linguistic formulation is disturbed in fluent aphasia and relatively preserved in early AD (Bayles et al., 1989; Blanken et al., 1987; Cardebat et al., 1993; Ehrlich et al., 1997). The evidence of a linguistic formulation deficit in mild fluent aphasics is not surprising given linguistic deficits are criteria in the definition of aphasia. The patients with mild fluent aphasia had difficulty across most verbal tasks varying from retelling and interpreting a fable (in the form of a lesson) to explaining the meaning of proverbs. The aphasic group performed lower on the linguistic formulation domain than the patients with AD and the normal control group.

Additionally, the present findings support the conclusion of Blanken et al. (1987) that the linguistic disturbances in AD do not represent an impairment of lexicalization, but rather an inadequacy in accessing the appropriate meaning through impaired inferencing. The AD group received ratings comparable to the normal control group in the linguistic formulation domain. We had anticipated that the AD group would display some mild to moderate formulation difficulties on the conceptually more complex discourse tasks such as constructing a fable lesson or explaining the meaning for a proverb since anomia is a well-documented deficit in the early stages of AD. Perhaps the very mild degree of impairment in our patients with AD contributed to our failure to find a difference. Alternatively, it may be that even normal controls have some formulation difficulties as shown in the normal example and thus more variability in linguistic formulation is acceptable.

The most prominent difficulties for the AD group were in the pragmatic domain of drawing inferences. The patients with AD performed lower on the fable retell task, the picture generation story task, abstracting the central meaning in the form of a gist, and deriving the didactic meaning in the form of a lesson. The ability to draw inferences between textual content and real world knowledge tends to be relatively preserved in patients with fluent aphasia and impaired in early AD. Cardebat et al. (1993) reported a similar difficulty with inferencing in AD using a different narrative methodology. Specifically, these researchers found that patients with AD consistently failed to mention the complicating action which represents the most important story event. In another study, Chapman et al. (1995) found that patients with early stage AD had difficulty drawing the necessary inferences between a static picture and real world knowledge to allow them to produce a dynamic narrative. Thus, disturbances in the pragmatic domain of drawing inferences between textual knowledge and real world knowledge were manifested independent of linguistic formulation difficulties in patients in the early stages of AD.

In addition to linguistic and pragmatic factors, our patients with AD may have been unable to produce a dynamic narrative from the single frame picture because of cognitive deficits. For example, they may have had memory problems that precluded storing the information long enough to draw the necessary inferences. Another possible explanation is that the subject did not attend to the most salient aspects of the picture.

The proverb explanation task was the least informative task in discriminating between the two patient populations. However, when used in combination with the multiple choice task, the informative value increased. Both patients with fluent aphasia and patients with AD had difficulty spontaneously producing the nonliteral meaning for proverbs. The similarities between groups were underscored by the fact that five patients with fluent aphasia were subjectively misclassified as having dementia based on their proverb responses due to the difficulty in providing nonliteral explanations. In contrast, only the patients with AD had difficulty recognizing the nonliteral meaning for proverbs given choices. These findings are consistent with previous data that patients with fluent aphasia had minimal difficulty identifying the nonliteral meaning of proverbs when they did not have to propositionalize the meaning in language form (Chapman et al., 1997; Kempler, Van Lancker, & Read, 1988).

Previous evidence suggested that patients with early AD had little difficulty with familiar proverbs, but marked difficulty with unfamiliar ones (Chapman et al., 1997). In the present study, we had too few items to determine differences between familiar and unfamiliar proverb interpretation. Clearly, unfamiliar proverbs require greater inferencing than familiar ones. Familiar proverbs may be retrieved much like the meaning of single words, whereas the meaning of unfamiliar proverbs must be resolved through inferencing between the text and some plausible meaning from real life (Van Lancker, 1990).

It is important to note that not all subjects within groups conformed to a disease-distinct pattern, despite the relatively consistent findings confirming a discrepancy between patient groups in the linguistic domain and the pragmatic domain of drawing inferences. The exceptions were manifested in a number of patients with fluent aphasia who showed marked difficulties in drawing inferences and in some patients with AD who exhibited notable linguistic formulation deficits. The diagnostic category alone will not define the pattern. Clearly, some patients with aphasia show inferencing deficits and some patients with AD show early linguistic formulation deficits (Chapman et al., 1997). Knowledge of an individual's pattern of strengths and weakness in the linguistic and pragmatic domains is the key to adequate diagnosis and treatment for both populations.

The ability to interpret the speaker's intentions for communicative purposes may be intact at more severe levels of aphasia and later disease progression in AD (Bond et al., 1983; Hamilton, 1994). Minimal differences were found in the ability to interpret the intentions of the speaker, i.e., communicative intentions, across all three groups despite the disparate linguistic and pragmatic inferencing performance in aphasia and in AD. The ability to recognize a request made for information may be preserved late into the progression of AD (Hamilton, 1994). The only exception to this pattern was identified in the AD group who failed to respond appropriately on the picture story generation. Instead, they tended to give a picture description instead of a narrative with a sequence of events. Cardebat and colleagues (1993) described a similar disturbance in patients with AD.

Clinical implications

The conceptual framework described herein offers an informative method for parsing discourse into component domains relevant to functional communication for assessment and treatment purposes in adult neurogenic populations. The most important principle to be derived from our research and others (Blanken et al., 1987; Ehrlich, 1994; Hamilton, 1994; Cardebat et al., 1993) is that discourse tasks are more revealing than isolated language measures in characterizing the full scope of communication breakdown. In particular, discourse tasks that vary in length and conceptual complexity provide a unique window to view both linguistic planning processes and pragmatic phenomena such as communicative intentions, drawing inferences between different types of knowledge, and using the appropriate amount of language to convey information. Such an approach allows us to broaden the focus of our clinical descriptions and to examine co-occurrence of linguistic difficulties and pragmatic impairments regardless of the patient's diagnostic category. If we take a narrow a path in clinical assessment and treatment because of preconceived notions of expected symptomatology, we are likely to misdiagnose and mismanage the communication breakdown in a large number of patients.

The present methodology has limitations as to how much performance on story retell and gist construction tasks may actually reflect what truly goes on in a communicative exchange. Some insights can be offered regarding the type of tasks that are the most illuminating when attempting to characterize communicative function in the linguistic and pragmatic domains. Based on our work and others, we propose that the most instructive methods would be tasks that meet some of the following criteria:

(a) tasks that require holistic processing rather than isolated observations about pictures. For example, picture descriptions do not require that an individual go beyond a primary analysis of the depicted elements (Cardebat et al., 1993; Chapman & Ulatowska, 1993). A task that requires the individual to construct a narrative based on the pictured information forces drawing inferences. The most effective pictures for eliciting narrative discourse are those that are dynamic and contextually rich, not pictures depicting isolated people performing unrelated activities.

(b) tasks that involve reformulating the text information at different levels of generalization. For example, the various discourse tasks used in the present study such as story gist and story lesson place different demands on drawing inferences between the textual information and real world knowledge. Thus, as the subject proceeds from a retell or story generation, to a gist, and finally to a lesson, there exist greater and greater demands to condense the original story information, paraphrase it using his/her own words, and integrate the text information with real world knowledge through drawing inferences.

(c) tasks that elicit linguistic formulation for texts of varying lengths. For example, longer texts are required for retells and shorter responses are possible with gist and lesson interpretations. Difficulties in accessing the language to convey one's thoughts may be more transparent when formulating longer responses and spontaneously generated responses rather than retells.

(d) tasks that involve encoding the information in short term memory storage rather than viewing the stimulus while producing the response. This is recommended because viewing the stimulus tends to encourage less paraphrasing and less inferencing.

Theoretical implications

Discourse studies in neurolinguistics help to elucidate the nature of the organization of the complex processes underlying human communication as well as the neurobiology of discourse. This study advances the theoretical notion that linguistic and pragmatic phenomena can be dissociated in patients with primarily linguistic deficits and those with more conceptual impairments. The disparity found between patients with fluent aphasia and patients with AD supports Caramazza's (1984) two postulations about the informative value of contrastive studies of brain-damaged populations. First, he claimed that the information from behaviors in brain-damaged patients could reveal components of complex behavior that are not apparent in normal behavior (i.e., fractionation). Clearly, in normals the role of linguistic formulation and pragmatic inferencing operates effectively and interdependently in discourse production such that two processes are not readily separated. We also speculate that in severely impaired patients, both systems break down such that it becomes problematic to identify the separate components. In our study, the components of linguistic formulation and pragmatic ability of drawing inferences were dissociated across a variety of discourse production tasks in two different brain damaged groups. Second, Caramazza proposed that brain damage affects certain functions and spares others to reveal potential associations and dissociations (i.e., transparency). This latter pattern was evident in the dissociations found in aphasia and AD at mild/early levels of impairment between the linguistic domain and the pragmatic domain of drawing inferences, even though discourse output in the two clinical populations showed similar changes in reduced information, verbosity, and incoherence.

The discourse differences between patients with fluent aphasia and patients with AD have implications for the neurobiological organization of discourse functions. The inferencing deficits in the patients with early AD may be due to involvement of the right hemisphere. Evidence from studies of patients with right hemispheric damage has suggested that these patients have difficulty drawing inferences and deriving nonliteral meanings (Wapner, Hamby and Gardner, 1981). Perhaps the relatively intact right hemisphere in our aphasic population allowed them to draw inferences and interpret proverbial meaning at a nonliteral level when they did not have to simultaneously retrieve the words to express the information.

Moreover, the evidence that a number of patients did not conform to the classification schema for their disease type (i.e., APH or AD) warrants further examination into the neurobiology of their brain disease. One might speculate that patients with mild aphasia who show obvious difficulties in the pragmatic domain of drawing inferences may have experienced a more diffuse brain injury due to more widespread ischemia that did not resolve. For patients with early stage AD who show early prominent language formulation symptoms, it would be interesting to determine whether these patients showed more focal brain disease in the left Perisylvian region in the early stage. Additionally, comparing patients with AD who have language deficits as a prominent early symptom may be an intriguing contrastive group to patients with primary progressive aphasia.

From a humanitarian sentiment, the differences in discourse components between patients with aphasia and patients with AD indicate that the similarities may well reflect different underlying mechanisms. Identifying the individual's strengths and weaknesses along the component dimensions of discourse function will guide more detailed diagnosis leading to improved intervention. In aphasia, we often fail to go beyond the linguistic deficits even though a number of patients show deficits in pragmatic domains such as drawing inferences. This domain needs to be assessed in aphasia. While intervention is the rule rather than the exception in aphasia, there is considerable controversy as to whether some form of intervention be implemented with early stage AD patients. At a general level, the speech-language pathologist's role is quite similar to the widely accepted practices of aphasia management. That is, it is necessary to help the patient to be communicatively viable for as long as possible. As evident from this study, our patients with aphasia and dementia have revealed an incredible array of strategies that allow them to achieve some level of successful communication. Thus, the identified strategies provide empirical data to guide professionals to the development of more appropriate methods for enhancing communication in these populations.

Acknowledgement—This investigation was supported by grants for the National Institute of Aging/National Institutes of Health (AG09486 and S-P30-AG12300-02) for the Alzheimer's Disease Center. We express our gratitude to Hanna K. Ulatowska who designed the tasks and Jacqueline A. Prince for help in data management.

References

Adams, C., Labouvie-Vief, G., Hobart, C., & Corosz, M. (1990). Adult age group differences in story recall style. *Journal of Gerontology: Psychological Sciences, 45,* 17-27.

Adams, C. (1991). Qualitative age differences in memory for text: A life-span developmental perspective. *Psychology and Aging, 6,* 323-336.

Adams, C., Smith, M. C., Nyquist, L., & Perlmutter, M. (1997). Adult age-group differences in recall for the literal and interpretive meanings of narrative text. *Journal of Gerontology: Psychological Sciences, 52B,* 187-195.

Bayles, K. A., (1982). Language function in senile dementia. *Brain and Language, 16,* 265-280.

Bayles, K. A., Boone, D. R., Tomoeda, C. K., Slauson, T. J., & Kaszniak, A. W. (1989). Differentiating Alzheimer's patients from the normal elderly and stroke patients with aphasia. *Journal of Speech and Hearing Disorders, 54,* 74-87.

Blanken, G., Dittmann, J., Haas, J., & Wallesch, C.-W. (1987). Spontaneous speech in senile dementia and aphasia: Implications for a neurolinguistic model of language production. *Cognition, 27,* 247-274.

Bond, S. L., Ulatowska, H. K., Macaluso-Haynes, S., & May, E. B. (1983). Discourse production in aphasia: Relationship to severity of impairment. In R. H. Brookshire (Ed.), *Proceedings of the clinical aphasiology conference* (pp. 202-210). Minneapolis: BRK Publishers.

Caramazza, A. (1984). The logic of neuropsychological research and the problem of patient classification in aphasia. *Brain and Language, 21,* 9-20.

Cardebat, D., Demonet, J.-F., & Doyon, B. (1993). Narrative discourse in dementia. In H. H. Brownell and Y. Joanette (Eds.), *Narrative discourse in neurologically impaired and normal aging adults* (pp. 317-332). San Diego: Singular Publishing Group, Inc.

Carlomagno, A. (1994). *Pragmatics in Aphasia.* (G. Hodgkinson, Trans.). San Diego: Singular Publishing Group, Inc. (Original Work published 1989).

Chapman, S. B., & Ulatowska, H. K. (1991). Nature of language impairment in dementia: Is it aphasia? *Texas Journal of Audiology and Speech Pathology, 17,* 3-9.

Chapman, S. B., & Ulatowska, H. K. (1992). Methodology for discourse management in the treatment of aphasia. *Clinics in Communication Disorders, 2,* 64-81.

Chapman, S. B., & Ulatowska, H. K. (1994). Differential diagnosis in aphasia. In R. Chapey (Ed.), *Language intervention strategies in adult aphasia* (pp. 121-131). Baltimore: Williams & Wilkins.

Chapman, S. B., Ulatowska, H. K., King, K., Johnson, J., & McIntire, D. D. (1995). Discourse in early Alzheimer's disease versus normal advanced aging. *American Journal of Speech-Language Pathology, 4,* 125-129.

Chapman, S. B., Ulatowska, H. K., Franklin, L. R., Shobe, A. E., Thompson, J. L., & McIntire, D. D. (1997). Proverb interpretation in fluent aphasia and Alzheimer's disease: Implications beyond abstract thinking. *Aphasiology, 11,* 337-350.

Chapman, S. B., & Ulatowska, H. K. (1997). Discourse in dementia: Consideration of consciousness. In M. Stamenov (Ed.), *Language structure, discourse and the access to consciousness* (pp. 155-188). Amsterdam: John Benjamins Publishing Co.

Davis, G. A. (1986). Pragmatics and treatment. In R. Chapey (Ed.), *Language interventions strategies in adult aphasia* (pp. 169-193). Baltimore: Williams & Wilkins.

Delis, D. C., Kramer, J., & Kaplan, E. (1984). *The California proverb test.* Unpublished protocol.

DeSanti, S., Koenig, L., Obler, L. K., & Goldberger, J. (1994). Cohesive devices and conversational discourse in Alzheimer's disease. In R. L. Bloom, L. K. Obler, S. De Santi, and J. S. Ehrlich (Eds.), *Discourse analysis and applications: Studies in adult clinical populations* (pp. 201-215). Hillsdale: Lawrence Erlbaum Associates.

Ehrlich, J. S. (1994). Studies of discourse production in adults with Alzheimer's disease. In R. L. Bloom, L. K. Obler, S. De Santi, and J. S. Ehrlich (Eds.), *Discourse analysis and applications: Studies in adult clinical populations* (pp. 149-160). Hillsdale: Lawrence Erlbaum Associates.

Ehrlich, J. S., Obler, L. K., & Clark, L. (1997). Ideational and semantic contributions to narrative production in adults with dementia of the Alzheimer's type. *Journal of Communication Disorders, 30*, 79-99.

Folstein, M. F., Folstein, S. E., & McHugh, P. R. (1975). Mini-Mental State: A practical method for grading the cognitive state of patients for the clinician. *Journal of Psychiatric Research, 12*, 189-198.

Garcia, L. J., & Joanette, Y. (1994). Conversational topic-shifting analysis in dementia. In R. L. Bloom, L. K. Obler, S. DeSanti, and J. S. Ehrlich (Eds.), *Discourse analysis and applications: Studies in adult clinical populations* (pp. 161-183). Hillsdale: Lawrence Erlbaum Associates.

Goffman, E. (1974). *Frame analysis: An essay on the organization of experience.* Cambridge, MA: Harvard University Press.

Goodglass, H., & Kaplan, E. (1983). *The assessment of aphasia and related disorders* (2nd ed.). Philadelphia: Lea & Febiger.

Grice, P. (1975). Logic and conversation. In P. Cole and J. Morgan (Eds.), *Syntax and semantics 9: Pragmatics* (pp. 41-58). New York: Academic Press.

Hamilton, H. (1994). *Conversations with an Alzheimer's patient.* Cambridge: Cambridge University Press.

Henderson, V. W. (1996). The investigation of lexical semantic representation in Alzheimer's disease. *Brain and Language, 54*, 177-178.

Hier, D., Hagenlocker, K., & Shindler, A. (1985). Language disintegration in dementia on a picture description task. *Brain and Language, 25*, 117-133.

Hodges, J. R., Patterson, K., Graham, N., & Dawson, K. (1996). Naming and knowing in dementia of Alzheimer's type. *Brain and Language, 54*, 302-325.

Holland, A. (1982). Observing functional communication of aphasic adults. *Journal of Speech and Hearing Disorders, 47*, 50-56.

Hughes, C. P., Berg, L., & Danziger, W. L. (1982). A new clinical scale for the staging of dementia. *British Journal of Psychiatry, 140*, 566-572.

Kempler, D., & Zelinski, E. M. (1994). Language in dementia and normal aging. In F. A. Huppert, C. Brayne, and D. W. O'Connor (Eds.), *Dementia and normal aging* (pp. 331-365). New York: Cambridge University Press.

Kempler, D., Van Lancker, D., & Read, S. (1988). Proverb and idiom comprehension in Alzheimer disease. *Alzheimer Disease and Associated Disorders, 2*, 38-49.

Kontioloa, P., Laaksonen, R., Sulkava, R., & Erkinjuntti, T. (1990). Pattern of language impairment is different in Alzheimer's disease and multi-infarct dementia. *Brain and Language, 38*, 364-383.

Levinson, S. C. (1983). *Pragmatics.* New York: Cambridge University Press.

Martin, A., & Fedia, P. (1983). Word production and comprehension in Alzheimer's disease: The breakdown of semantic knowledge. *Brain and Language, 19*, 124-141.

Matthews, P. J., Obler, L. K., & Albert, M. L. (1994). Wernicke and Alzheimer on the language disturbances of dementia and aphasia. *Brain and Language, 46*, 439-462.

McKhann, G., Drachman, D., Folstien, M., Katzman, R., Price, D., & Stadlan, E. M. (1984). Clinical diagnosis of Alzheimer's disease. *Neurology, 34*, 939-944.

Nicholas, M., Obler, L. K., Albert, M. L., & Helm-Estabrooks, N. (1985). Empty speech in Alzheimer's disease and fluent aphasia. *Journal of Speech and Hearing Research, 28*, 405-410.

Obler, L. K. (1981). Review of *Le language des déments. Brain and Language, 12*, 375-386.

Obler, L. K. (1983). Language and brain dysfunction in dementia. In S. Segalowitz (Ed.), *Language functions and brain organization* (pp. 267-282). New York: Academic Press.

Obler, L. K., & Albert, M. L. (1984). Language in aging. In M. L. Albert (Ed.), *Clinical neurology of aging* (pp. 245-253). New York: Oxford University Press.

Ska, B., & Guenard, D. (1993). Narrative schema in dementia of the Alzheimer's type. In H. H. Brownell and Y. Joanette (Eds.), *Narrative discourse in neurologically impaired and normal aging adults* (pp. 299-316). San Diego: Singular Publishing Group, Inc.

Smith, B. R., & Leinonen, E. (1992). *Clinical Pragmatics.* New York: Chapman & Hall.

Ulatowska, H. K., Allard, L., Donnell, A., Bristow, J., Haynes, S. M., Flower, A., & North, A. J. (1988). Discourse performance in subjects with dementia of the Alzheimer's disease type. In H. Whitaker (Ed.), *Neuropsychological studies in nonfocal brain damage* (Vol. 2, pp. 108-131). New York: Springer-Verlag.

Ulatowska, H. K., & Chapman, S. B. (1991). Neurolinguistics and aging. In D. N. Ripich (Ed.), *Geriatric Communication Disorders* (pp. 21-38). Austin: Pro-Ed.

Ulatowska, H. K., & Chapman, S. B. (1994). Discourse macrostructure in aphasia. In R. L. Bloom, L. K. Obler, S. DeSanti, and J. Ehrlich (Eds.), *Discourse in adult clinical populations* (pp. 19-46). New York: Lawrence Erlbaum Associates.

Ulatowska, H. K., & Chapman, S. B. (1995). Discourse studies. In R. Lubinski (Ed.), *Dementia and communication* (pp. 115-132) San Diego: Singular Publishing Group.

Ulatowska, H. K., Chapman, S. B., & Johnson, J. (1995). Processing of proverbs in aphasics and old-elderly. *Clinical Aphasiology, 23*, 179-193.

Ulatowska, H. K., Chapman, S. B., & Johnson, J. (in press). In H. H. Hamilton (Ed.), Inferencing in processing of text in elderly populations. *Anthology on old age and language.* Dallas: Garland Publishing.

Ulatowska, H. K., Chapman, S. B., Highley, A. P., & Prince, J. A., (in press). Discourse in healthy old-elderly adults: A longitudinal study. *Aphasiology.*

Ulatowska, H. K., Chapman, S. B., Johnson, J., & Branch, M. S., (submitted). Macrostructure and inferential processing in discourse of aphasic patients.

Van Lancker, D. (1990). The neurology of proverbs. *Behavioral Neurology, 3,* 169-187.

Wapner, W. A., Hamby, S., & Gardner, H. (1981). The role of the right hemisphere in the apprehension of complex linguistic materials. *Brain and Language, 14,* 15-33.

Appendix

Fables

Farmer and Sons

A farmer worked in a vineyard and became rich. He wanted his sons to be just like him. On his deathbed the farmer told his sons that there was a great treasure buried in the vineyard. After the farmer died, the sons went to the vineyard and dug up the soil. They could not find a buried treasure. At harvest time, the vineyard produced the best grapes ever. Now the sons understood the meaning of the treasure.

Raven and Pigeons

A hungry raven saw that pigeons in the pigeon coop had a lot of food. He painted his feathers white to look like them. But when he started to crow, they realized that he was a raven and chased him away. So he returned to his own kind. But the other ravens did not recognize him because he had his feathers painted white, so they also chased him away.

Two Roosters

Two roosters were fighting over the chicken yard. The one who was defeated hid in the corner. The other rooster flew to the top of the roost and began crowing and flapping his wings to boast of his victory. Suddenly, an eagle swooped down, grabbed the rooster and carried him away. This was good luck for the defeated rooster. Now he could rule over the roost and have all the hens that he desired.

Single frame picture

The picture is a Norman Rockwell print depicting a rural scene around the 1930's. The characters in the picture (going from left to right) are an older man dressed in workclothes, looking down at the ground; a young man dressed in a suit, looking off to the left side of the picture; and a sheep dog with his head on the young man's lap, looking at the ground. The older man has two hats in his hand and is smoking a rolled cigarette. The young man has a suitcase beside him with a State U. sticker on it, books on the suitcase, and a small wrapped package in his lap. They are sitting next to each other on the running board of an old truck. The truck says "Bar W Ranch" on the door and has built up wooden sides on the truck bed. In the foreground of the picture is a bare dirt ground, and the side of a railroad track. In the left hand corner of the picture there is a red lantern and a red flag sitting on a box.

Proverbs

Spontaneous

Familiar
1. Too many cooks spoil the broth.
2. Don't count your chickens before they are hatched.

Unfamiliar
3. One swallow doesn't make a summer.
4. The used key is always bright.

Multiple choice

1. Too many cooks spoil the broth.
   A) One person can make soup better than ten.
   B) Too many cooks, the broth is the first course.
   C) A penny saved is a penny earned.

D) A task is at risk when more people are involved than are needed.

2.   Don't count your chickens before they are hatched.
        A) One shouldn't always assume that things will turn out the way one expects.
        B) The good is the enemy of the best.
        C) Chickens don't continue to sit on eggs after they have hatched.
        D) There may be fewer chicks than there were eggs.

3.   One swallow doesn't make a summer.
        A) It's not wise to draw a conclusion based on a single example.
        B) One shouldn't think that winter is over just because the first bird has arrived.
        C) People who live in glass houses shouldn't throw stones.
        D) One swallow of a cold drink will only make you want more.

4.   The used key is always bright.
        A) The key to success is to always use bright ideas.
        B) Do unto others as you would have others do unto you.
        C) Regular practice results in the best performance.
        D) Tools stay shiny when one frequently works with them.

 **Pergamon**

J. Neurolinguistics, Vol. 11, Nos 1–2, p. 79–87, 1998
© 1998 Published by Elsevier Science Ltd. All rights reserved
Printed in Great Britain
0911-6044/98 $19.00 + 0.00

PII: S0911-6044(98)00006-2

Coherence and informativeness of discourse in two dementia types

Matti Laine*, Minna Laakso†, Elina Vuorinen*, and Juha Rinne*

*University of Turku, Finland; †University of Helsinki, Finland.

Abstract—We examined the coherence and informativeness of discourse in vascular dementia (VaD) and probable Alzheimer's disease (AD) by analyzing work history interviews in 8 patients with VaD, 11 patients with AD, and 19 age- and education-matched normal controls. The patient groups had comparable levels of cognitive impairment (mild-to-moderate dementia). The results show that both VaD and AD patients exhibited impaired global thematic coherence and reduced informativeness in their discourse. By contrast, the degree of local coherence between two successive utterances did not reliably differentiate the patient groups from the controls. A case-by-case analysis indicated that severely impaired global coherence was found among the AD patients only. Correlational analyses showed that global coherence was the only discourse variable related to conceptual/semantic impairment.

Introduction

This study focuses on certain aspects of discourse in the two most common dementia types, Alzheimer's dementia (AD) and vascular dementia (VaD). Comparison of these dementia types is of interest as the few relevant studies have dealt with AD patients and discourse analyses of VaD patients are lacking, even though VaD accounts for about one third of dementia cases.

Studies of language-related functions in AD have revealed widespread changes, particularly in semantically mediated linguistic tasks. Such tasks include language comprehension, picture naming, word-picture matching, and word classification. As regards spontaneous speech, fluent and grammatical but empty output has been considered as most typical in early AD (Obler & Albert, 1981). There are fewer studies on language functions in VaD but some authors at least have suggested a pattern somewhat different from AD: less fluent, dysarthric output is considered more common. Anomia, a cardinal feature of acquired language disorders, has been noted in VaD as well (Villardita, 1993).

Extensive cognitive deterioration in dementia, affecting both linguistic and non-linguistic functions, should affect the pragmatic aspects of communication as well. In particular, one would expect that semantic breakdown and attentional/memory disorders would impair the maintenance of a coherent discourse. Indeed, it has been reported that in conversations and narratives, AD patients show less cohesion and coherence (Appell, Kertesz & Fisman 1982; Ripich & Terrell, 1988; Glosser & Deser, 1990) and appear less aware of their errors (McNamara et al., 1992). In addition, noninformative, empty speech output has been connected to AD (Nicholas et al., 1985). We do not know of any similar analyses with VaD patients.

We analyzed two discourse phenomena, coherence and informativeness, in patients with VaD and AD. These were chosen because they are crucial for the integrity of topical content in discourse and previous studies have suggested that the discourse of AD patients is characterized by decreased coherence and 'emptiness'. To the best of our knowledge, this is the first study that extends systematic discourse analysis to VaD as well. Moreover, in earlier studies the cognitive correlates of impaired discourse in dementia received only scant attention. Our previous lexical-semantic analyses of the present patient samples (Laine, Vuorinen & Rinne 1997) allow us to determine whether these changes in discourse are related to concomitant deficits in semantic function and lexical retrieval.

Materials and Methods

Subjects

All subjects were tested at the Department of Neurology, University of Turku. Informed consent was obtained prior to testing. Details of the diagnostic procedures are presented in Laine et al. (1997). Background information is summarized in Table 1. The VaD patients were demented and showed clinical and neuroradiological evidence of cerebrovascular disease which was temporally related to the onset of cognitive impairment showing stepwise progression. Thus, they fulfilled the recently proposed criteria for probable vascular dementia (Román et al., 1993). The patients with AD met the NINCDS-ADRDA clinical criteria for probable AD (McKhann et al., 1984). Severely demented patients were excluded from both groups. In the present study, the number of participants was slightly smaller than in Laine et al. (1997): taped conversation was missing for three subjects, one VaD patient was too dysarthric for discourse analysis, and one AD patient produced too little spontaneous speech.

As Table 1 indicates, all groups were matched by age and educational level. In addition, the two patient groups were equated on average Mini-Mental State Examination (MMSE) (Folstein, Folstein & McHugh, 1975) performance and average Clinical Dementia Rating (Hughes et al., 1982). Approximate disease duration was also comparable in the dementia groups.

Analysis of conversational speech

All subjects were interviewed concerning their work history and this interview was recorded. Tapes were transcribed and the protocols were analyzed. To ensure comparable speech samples for the subjects, their first twenty utterances in the interview were identified. 'Utterance' was defined by syntactic and semantic criteria, as well as by connectives and/or pauses (for comparable criteria, see e.g., Glosser, Wiener & Kaplan (1988)). They ranged from complete clause structures to syntactically incomplete utterances. However, as one of our objectives was to study the topical coherence of discourse, minimal verbal responses and interactional phrases not maintaining the topic were excluded. In addition, occasional unclear utterances (due to poor recording and/or dysarthria) were left out of the analysis.

	AD (n=11)	VaD (n=8)	Controls (n=19)	Group difference tested by one-way ANOVA
Age	67.4 (8.2) 52-80	67.1 (6.5) 57-74	67.4 (6.3) 58-79	F=0.0; p=n.s.
Education[1]	1.0 (0.0) -	1.25 (0.7) 1-3	1.21 (0.4) 1-2	F=1.04; p=n.s.
Disease duration	30.6 (14.8) 12-54	24.0 (12.4) 12-48	-	F=1.01; p=n.s.
Mini-Mental score	18.7 (3.4) 13-25	20.3 (4.7) 12-25	28.3 (1.2) 26-30	F=45.79; p<.0001[3]
Clinical Dementia Rating[2]	2.3 (0.5) 1-3	2.0 (0.5) 2-3	-	F=1.40; p=n.s.

[1] Education was defined by a three-level scale (1 = primary school level, 2 = secondary school level, 3 = high school/college level).

[2] A modified Clinical Dementia Rating (Hughes et al. 1982) was employed: 0 = no dementia, 1 = questionable dementia, 2 = mild dementia, 3 = moderate dementia.

[3] Group comparisons (Tukey's method) showed that both patient groups differed significant-ly (p < .01) from the controls but not from each other.

Table 1. Background information on the subject groups: means, standard deviations and ranges for age, education, disease duration in months, Mini-Mental State score, and Clinical Dementia Rating

Two independent judges scored each utterance for four different features. The first two features relate to coherence. Coherence refers to the appropriate maintenance of topic in discourse (Halliday & Hasan, 1976). Two aspects of coherence were differentiated: (a) local coherence is an estimate of how closely an utterance is thematically related to the immediately preceding utterance (or, in the case of the first utterance, to the question posed by the examiner). The scale ranged from 1 to 5 where 1 = very low and 5 = very high local coherence (cf. Glosser & Deser, 1990). (b) The global coherence measure evaluates the relatedness of the utterance to the general topic of the examiner's question. This scale also ranged from 1 to 5 where 1 = very low and 5 = very high global coherence.

The latter two discourse features relate to the informative content conveyed by an utterance: (c) use of non-referential lexical items. These were instances where a referring lexical item was used but its referent was not specified or evident in the immediate context. Each utterance was scored for the existence of non-referential items (0 = no, 1 = yes). (d) The informativeness scale evaluates the existence and extent of new information in an utterance. A three-point scale was employed where 0 = no new information, 1 = partially new information, and 2 = new information conveyed.

Interrater reliability for the four variables was satisfactory. Product-moment correlations between the two independent scorings were as follows: local coherence .65, global coherence .73, non-referential lexical items .81, and informativeness .65. The final data matrix was prepared by averaging the two ratings.

Utterance	Local coherence	Global coherence	Non-ref. items	Informativeness
Patient with VaD				
Examiner: tell me a little about the work you had				
Patient: now I haven't had for a long time well. I haven't had any permanent ones anymore anymore	3	3	0	2
I sell sold well my son [has] them or I gave them to him	3	2	1	2
and he has just been using them for a longer time	4	2	1	2
Examiner: uhm				
Patient: so [they] are own indeed	2	1	0	1
Patient with AD				
Examiner: so tell me about your work history				
Patient: well we have been fixing that new building	1	1	0	2
in March the top floor should be ready	4	1	0	2
Examiner: oh				
Patient: yeah freezers were already taken to there downstairs	4	1	0	2
then when /ver/ son stays there as the head of that old house	3	1	0	2
Control subject				
Examiner: [...] and could you tell a bit about that profession, your profession				
Subject: well...34 years I was... working in a prison	5	5	0	2
first 17 years at the N.N. local prison	5	5	0	2
and after that another 17 years at the N.N. regional prison	5	5	0	2
the work was interesting	4	4	0	2

Table 2. Discourse extracts from AD, VaD, and control subjects

Examples of the scoring of coherence and informativeness in the subject groups

In order to give a better idea of the scoring method we employed, Table 2 includes samples of discourse from an AD patient, a VaD patient, and a normal control. These examples also highlight some of the difficulties our dementia patients had in the interviews.

Measures related to lexical retrieval

Our previous analysis of the present subjects provided us with several variables that reflect lexical retrieval. These will be only summarized here as they are described in detail in Laine et al. (1997). By correlating these variables with the coherence and informativeness measures, we wished to study to what extent disorders of lexical function are related to discourse impairments (coherence and informativeness) in our dementia patients.

(a) Overall naming performance was assessed by the Finnish version of the Boston Naming Test (BNT: Laine et al., 1994). The naming score was the number of items that were spontaneously named correctly within 25 seconds.

(b) Semantic abilities related to lexical retrieval were measured by three tasks. Two of them were presented during naming of 30 BNT items (see Laine et al. (1997) for details). The first was a multiple-choice task where the subject was to choose the specific semantic feature related to the BNT item to be named from among four alternatives. For example, the following alternatives were offered for the target 'helicopter': one moves with it in the water, one flies with it, it moves on tracks, it moves in the snow. In the second semantic multiple-choice task, the subject had to find the correct superordinate for the BNT item to be named from among eight written category names (furniture, building, plant, animal, vehicle, tool, musical instrument, and game accessory). The third semantic task was administered separately: the odd-one-out task required the subject to decide which picture in a group of five does not go with the others. Because all of the pictures are semantically related, successful performance required fine-grained semantic analysis (e.g., desk, chair, sofa, rocking chair, stool).

According to the statistical analysis reported by Laine et al. (1997), the two dementia groups were inferior to the control group on all four of these variables.

Results

Performance of the subject groups on the discourse-related measures

Group comparisons were performed by one-way ANOVAs.

(a) Local coherence (Figure 1): For this variable, only a marginally significant effect was present ($F(2,35) = 4.68$; Levene's test p=.03; Brown-Forsythe equality of means test p = .07). Figure 1 shows that both patient groups tended to have lower local coherence values than the controls did.

(b) Global coherence (Figure 2). The one-way ANOVA revealed a significant main effect ($F(2,35) = 7.55$; $p = .002$). Group comparisons (Tukey's test) showed that both the VaD and the AD patients were significantly inferior to the controls (p values <. 05. and < .01, respectively).

(c) Use of non-referential nouns/pronouns (Figure 3) also showed a significant main effect in the one-way ANOVA ($F(2,35) = 7.29$; Levene's test $p < .0001$; Brown-Forsythe equality of means test $p = .03$). In group comparisons, both the VaD and AD groups produced significantly more non-referential items than the controls.

(d) Informativeness (Figure 4). A significant main effect was obtained on this variable as well ($F(2,35) = 13.26$; Levene's test $p < .0001$; Brown-Forsythe equality of means test $p = .003$). Group comparisons confirmed that both patient groups produced significantly less informative output than the controls did.

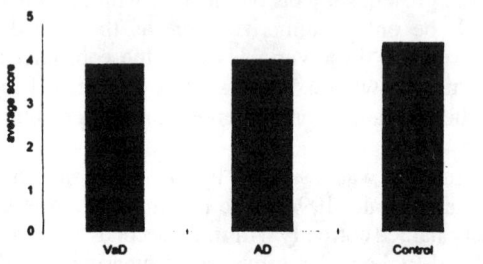

Figure 1.   Local coherence ratings for the VaD, AD, and control subjects.

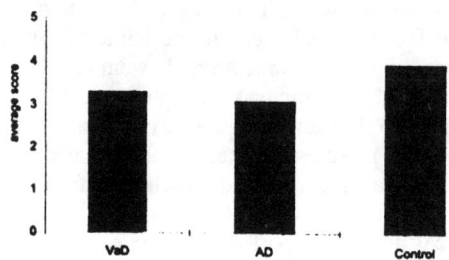

Figure 2.   Global coherence ratings for the VaD, AD, and control subjects.

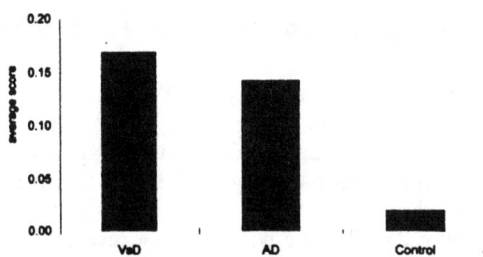

Figure 3.   Proportion of utterances including non-referential lexical items for the VaD, AD, and control subjects.

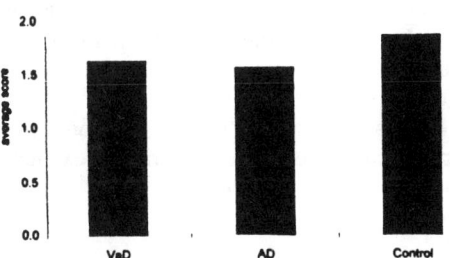

Figure 4.   Informativeness ratings for the VaD, AD, and control subjects.

In addition to the group analysis, we searched individual protocols for sequences of severely disrupted global coherence. This analysis suggests that severely impaired global coherence may be more common in AD than in VaD: we found three AD patients where both raters had scored a sequence of four or more utterances with the lowest possible global coherence score (= 1). In the VaD group, a maximum of two sequential utterances received this score.

Relationships between lexical-semantic and discourse variables

We explored the interrelationships between lexical and discourse variables through correlation coefficients. In this analysis, the two patients groups were combined. Pearson's product-moment correlations were calculated between the four discourse variables (local coherence, global coherence, use of non-referential lexical items, and informativeness) and the four lexical-semantic measures (BNT summative score, choice of specific semantic feature, choice of superordinate, odd-one-out task performance). The correlation coefficients are given in Table 3.

	Local coherence	Global coherence	Use of non-referential lexical items	Informativeness
BNT # correct	.49*	.62**	-.43	.49*
Choice of specific semantic feature	.25	.38	-.29	.42
Choice of superordinate	.24	.48*	-.31	.26
Odd-one-out task	.31	.48*	-.31	.22

*p < .05
**p < .01

Table 3.   Correlation coefficients between lexical-semantic and discourse-related variables in the combined group of VaD and AD patients (n = 19)

Table 3 indicates that only global coherence correlates significantly with semantic abilities (choice of superordinate, odd-one-out task performance) in our dementia patients. Both local and global coherence correlate with overall naming but those relationships could be caused in part by the common variance BNT performance has with general cognitive impairment (correlation between BNT and MMSE score = .52; $p < .05$). Informativeness correlates significantly with BNT score whereas the use of non-referential lexical items does not show significant correlations with any of the variables examined here.

Discussion

This study had two main purposes. The first was to confirm the existence of certain pragmatic difficulties in dementia patients. We also wanted to see whether the pragmatic difficulties reported in AD patients apply to VaD patients as well. On the basis of our results, both hypotheses were confirmed. Our study replicated previous findings with regard to impaired global but not local coherence (Glosser & Deser, 1990), abnormally frequent use of non-referential lexical items (Ripich & Terrell, 1988), and decreased informativeness (Nicholas et al., 1985) in AD patients. Moreover, with our limited patient samples, we were able to show that these findings apply to VaD patients as well. On the basis of the discourse measures used here, the two patient groups behaved quite similarly fashion. The only possible difference, worth exploring in further studies, relates to global coherence where case-by-case analyses suggested that AD patients may be somewhat more impaired than VaD patients.

The correlation analyses revealed an interesting clue to the underlying mechanisms of impaired global coherence. Global coherence was the only discourse variable that correlated significantly to performance level at semantic processing tasks. Thus global coherence in discourse may have some of its roots in 'deep', conceptual/semantic structures. As was noted in the introduction, semantic impairment is very often present even at the early stages of dementia. Local coherence, on the other hand, reflects the relationship of an utterance to the preceding one, and it can remain quite good even when the overall topic of conversation becomes lost.

In our study, informativeness of discourse correlated significantly with overall BNT performance. It is quite understandable that the word-retrieval difficulties many of our dementia patients suffered from hampered their ability to convey new information during conversation. On the other hand, the use of non-referential lexical items did not show statistically significant correlations to semantic-lexical variables. One might speculate that the use of such non-informative items is a strategic choice which only indirectly reflects word-finding difficulty. Moreover, the strict and dichotomous scoring (existent vs. non-existent) for non-referential lexical items may have affected the magnitude of the correlation coefficients.

In conclusion, the present study shows that both VaD and AD patients exhibit impaired global coherence and informativeness in their discourse. By contrast, the local coherence measure did not reliably differentiate the patient groups from the matched controls. Global coherence appeared to be most impaired in the AD group, but lowered scores for this measure were observed in the VaD group too. Correlational analyses showed that among the discourse variables studied, global thematic coherence was the only one that was significantly related to the degree of conceptual/semantic impairment in dementia.

References

Appell, J., Kertesz, A., & Fisman, M. (1982). A study of language functioning in Alzheimer patients. *Brain and Language, 17,* 73-91.

Folstein, M. F., Folstein, S. E., & McHugh, P. R. (1975). "Mini-Mental State": A practical method for grading the cognitive state of patients for the clinician. *Journal of Psychiatric Research, 12,* 189-198.

Glosser, G., & Deser, T. (1990). Patterns of discourse production among neurological patients with fluent language disorders. *Brain and Language, 40,* 67-88.

Glosser, G., Wiener, M., & Kaplan, E. (1988). Variations in aphasic language behaviors. *Journal of Speech and Hearing Disorders, 53,* 115-124.

Halliday, M. A. K., & Hasan, R. (1976). Cohesion in English. London: Longmans.

Hughes, C. P., Berg, L., Danziger, W. L., Coben, L. A., & Martin, R. L. (1982). A new clinical scale for staging of dementia. *British Journal of Psychiatry, 140,* 566-572.

Laine, M., Goodglass, H., Niemi, J., Koivuselkä-Sallinen, P., Tuomainen, J., & Marttila, R. (1994). Adaptation of the Boston Diagnostic Aphasia Examination and the Boston Naming Test into Finnish. *Scandinavian Journal of Logopedics and Phoniatrics, 18,* 83-92.

Laine, M., Vuorinen, E., & Rinne, J. (1997). Picture naming deficits in vascular dementia and Alzheimer's disease. *Journal of Clinical and Experimental Neuropsychology, 19,* 126-140.

McKhann, G., Drachman, D., Folstein, M., Katzman, R., Price, D., & Stadlan, E. M. (1984). Clinical diagnosis of Alzheimer's disease. Report of the NINCDS-ADRDA Work Group under the auspices of Department of Health and Human Services Task Force on Alzheimer's disease. *Neurology, 34,* 939-944.

McNamara, P., Obler, L. K., Au, R., Durso, R., & Albert, M. L. (1992). Speech monitoring skills in Alzheimer's disease, Parkinson's disease, and normal aging. *Brain and Language, 42,* 38-51.

Nicholas M., Obler, L., Albert, M., & Helm-Estabrooks, N. (1985). Empty speech in Alzheimer's disease and fluent aphasia. *Journal of Speech and Hearing Research, 28,* 405-410.

Obler, L. K., & Albert, M. L. (1981). Language in the elderly aphasic and dementing patient. In M.T. Sarno (Ed.), *Acquired Aphasia* (pp. 385-398). New York: Academic Press.

Ripich, D. N., & Terrell, B. Y. (1988). Patterns of discourse cohesion and coherence in Alzheimer's disease. *Journal of Speech and Hearing Disorders, 53,* 8-15.

Román, G. C., Tatemichi, T. K., Erkinjuntti, T., Cummings, J. C., Masdeu, J. C., Garcia, J. H., Amaducci, L., Orgogozo, J.-M., Brun, A., Hofman, A., Moody, D. M., O'Brien, M. D., Yamaguchi, T., Grafman, J., Drayer, B. P., Bennett, D. A., Fisher, M., Ogata, J., Kokmen, E., Bermejo, F., Wolf, P. A., Gorelick, P. B., Bick, K. L., Pajeau, A. K., Bell, M. A., DeCarli, C., Culebras, A., Korczyn, A. D., Bogousslavsky, J., Hartmann, A., & Scheinberg, P. (1993). Vascular dementia: Diagnostic criteria for research studies. Report of the NINDS-AIREN International Workshop. *Neurology, 43,* 250-260.

Villardita, C. (1993). Alzheimer's disease compared with cerebrovascular dementia. Neuropsychological similarities and differences. *Acta Neurologica Scandinavica, 87,* 299-308.

Pergamon

J. Neurolinguistics, Vol. 11, Nos 1–2, p. 89–102, 1998
© 1998 Published by Elsevier Science Ltd. All rights reserved
Printed in Great Britain
0911-6044/98 $19.00 + 0.00

PII: S0911-6044(98)00007-4

Affective prosodic disturbance subsequent to right hemisphere stroke: A clinical application

Robert T. Wertz[*†], *Constance R. Henschel*[†], *Linda L. Auther*[*], *John R. Ashford*[*], *and Howard S. Kirshner*[†]

[*]Veterans Administration Medical Center, Nashville, Tennessee
[†]Vanderbilt University School of Medicine, Nashville, Tennessee

Abstract—We examined Ross' (1981) hypothesis regarding the disruption of affective prosody subsequent to right hemisphere brain damage (RHD). Twenty patients who had suffered a right hemisphere stroke were compared with 18 normal, non-brain-damaged subjects for affective prosody in spontaneous speech, gesturing accompanying spontaneous speech, repetition of affective prosody, comprehension of affective prosody, and comprehension of affective gestures. In addition, we attempted to classify our RHD subjects with Ross' prosodic taxonomy, determine the relationship between classification and site of lesion, and explore the contribution of coexisting dysarthria to prosodic disturbance. All RHD subjects displayed affective prosodic disturbance in spontaneous speech. Only one normal subject was judged mildly dysprosodic. The RHD group had significantly more difficulty in repeating affective prosody and comprehending affective gestures. There were no significant group differences in gestures accompanying spontaneous speech or comprehending affective prosody. Eighty percent of the RHD subjects were classified with the Ross taxonomy, however there was no systematic relationship between classification and site of lesion. And, while dysprosody without a coexisting dysarthria was present in three RHD subjects, 17 displayed a coexisting dysarthria. We conclude that affective dysprosody is common subsequent to RHD, however the relationship between classification of dysprosody and site of lesion or the contribution of coexisting dysarthria to dysprosody is not clear.

Introduction

The perception and production of linguistic and affective prosody and their disruption subsequent to brain damage have received considerable debate in the literature. Currently, it is not clear whether brain damage does or does not disrupt prosody (Baum & Pell, 1997); whether linguistic prosody is disrupted by a left hemisphere lesion and affective prosody is disrupted by a right hemisphere lesion (Behrens, 1988; Bradvik et al., 1991; Bryan, 1989; Cancelliere & Kertesz, 1990; Emmory, 1987; Gorelick & Ross, 1987; Shapiro & Danley, 1985; Weintraub, Mesulam, & Kramer, 1981); whether right hemisphere lesions disrupt both the perception and production of affective prosody, only perception of affective prosody, or only the production of affective prosody (Gorelick & Ross, 1987; Heilman et al., 1984; Lalande et al., 1992); or whether prosodic disturbance results from dysarthria (Duffy & Folger, 1996; Kent & Rosenbek, 1982; Ropper, 1987) rather than another process independent from involvement of the muscles utilized in speech.

Tompkins (1995) has summarized the data regarding prosodic disturbance subsequent to right hemisphere brain damage. She observes: "... some right hemisphere damaged adults have speech that is less varied intonationally than that of non-brain-damaged adults ...

some right hemisphere damaged adults sound hypermelodic ..." and "whether linguistic or emotional prosody is the subject of concern, the evidence on prosodic production problems after right hemisphere damage is mixed" (pp. 21-22).

Prosody defined

Several definitions of prosody exist, but all are similar. Monrad-Krohn (1947) defined prosody as the melodic line of speech produced by variations of pitch, rhythm, and stress of pronunciation. For Hargrove and McGarr (1994), prosody represents the linguistic use of vocal aspects of speech without consideration of the segmental aspects (speech sounds or phonemes). They observe that the rules of prosody are systematic, bound by conventions, and convey important information to listeners. Tompkins (1995) says prosody includes elements of speech melody, rate, stress, juncture, and duration; has a pervasive influence on communication; and is critical in speech production and perception. Prosody can convey linguistic meaning—whether the utterance is a statement or a question or whether a person lives in the "white house" or the "White House"—or emotional or affective meaning—whether the speaker is happy, sad, angry, surprised, etc.

Affective prosody

Ross (1981) has popularized the notion that the right hemisphere is responsible for producing and perceiving affective prosody. Specifically, he suggests that ". . . affective components of language, encompassing prosody and emotional gesturing, are a dominant function of the right hemisphere, and their functional-anatomical organization in the right hemisphere mirrors that of propositional language in the left hemisphere" (p. 561). Data to support this position are provided by ten patients (Ross, 1981); an additional 14 patients (Gorelick & Ross, 1987); and a case report (Wolfe & Ross, 1987).

Explorations of Ross' hypothesis indicate mixed results. There is considerable evidence that right hemisphere brain damage disrupts perception of affect (Blonder, Bowers, & Heilman, 1991; Bryan, 1989; Darby, 1993; Heilman et al., 1984; Lalande et al., 1992). However, Van Lancker and Sidtis (1992) and Cancelliere and Kertesz (1990) report that either a left or right hemisphere lesion may result in problems perceiving prosody. And, Tompkins and Flowers (1985) suggest differences in the perception of emotional intonation between left and right brain damaged people may depend on the difficulty of the task. Support for deficits in the production of affective prosody subsequent to right hemisphere brain damage, other than reports by Ross and colleagues, must be qualified. It ranges from Baum and Pell's (1997) observation that the right hemisphere is not specifically engaged in the production of affective prosody to Weintraub, Mesulam, and Kramer's (1981) suggestion that right hemisphere damage may affect both linguistic and affective prosody. In between is Bradvik et al.'s (1991) report that prosodic impairment following a right hemisphere lesion could be linguistic in nature and not secondary to an affective disorder; Cancelliere and Kertesz's (1990) results that indicate both left and right hemisphere lesions disrupt expression of emotional prosody; and Shapiro and Danley's (1985) observation that right anterior and central lesions disrupt emotional and nonemotional prosody, and right posterior lesions result in exaggerated pitch variation and intonational range in producing both emotional and nonemotional prosody.

Nevertheless, Ross and colleagues have provided a testable hypothesis. Specifically, they contend that right hemisphere brain damage results in impaired spontaneous, affective prosody and gesturing; impaired repetition of affective prosody; impaired comprehension of affective prosody; and impaired comprehension of emotional gesturing. Moreover, Ross (1981) has provided a taxonomy for classifying what he calls the "aprosodias" into types similar to those seen in aphasia subsequent to left hemisphere brain damage. These include: motor, sensory, global, conduction, transcortical motor, transcortical sensory, mixed transcortical, and anomic. And, as in the classification of aphasic types, Ross provides an anatomical relationship in the right hemisphere for each aprosodic type.

Dysarthria

Impaired speech motor control associated with dysarthria can impair prosody (Darley, Aronson, & Brown, 1975; Duffy, 1995). And, Tompkins (1995) indicates that dysarthria may contribute to the perception of restricted intonational variability in some right hemisphere damaged adults. Thus, we need to ask whether disrupted linguistic or affective prosody subsequent to right hemisphere brain damage results from some prosodic mechanism that resides in the right hemisphere or dysarthria, for example, associated with a unilateral upper motor neuron lesion.

Kent and Rosenbek (1982), discussing prosodic disturbance and neurologic lesion, observed, "The right cerebral hemisphere . . . appears to have a privileged role in the processing of prosodic and affective information" (p. 285). Tentatively, they called the speech pattern observed in their patients with a right-cerebral lesion a dysarthria. While they recognized that there may be some dispute about the suitability of this terminology, they point out several similarities between the speech patterns of patients with a right-cerebral lesion and patients who display a hypokinetic dysarthria associated with Parkinson's disease.

Similarly, Ropper's (1987) patients with a right hemisphere stroke displayed severe dysarthria characterized by slowness of speech; incomplete pronunciation of each syllable; hypophonic, slightly harsh voice quality; and restricted amplitude that resulted in monotonous speech. Duffy and Folger (1996) report that some of their patients with right upper motor neuron lesions produced prosodic abnormalities that were characterized by slow rate, increased rate in segments, excessive and equal stress, or combinations of these signs. And, Hird and Kirsner's (1993) comparison of patients with right hemisphere lesions, hypokinetic dysarthria, or ataxic dysarthria failed to reveal significant acoustic differences among groups in the production of linguistic or emotional prosody.

Clinical questions

Several generations of speech-language pathologists have been taught that a right hemisphere lesion will impair production and perception of affective prosody, and these deficits can be observed by clinical, bedside appraisal. We examined this premise by evaluating spontaneous speech, spontaneous gesturing, repetition of affective prosody, comprehension of affective prosody, and comprehension of affective gesturing in patients who had suffered a right hemisphere stroke and in normal, non-brain-damaged adults. Specifically, we asked: does perception and production of affective prosody and gesturing differ between right hemisphere damaged patients and non-brain-damaged adults, can prosodic disturbance subsequent to right hemisphere damage be classified according to the Ross (1981) taxonomy, what is the anatomical relationship to prosodic classification in right hemisphere damaged patients, and what is the relationship of dysarthria to prosodic disturbance in right hemisphere damaged patients?

Methods

Subjects

Twenty consecutive patients who had suffered a right hemisphere stroke and who were admitted to an inpatient rehabilitation program were evaluated. All met the following selection criteria: had suffered a first, single, right hemisphere stroke; displayed no significant aphasia; and had no other, coexisting neurological disease. Eighteen normal, non-brain-damaged volunteers from the community were matched with the right hemisphere sample for age and education. All had no history of stroke or other neurological disease.

Table 1 shows demographic information for the two groups. Right hemisphere damaged (RHD) subjects,.11 males and nine females, had a mean age of 64.75 years and a mean education of 13.00 years. Normal subjects, ten males and eight females, had a mean age of 59.94 years and a mean education of 14.72 years. There was no significant difference ($p < .05$) in age or years of education between groups. Fifteen of the RHD subjects had suffered an ischemic stroke, and five had suffered a hemorrhagic stroke. Seventeen of the 20 RHD subjects displayed right hemiplegia. The RHD group was an average of 5.2 weeks postonset, with a range of one to 12 weeks postonset.

Variable	Group					
	RHD (n = 20)			Normal (n = 18)		
	n	$\bar{x}$	Range	n	$\bar{x}$	Range
Age in Years	20	64.75	22-86	18	59.94	22-73
Education in Years	20	3.00	6-20	18	14.72	7-20
Weeks Postonset	20	5.20	1-12			
Gender						
Male	11			10		
Female	9			8		
Stroke						
Ischemic	15					
Hemorrhagic	5					
Right Hemiplegia	17					

Table 1. Demographic data for the right hemisphere-damaged (RHD) and normal groups.

Measures

We followed the clinical protocol provided by Ross (1981), as closely as possible, to evaluate prosody, dysarthria, and gesturing in spontaneous speech; repetition of affective prosody; comprehension of affective prosody; and comprehension of affective gesturing. All tasks were presented by the same speech-language pathologist to all RHD and normal subjects. Scoring was done "on-line" at the time of evaluation. Audio recordings were made of subjects' spontaneous speech samples and repetitions of affective prosody for later intra- and interjudge reliability. All RHD subjects were evaluated on the rehabilitation ward, and all normal subjects were evaluated in a speech clinic or their homes.

Spontaneous prosody, dysarthria, and gesturing

To evaluate prosody, dysarthria, and gesturing in spontaneous speech, each RHD subject was asked, "Tell me how you feel about your illness and being in the hospital?" Each normal subject was asked, "Have you ever been really sick or know someone who was? Tell me about it." A minimum of two minutes of spontaneous speech was obtained from each RHD and normal subject.

Prosody in spontaneous speech was rated on an eight-point, equally-appearing interval scale, ranging from "0," no affective prosodic disturbance, to "7," most severe prosodic disturbance. Prosody judgements were based on each subject's speech melody, rate, stress, juncture, and duration and, specifically, whether the prosody was affectively appropriate for the content of the subject's response. Similarly, dysarthria was rated on an eight-point scale, ranging from "0," no dysarthria, to "7," most severe dysarthria. The presence and severity of dysarthria was based on the subject's use of respiration, phonation, articulation, and resonance to produce intelligible and naturally sounding speech (Wertz & Rosenbek, 1992). Gesturing during spontaneous speech was rated on a three-point scale, ranging from "0," normal; through "1," reduced; to "2," absent.

Repetition of affective prosody

To evaluate repetition of affective prosody, each subject was instructed, "I'll say a sentence. You say the sentence back to me exactly as I said it to you." Six sentences, each conveying a different emotion—happy, sad, anger, surprise, tearful, disinterest—were presented, live voice, with the examiner facing the subject. Sentence stimuli were:

"You bought a new car?"
"We'll be back tomorrow."
"I've got to go to Chicago."
"It's fine with me."
"We're going to the movies."
"It's time to leave now."

Each subject's repetition of each sentence was scored "+," appropriate prosody that matched the examiner's production, or "-," inappropriate, reduced, or absent prosody compared with the examiner's production. Thus, performance could range from zero to six correct.

Comprehension of affective prosody

The same sentences and emotions were used to assess comprehension of affective prosody. Each subject was instructed, "I'll say a sentence, and you tell me how I feel or point to the word that indicates how I feel." A different randomization of words, in large type, indicating the six emotions—happy, sad, anger, surprise, tearful, disinterest—was placed before the subject for each stimulus presentation. Verbal or gestural responses—pointing to a word—were accepted. The examiner stood behind the patient and spoke each sentence conveying a specific emotion. Plus or minus scoring was used, thus performance could range from zero to six correct.

Comprehension of affective gesturing

The same emotions were used to assess comprehension of affective gestures. Each subject was instructed, "I'll show you a gesture, and you tell me how I feel or point to the word that indicates how I feel." As in the comprehension of affective prosody task, a different randomization of words, in large type, indicating the six emotions was placed before the subject for each stimulus presentation. Verbal or gestural responses were accepted. The examiner stood in front of the subject and conveyed each emotion with the appropriate gesture and facial expression. Plus or minus scoring was used, thus performance could range from zero to six correct.

Reliability

As indicated above, subject's spontaneous speech and repetition of sentences were audio tape recorded for reliability analysis. Table 2 shows intra- and interjudge reliability. Intrajudge reliability results from the examiner listening to the audio tape recordings, approximately one week after the examination, and re-rating the subject's performance. Interjudge reliability results from a second speech-language pathologist, "blinded" to the subject's group assignment (RHD or normal), listening to the audio tape recordings and rating the subject's performance.

A 47% sample was used for interjudge reliability. As shown in Table 2, kappa values (Cohen, 1960) indicate significant interjudge reliability ($p < .0005$) for the presence of dysprosody and dysarthria in spontaneous speech, severity of dysarthria in spontaneous speech, and severity of dysprosody on the affective repetition task. Interjudge agreement was significant ($p < .004$) for severity of dysprosody in spontaneous speech. An 89% sample was used for intrajudge reliability. As shown in Table 2, intrajudge reliability was significant ($p < .0005$) for all presence and severity measures in spontaneous speech and for severity of dysprosody on the affective repetition task.

| Measure | Kappa | |
	Interjudge	Intrajudge
Spontaneous Speech		
Presence of Dysarthria	.89	.76
Presence of Dysprosody	1.00	.94
Severity of Dysarthria	.42	.53
Severity of Dysprosody	.37*	.64
Affective Repetition	.73	.49

*Significant at $p < .004$
All other Kappas significant at $p < .0005$

Table 2.   Kappa statistic for interjudge (47% sample) and intrajudge (89% sample) scorer reliability.

Classification and localization

The Ross (1981) taxonomy, shown in Table 3, was used to classify subjects into an aprosodic type based on their performance in the spontaneous prosody and gesturing, affective repetition, comprehension of affective prosody, and comprehension of affective gesturing tasks. Ratings of 1 through 7 on the spontaneous prosody task; 1 or 2 on the spontaneous gesturing task; or one or more errors on the affective repetition, comprehension of affective prosody, and comprehension of affective gesturing tasks were used to signify impairment. Thus, for example, as shown in Table 3, a subject with a rating of 1-7 on the spontaneous prosody task; 1-2 on the spontaneous gesturing task; one or more errors on the affective repetition task, and no errors on the comprehension of affective prosody or comprehension of affective gesturing tasks was classified as demonstrating "motor" aprosodia.

Similarly, we employed Ross' (1981) suggested site of lesion relationships with the aprosodic types to determine brain-behavior relationships in our RHD subjects. For example, a "motor" aprosodia is predicted to result from a lesion in the right frontal or frontal-parietal areas, and a "sensory" aprosodia is predicted to result from a lesion in the right posterosuperior temporal and posteroinferior parietal lobes. Site of lesion data were available from CT or MRI scans for 16 of our RHD subjects.

Type	Signs			
	SPG	R	CP	CG
Motor	P	P	G	G
Sensory	G	P	P	P
Global	P	P	P	P
Conduction	G	P	G	G
Transcortical Motor	P	G	G	G
Transcortical Sensory	G	G	P	P
Mixed Transcortical	P	G	P	P
Anomic	G	G	G	P

Table 3. Taxonomy for classifying the "aprosodias" from spontaneous speech and gesturing (SPG), repetition of affective sentences (R), comprehension of affective prosody (CP), and comprehension of affective gestures (CG). P = poor and G = good. (After Ross, 1981)

Results

Table 4 shows a comparison of the presence of dysprosody, dysarthria, and reduced or absent gesturing in the spontaneous speech task between the RHD and normal subjects. None of the normal subjects displayed dysarthria. Conversely, 17 of the 20 RHD subjects were rated dysarthric. A Chi Square analysis indicated significantly more RHD subjects were dysarthric (p < .0001). Similarly, only one of the normal subjects was rated dysprosodic, however all RHD subjects were rated dysprosodic. A Chi Square analysis indicated significantly more RHD subjects were dysprosodic (p < .0001). Nine of the normal subjects and 13 of the RHD subjects displayed reduced or absent gestures accompanying spontaneous speech. A Chi Square analysis indicated no significant difference (p < .05) between groups in gestural performance accompanying spontaneous speech.

Measure	Group			
	RHD (n = 20)		Normal (n = 18)	
	+	-	+	-
Dysarthria	17*	3	0	18
Dysprosody	20*	0	1	17
Gestures	13	7	9	9

*Significant at p < .0001

Table 4.   Number of subjects in the right hemisphere damaged (RHD) and normal groups with (+) and without (-) dysarthria, dysprosody, and diminished or absent gestures in spontaneous speech.

Rated performance on all tasks for each group is shown in Table 5. Higher ratings on the spontaneous speech task (dysprosody and dysarthria) indicate more impairment. Higher scores on the spontaneous gesturing, affective repetition, comprehension of affective prosody, and comprehension of affective gesturing tasks indicate less impairment.

As implied by the Chi Square analyses, t tests indicated the RHD group displayed significantly more severe dysarthria (p < .001) and significantly more severe dysprosody (p < .001) in spontaneous speech than the normal group. There was no significant difference (p < .05) between groups for gestures accompanying spontaneous speech. The RHD group made significantly more errors than the normal group in repetition of affective prosody (p < .001) and comprehension of affective gestures (p < .001). RHD subjects made more errors in comprehension of affective prosody than normal subjects, however this difference approached but did not reach significance (p < .058).

Measure	Group				Group Difference
	Normal		RHD		
	$\bar{x}$	Range	$\bar{x}$	Range	$\bar{x}$
Spontaneous Speech					
Dysarthria	0.00	0-0	2.15	0-5	2.15*
Dysprosody	0.06	0-1	3.05	1-6	2.99*
Gestures	1.39	0-2	0.95	0-2	0.44
Affective Repetition	5.44	4-6	2.50	0-6	2.94*
Prosodic Comprehension	5.39	5-6	4.63	1-6	0.76
Gestural Comprehension	6.00	6-6	5.11	4-6	0.89*

*Significant at p < .001

Table 5. Comparison of the normal and right hemisphere damaged (RHD) groups for rated dysarthria and dysprosody (0-7) and gestures (0-2) in spontaneous speech and errors (0-6) in affective repetition, comprehension of affective prosody, and comprehension of affective gestures.

Table 6 shows the type of aprosodia for subjects in each group according to the Ross (1981) taxonomy. Two (11%) of the normal subjects were classified as demonstrating a conduction aprosodia, 11 (61%) were unclassified, and five (28%) were normal. Four (20%) of the RHD subjects were classified as demonstrating a motor aprosodia, 11 (55%) were classified as global aprosodia, one (5%) was classified as transcortical motor aprosodia, and four (20%) were unclassifiable.

Type	Group			
	Normal		RHD	
	n	%	n	%
Motor			4	20
Sensory				
Global			11	55
Conduction	2	11		
Transcortical Motor			1	5
Transcortical Sensory				
Mixed Transcortical				
Anomic				
Unclassified	11	61	4	20
Normal	5	28		

Table 6. Classification of the number (N) and percent of normal and right hemisphere damaged (RHD) subjects with Ross' (1981) aprosodic taxonomy.

Site of lesion data were available for 16 of our 20 RHD subjects. Table 7 shows a comparison of site of lesion with classification by the Ross (1981) taxonomy. Of the four RHD subjects classified as motor aprosodia, two had a temporal-parietal lesion, one had a subcortical lesion, and site of lesion for one patient was not known. Of the 11 patients classified as global aprosodia, two had a parietal lesion, two had a parietal-occipital lesion, one had a frontal-temporal-parietal lesion, one had a frontal-temporal-parietal-occipital lesion, two had a subcortical lesion, and the site of lesion was unknown for three subjects. The one RHD subject classified as transcortical motor aprosodia had a subcortical lesion. And, of the four RHD subjects who were unclassifiable, one had a frontal-parietal lesion, two had a frontal-temporal-parietal lesion, and one had a subcortical lesion.

Dysarthria coexisted with dysprosody in 17 of the 20 RHD subjects, and three displayed dysprosody with no coexisting dysarthria. None of the normal subjects was rated dysarthric.

Site	Type			
	M	G	TM	UN
Parietal		2		
Frontal-Parietal				1
Temporal-Parietal	2			
Parietal-Occipital		2		
Frontal-Temporal-Parietal		1		2
Frontal-Temporal-Parietal-Occipital		1		
Subcortical	1	2	1	1
Undetermined	1	3		

M = Motor
G = Global
TM = Transcortical Motor
UN = Unclassified

Table 7. Relationship between aprosodic type and site of lesion for the right hemisphere damaged subjects.

Discussion

Our clinical test of Ross' (1981) hypothesis about affective components of language, encompassing prosody and emotional gesturing, being a dominant function of the right hemisphere and the functional-anatomical organization of affective prosody in the right hemisphere mirroring that of propositional language in the left hemisphere provides mixed support. We asked: does perception and production of affective prosody and gesturing differ between RHD patients and non-brain damaged adults, can prosodic disturbance subsequent to RHD be classified according to the Ross (1981) taxonomy, what is the anatomical relationship to prosodic classification in RHD patients, and what is the relationship of dysarthria to prosodic disturbance in RHD patients?

Reliability

It appears clinicians can rate prosodic performance in spontaneous speech and repetition of affective prosody reliably. Intra- and interjudge agreement was significant. However, there is a need to determine intra- and interjudge reliability in rating gestures accompanying spontaneous speech, comprehension of affective prosody, and comprehension of emotional gestures. Moreover, there is a need to determine the consistency of subject's performance on all tasks by comparing test and retest performance.

Prosodic disturbance subsequent to RHD

Dysprosody appears to be a common occurrence subsequent to RHD. All of our RHD subjects displayed mild to severe prosodic disturbance in spontaneous speech. Only one normal subject was rated dysprosodic, and the rating was mild—"1" on the 0-7 rating scale. The influence of RHD on gesturing during spontaneous speech is debatable. Thirteen (65%) RHD subjects displayed reduced or absent gesturing during spontaneous speech. However, nine (50%) of the normal subjects also displayed reduced or absent gesturing during spontaneous speech, and there was no significant difference in the presence or severity, when reduced or absent, between groups. Thus, our results do not indicate reduction in gestures accompanying spontaneous speech can be attributed solely to RHD.

RHD does influence repetition of affective prosody and comprehension of affective gesturing. Normal subjects repeated affective prosody significantly more correctly than RHD subjects. However, eight (45%) of the normal subjects failed on one or two of the affective repetition stimuli. Conversely, 19 (95%) of the RHD subjects failed on one to six of the affective repetition stimuli, and mean performance, 2.50, was significantly lower than normal mean performance, 5.44. Of course, poor repetition of affective prosody may result from failure to comprehend the affect to be repeated, failure to produce the requested affect, or both. Of the eight normal subjects with impaired affective repetition, five also had impaired comprehension of affective prosody. Of the 19 RHD subjects with impaired affective repetition, 12 had impaired comprehension of affective prosody. None of the normals displayed impaired comprehension of affective gestures. Conversely, 14 (70%) of the RHD subjects displayed impaired comprehension of affective gestures, however severity was mild—one or two errors—in 13 RHD subjects.

Comprehension of affective prosody approached (p < .058), but did not reach, significance between groups. Eleven (61%) of the normal subjects displayed problems in comprehending affective prosody, however none missed more than one of the six stimuli. Conversely, 12 (60%) of the RHD subjects displayed problems in comprehending affective prosody, and performance ranged from zero to five correct for the six stimuli. Ross' (1981) taxonomy predicts that problems comprehending affective prosody is not universal across RHD patients. Those with a motor, conduction, transcortical motor, or anomic aprosodia are expected to have good comprehension of emotional prosody. Thus, group comparison of comprehension performance between RHD subjects and normals may be inappropriate.

Classification

Sixteen (80%) of our RHD subjects could be classified in the Ross (1981) taxonomy. Moreover, two (11%) of the normal subjects could be classified. Both were classified as demonstrating conduction aprosodia. An additional 11 normal subjects who made errors on the prosodic tasks were unclassifiable in the Ross taxonomy, and five were not classified, because they made no errors on any task.

For the RHD group, 11 of the classified subjects demonstrated global aprosodia, four classified as motor aprosodia, and one classified as transcortical aprosodia. All of the four unclassified subjects would have demonstrated motor aprosodia, however three had mildly impaired comprehension of affective gestures, and one had mildly impaired comprehension of affective prosody.

Classification of brain damaged patients is problematic. In aphasia, two approaches have been employed. The Western Aphasia Battery (Kertesz, 1982) uses discrete cut-off scores for fluency, auditory comprehension, repetition, and naming. This approach results in classifying almost all aphasic people. The Boston Diagnostic Aphasia Examination (Goodglass and Kaplan, 1983) uses a range of performance in specific behavioral characteristics—melodic line, phrase length, articulatory agility, grammatical form, paraphasia in running speech, repetition, word finding, and auditory comprehension. This approach results in classifying 40 to 60 percent of aphasic people. Our application of the Ross taxonomy to classify our sample of RHD subjects employed discrete cut-off scores. Any rating other than "0" (normal) on the spontaneous speech and gesturing task and any error on the affective repetition, affective comprehension, and comprehension of affective gestures tasks signified impairment. This approach resulted in our failure to classify 20% of our RHD subjects and failure to differentiate RHD patients from 72% of the normal subjects with the classification system.

Functional-anatomical relationships

One of the purposes in classifying brain damaged patients is to explore brain-behavior relationships. This has been the traditional approach in aphasia, and it appears to be Ross' (1981) approach in classifying the aprosodias subsequent to RHD. Specifically, aprosodic types should relate with lesion localization. Our results imply the relationships between lesion localization and type of aprosodia are not precise. For example, none of the three patients for whom we had localization data and who classified as motor aprosodia had a lesion in the right hemisphere analogous to left hemisphere lesions that cause Broca's aphasia. Of the eight patients on whom we had localization data and who classified as global aprosodia, only two had the large supra- and infra-Sylvian infarction predicted by Ross to result in global aprosodia. Our one patient who classified as transcortical motor aprosodia had a subcortical lesion similar to that of Ross' (1981) transcortical motor aprosodic patient whose lesion involved the anterior limb of the internal capsule, head of the caudate nucleus, and putamen, without evidence of cortical involvement. This localization in the right hemisphere, as Ross observed, is different from the left medial frontal lesions believed to result in transcortical motor aphasia. Thus, lesion localization in our sample of RHD aprosodic patients does not demonstrate that "functional-anatomic organization in the right hemisphere mirrors that of propositional language in the left hemisphere" (Ross, 1981, p. 561).

Dysarthria

Kent and Rosenbek (1982) elected to call their patients with dysprosody subsequent to RHD dysarthric. Similarly, Ropper (1987) and Duffy and Folger (1996) report prosodic disturbance in dysarthric patients who have sustained RHD. Thus, it is not clear whether dysprosody subsequent to RHD results from dysarthria or some prosodic mechanism that resides in the right hemisphere.

All of our 20 RHD patients displayed prosodic disturbance. Seventeen displayed a coexisting dysarthria, and three did not. Prosodic disturbance in the three patients without dysarthria was mild, "1" or "2" on the seven-point rating scale, in spontaneous speech. However repetition of affective prosody, comprehension of affective prosody, and comprehension of affective gestures ranged from mild to severe in these patients. In the 17 patients with coexisting dysarthria, the severity of dysarthria ranged from mild, "1" on the seven-point rating scale, to moderately severe, "5." Dysprosody ratings in these patients' spontaneous speech ranged from mild, "2," to severe, "6." Similarly, their impairment on the repetition of affective prosody, comprehension of affective prosody, and comprehension of affective gestures ranged from mild to severe. Thus, except for dysprosody in spontaneous speech, there was no apparent difference between RHD patients who had coexisting dysarthria and those who were not dysarthric.

The type of dysarthria (Duffy, 1995) present in our RHD patients was either unilateral upper motor neuron or hypokinetic. Both disrupt prosody. However, the presence of dysarthria does not explain deficits in comprehension of affective prosody or comprehension of affective gestures. Thus, our three patients who were dysprosodic and not dysarthric imply prosodic disturbance can result from RHD that is not attributable to dysarthria. In the majority of cases, however, it is not clear whether prosodic disturbance in spontaneous speech and repetition of affective prosody results from dysarthria, a "pure" dysprosody, or both.

Conclusions

Our data provide answers to the questions we posed. First, RHD patients displayed a significant disruption in prosody in spontaneous speech, repetition of affective prosody, and comprehension of affective gestures. RHD patients did not differ significantly from normals in gestures accompanying spontaneous speech or comprehension of affective prosody. Second, dysprosody in RHD patients could be classified by Ross' (1981) taxonomy in 80 percent of the cases. Third, there was no systematic relationship between site of lesion and type of aprosodia in our RHD patients. And, fourth, while three of our RHD patients displayed dysprosody with no coexisting dysarthria, most RHD patients had coexisting dysarthria. Thus, it is not clear what contributes to the perception of dysprosody in spontaneous speech and repetition of affective prosody in RHD patients who have a coexisting dysarthria.

Additional tests of Ross' (1981) hypothesis about dysprosody subsequent to RHD are necessary. These may want to employ additional stimuli or trials to establish the range of normal performance. The number of trials and stimuli we employed resulted in considerable overlap between RHD and normal subjects on the repetition of affective prosody and comprehension of affective prosody tasks. Thus, the sensitivity and specificity of the measures we employed is unacceptable. In addition, Ross' (1981) taxonomy and its relationship to site of lesion requires additional exploration with larger samples. Moreover, there is a need to establish test-retest reliability on all tasks and inter- and intrajudge reliability for observation of gestures accompanying spontaneous speech, comprehension of affective prosody, and comprehension of affective gestures. Finally, there is a need to differentiate dysprosody resulting from dysarthria from dysprosody resulting from RHD. This may require perceptual and acoustic comparison of dysarthric subjects without RHD, subjects with RHD and no dysarthria, and subjects with RHD and coexisting dysarthria.

References

Baum, S. R., & Pell, M. D. (1997). Production of affective and linguistic prosody by brain-damaged patients. *Aphasiology, 11*, 177-198.

Behrens, S. J. (1988). The role of the right hemisphere in the production of linguistic stress. *Brain and Language, 33*, 104-127.

Blonder, L. X., Bowers, D., & Heilman, K. M. (1991). The role of the right hemisphere in emotional communication. *Brain, 114*, 1115-1127.

Bradvik, B. B., Dravins, C., Holtas, S., Rosen, I., Ryding, E., & Ingvar, D. H. (1991). Disturbance of speech prosody following right hemisphere infarcts. *Acta Neurologica Scandanavia, 84*, 114-126.

Bryan, K. L. (1989). Language prosody and the right hemisphere. *Aphasiology, 3*, 285-299.

Cancelliere, A. E., & Kertesz, A. (1990). Lesion localization in acquired deficits of emotional expression and comprehension. *Brain and Cognition, 13*, 133-147.

Cohen, J. (1960). A coefficient of agreement for nominal scales. *Educational and Psychological Measurement, 10*, 37-46.

Darby, D. G. (1993). Sensory aprosodia: A clinical clue to lesions of the inferior division of the right middle cerebral artery? *Neurology, 43*, 567-572.

Darley, F. L., Aronson, A. E., & Brown, J. R. (1975). *Motor speech disorders.* Philadelphia: W. B. Saunders.

Duffy, J. R. (1995). *Motor speech disorders: Substrates, differential diagnosis, and management.* St. Louis: Mosby.

Duffy, J. R., & Folger, W. N. (1996). Dysarthria associated with unilateral central nervous system lesions: A retrospective study. *Journal of Medical Speech-Language Pathology, 4*, 57-70.

Emmory, K. D. (1987). The neurological substrates for prosodic aspects of speech. *Brain and Language, 30*, 305-320.

Goodglass, H., & Kaplan, E. (1983). *Boston diagnostic aphasia examination.* Philadelphia: Lea and Febiger.

Gorelick, P. B., & Ross, E. D. (1987). The aprosodias: Further functional-anatomical evidence for the organization of affective language in the right hemisphere. *Journal of Neurology, Neurosurgery and Psychiatry, 50*, 553-560.

Hargrove, P. M., & McGarr, N. S. (1994). *Prosody management of communication disorders.* San Diego: Singular Publishing Group, Inc.

Heilman, K. M., Bowers, D., Speedie, L., & Coslett, H. B. (1984). Comprehension of affective and nonaffective prosody. *Neurology, 34*, 917-921.

Hird, K., & Kirsner, K. (1993). Dysprosody following neurogenic impairment. *Brain and Language, 45*, 46-60.

Kent, R. D., & Rosenbek, J. C. (1982). Prosodic disturbance and neurologic lesion. *Brain and Language, 15*, 259-291.

Kertesz, A. (1982). *Western aphasia battery.* New York: Grune & Stratton.

Lalande, S., Braun, C. M. J., Charlebois, N., & Whitaker, H. A. (1992). Effects of right and left cerebrovascular lesions on discrimination of prosodic and semantic aspects of affect in sentences. *Brain and Language, 42*, 165-186.

Monrad-Krohn, G. (1947). The prosodic quality of speech and its disorders. *Acta Psychologia Scandanavia, 22*, 225-265.

Ropper, A. H. (1987). Severe dysarthria with right hemisphere stroke. *Neurology, 37*, 1061-1063.

Ross, E. D. (1981). The aprosodias: Functional-anatomic organization of the affective components of language in the right hemisphere. *Archives of Neurology, 38,* 561-569.

Shapiro, B. E., & Danley, M. (1985). The role of the right hemisphere in the control of speech prosody in propositional and affective contexts. *Brain and Language, 25,* 19-36.

Tompkins, C. A. (1995). *Right hemisphere communication disorders: Theory and management.* San Diego: Singular Publishing Group, Inc.

Tompkins, C. A., & Flowers, C. R. (1985). Perception of emotional intonation by brain-damaged adults: The influence of task processing levels. *Journal of Speech and Hearing Research, 28,* 527-538.

Van Lancker, D., & Sidtis, J. J. (1992). The identification of affective-prosodic stimuli by left- and right-hemisphere damaged subjects: All errors are not created equal. *Journal of Speech and Hearing Research, 35,* 963-970.

Weintraub, S., Mesulam, M. M., & Kramer, L. (1981). Disturbances in prosody: A right hemisphere contribution to language. *Archives of Neurology, 38,* 742-744.

Wertz, R. T., & Rosenbek, J. C. (1992). Where the ear fits: Perceptual evaluation of motor speech disorders. *Seminars in Speech and Language, 13,* 39-54.

Wolf, G. I., & Ross, E. D. (1987). Sensory aprosodia with left hemiparesis from subcortical infarction: Right hemisphere analogue of sensory-type aphasia with right hemiplegia? *Archives of Neurology, 44,* 668-671.

Pergamon

J. Neurolinguistics, Vol. 11, Nos 1–2, p. 103–118, 1998
© 1998 Published by Elsevier Science Ltd. All rights reserved
Printed in Great Britain
0911-6044/98 $19.00 + 0.00

PII: S0911-6044(98)00008-6

The role of emotion in the linguistic and pragmatic aspects of aphasic performance

Marjorie Perlman Lorch[*], *Joan C. Borod*[†] *and Elissa Koff*[‡]

[*]Birkbeck College, University of London, London, UK; [†]Queens College and Mount Sinai Medical Center, City University of New York, New York, NY; [‡]Wellesley College, Wellesley, MA, USA.

Abstract—Considerations of aphasics' performance typically focus on aspects of linguistic impairment. Similarly, researchers tend to emphasize right brain-damaged subjects' relatively poor performance in response to emotional content or context. The spared or heightened emotional abilities of aphasic communication often go unnoticed. Research will be reviewed which suggests that aphasics have the ability to successfully utilize emotion in the comprehension and expression of both linguistic and pragmatic content and contexts. Evidence from a wide range of research on lexical processing, prosody, and discourse will be reviewed which indicates that emotion may play a facilitatory role in the comprehension and production of communication in language-impaired people. A large group study involving 15 left brain-damaged, 12 right brain-damaged and 16 normal controls was carried out to investigate posed and spontaneous emotional expression and perception, including the vocal and verbal, as well as facial, channels for spontaneous expression. Results will be considered with respect to the neuropsychological organization of linguistic and emotional cognitive systems.

Introduction

This paper is concerned with the communication of emotion in aphasic left hemisphere brain-damaged sufferers (LBDs) and right hemisphere brain-damaged sufferers (RBDs) compared with non-neurologically impaired normal controls (NC). The aim of this paper is to explore the relationship between linguistic and extralinguistic aspects of communicative behavior which comprise aphasic speech. Underlying this topic is the question of what constitutes pragmatically appropriate affective communication. General awareness of the social context and specific awareness of the interlocutors and their mental state/contents are key factors contributing to successful communication (Lesser & Milroy, 1993). Therefore, expression of emotion is a key component of communicative behavior.

The communication of emotion is multidimensional, involving several distinct modes or channels simultaneously—face, language, speech and voice, gesture, and posture. Further, it has been suggested that different channels of emotional expression are subserved by distinct neural systems, based on evidence regarding selective impairment in brain-damaged persons (Borod, 1993; Bowers, Bauer, & Heilman, 1993).

Considerations of the performance of LBD aphasic speakers typically focus on aspects of their linguistic impairment. Similarly, researchers tend to emphasize RBD persons' relatively poor performance in response to emotional content or pragmatic context. In the modern aphasia literature, there has been little attention paid to the other communicative aspects of affective behavior which occur together with language—i.e., the dynamic interplay of facial expression, vocal quality and body movement. However, in order to achieve effective communication, in its broadest sense, the expression of all of these aspects of higher cerebral function must be successfully integrated.

A historical perspective on considerations of emotional behavior in aphasic communication will be presented. Subsequently, a re-examination of data collected in an extensive group study on the effect of cortical brain damage on emotional expression will be carried out. These data were collected as part of an NIH-funded research project that tested hypotheses regarding the role of the right hemisphere in emotional cognition. RBDs were the experimental group, and LBDs were included, along with NCs, as a control group. In this study, focus on the performance of the LBDs on various tasks from this project will be provided to shed light on aphasic emotional communicative behavior.

The relation between propositional and emotional speech

The consideration of emotional speech alongside propositional aspects of speech has had a variable history. At various periods in aphasiological history, clinicians have been relatively more inclusive about what constitutes communication, including speech and voice, while in other periods, focus has been exclusively on linguistic capacities.

Broca was the first to observe that aphasics may retain the ability to express certain automatic and emotional aspects of speech. In 1861, he had noted how certain patients "have, in a certain way, two degrees of articulation. Under ordinary circumstances, they invariably pronounce their favorite word ('mot de prédilection'): but when they experience a fit of anger, they become capable of articulating a second word, most often a crude oath" (Broca, 1861, p. 333; cited in Harrington, 1987, p. 216).

Hughlings Jackson (1868) developed the notion that emotional speech can be selectively spared in aphasics with left hemisphere lesions, drawing the significant distinction between propositional and non-propositional speech. Aphasia was described as a loss of "higher" voluntary speech movements with sparing of "lower" involuntary, automatic speech, such as emotional exclamations:

> "The movements of speech are educated movements, and thus differ widely from those movements which may be said to be nearly perfect at birth, such as those for respiration, smiling, swallowing, etc. All the movements represented in the corpus striatum unilaterally require long education.... The muscles always acting bilaterally, and chiefly represented bilaterally in the corpus striata, are born with their centres for movements nearly perfect. Thus, then, the term "Intellectual Language" merges in the larger term "Special movements acquired by the Individual," and the term "Emotional Language" in the term "Inherited Movements" (Jackson, 1868, p. 275).

The conservation of emotional experience in aphasics was attested to by Pierre Marie (1906). He went further than Jackson in emphasizing the intact affective and pragmatic nature of their social interactions: "I would like to recall the voluntary politeness of these aphasics as additional proof of the conservation in their case of the affective and moral sphere" (translated in Lebrun, 1994, p. 235).

Further discussion of the issues of emotional behavior in the context of aphasia can be found in the writings of Goldstein (1948). Goldstein extended the Jacksonian distinction from propositional and non-propositional speech to what he termed "abstract language"– defined as "volitional, propositional and relational;" and "concrete language"—defined as including "the instrumentalities of speech" and emotional utterances. He noted that everyday language is a combination of both types of language and that if one of the two forms of language is particularly disturbed by pathology, the individual tries to overcome this by an increased use of the other form. Goldstein's notions of the concrete and abstract attitude are crucially involved in the emotional expression of the aphasic: "Because of the difference of the relationship of emotional and nonemotional language to the personality, there is a difference of impairment of both in aphasia"(Goldstein, 1948, p. 25).

Goldstein's notion of abstract attitude and the original Jacksonian distinction between propositional and nonpropositional speech, along with discussions of symbolic function developed by Head (1926), were reflected in the discussions of the left hemisphere in its role for language dominance in the second half of this century. This can be seen in a typical passage from Alajouanine and Lhermitte (1964):

> "The more the disorder affects the systems of symbolic formulation, the more the ability to deal verbally with ideas is impaired and the more prominent are dissociations between automatic and propositional language. In this latter state the ability to express and to understand appears not to depend on linguistic units *per se* but rather on psychological factors, among which...affective reactions...are noteworthy" (Alajouanine & Lhermitte, 1964, p. 216).

Arguments supporting a more holistic and integrated approach to the study of communication were put forth by an increasingly smaller minority with the rise of more componential approaches (e.g., Geschwind (1965) in neurology and Chomsky (1965) in linguistics–cf. Sarles, 1977). At this time, with the rise of the localizationist approach (e.g., Geschwind, 1973) and the notion of modularity of function (Fodor, 1983), neurolinguistic investigators pursued characterizations of aphasia which correlated the functional architecture of the left hemisphere with components of grammar (e.g., Caplan & Hildebrandt, 1988). However, with the growing interest in parallel and distributed processing models and connectionism, there has been a resurgence of interest in a Systems Approach to cognitive processing (e.g., Damasio, 1989).

Recently, the emphasis in aphasia research has broadened out from a focus on the clinical pathological correlation of linguistic functions to a more socially driven analysis of communicative interaction. This new interest in the consideration of pragmatic function, alongside linguistic function, is well represented by Lesser and Milroy (1993). Although there is a thorough discussion of gesture, body movement, and other non-verbal components of conversational communication in aphasia in this publication, no mention is made of emotional communication as part of the extralinguistic aspects of social interaction.

Neuropsychological research into emotional communication

Pragmatic and affective impairments are typically attributed to right hemisphere pathology. Many investigators have pursued the hypothesis of right hemisphere specialization for emotional processing (Borod, Koff, & Caron, 1983; Bryden & Ley, 1983; Buck, 1984; Heilman, Bowers, & Valenstein, 1985; Ross, 1985). The role of the right hemisphere in the higher-order organization of arousal, attention, intention, and spatial processing has, at various times, been implicated as the source of difficulties with various tasks involving emotional cognition (e.g., Borod, 1992; Gardner, Brownell, Wapner, & Michelow, 1983). The more specific hypothesis—that different emotional valences (positive and negative) are distinctly lateralized with positive emotions being a left hemisphere specialization and negative emotions being a right hemisphere specialization, has also received some support (e.g., Davidson, 1984; Sackeim, Greenberg, Weiman, Gur, Hungerbuhler, & Geschwind,

1982). Numerous studies have addressed these issues and tried to raise evidence in support of one of these two different formulations (see Borod, Caron, & Koff, 1981; Borod & Koff, 1984).

In terms of the neuropsychological organization of emotion and hemispheric processing, specific findings have been reported: a) the left side of the face (controlled predominantly by the right hemisphere) has been found to be more intense than the right side during both posed and spontaneous expressions for neurologically intact subjects (Borod, Koff, & White 1983; Dopson, Beckwith, Tucker, & Bullard-Bates, 1984; for review, see Borod, Santschi-Haywood, & Koff, 1997); b) a right hemisphere dominance for emotional prosody and impairments of these functions in RBD patients (Ross & Mesulam, 1979); c) spontaneous emotional expression occurs less often in RBDs than LBDs or NC (Blonder, Burns, Bowers, Moore, & Heilman, 1993; Buck & Duffy, 1980).

In addition to cortical right hemisphere control of emotional cognition, one must also include for consideration evidence of impairments in social or affectively appropriate behavior arising from a range of diffuse, bilateral and/or subcortical lesion sites. For example, pragmatically inappropriate language was found to be common in patients with diffuse or bilateral lesions in a detailed behavioral study of anosagnosic patients (Weinstein & Kahn, 1955). Referring to self in the second or third person, joking, slang or inappropriately informal language, euphemisms, cryptic speech, aphorisms, metaphors, or malapropisms reflecting pragmatic impairment were listed as typical features of these patients' speech. This pattern of language behavior (part of the larger syndrome of denial of illness) was only seen in lesions arising from a cerebral vascular accident if it involved deep subcortical structures or increase in cerebrospinal fluid pressure (which would give rise to diffuse and bilateral effects). Pragmatically, socially or affectively inappropriate language is generally an indication of pathology which extends beyond the language cortex of the dominant hemisphere (Lorch, 1995).

Given the current support for the view that there is a special role for the right hemisphere in the cognitive control of the expression and perception of emotion, this raises the question: What are the consequences for extralinguistic communication in aphasic persons with left brain damage and intact right hemisphere systems? When discussing the psychiatric aspects of aphasia, Benson (1973) states that appropriate expressions of frustration, anger and depression are quite common in aphasic sufferers. The notion that left hemisphere patients become depressed has been long-standing in the literature, first developed in depth by Goldstein (1948).

Research into the lateral differences in hemispheric valence of emotional expression, mentioned above, has in part been pursued in light of the incidence of depression following left but not right hemisphere stroke (Downhill & Robinson, 1994; Robinson & Szetela, 1981). A more complex picture has emerged in a longitudinal study of the emotional and social reactions of stroke sufferers. Nelson, Cicchetti, Satz, Sowa and Mitrushina (1994) found that differential recovery rates were dependent on side of lesion, with left hemisphere stroke sufferers having increased indifference, inappropriateness and depression as measured at 2 months post-onset. By 6 months post-onset, however, their emotional recovery seemed to stabilize, whereas the emotional functioning of right hemisphere stroke sufferers appeared to worsen at this point.

Paralinguistic features of voice

Determination of the significant behavioral features which comprise meaningful and appropriate displays of affective content in social communication is still limited. One aspect of communication which has received considerable attention is intonation. The communicative function of intonation is both emotional and linguistic, being used to convey emotional tone, linguistic (lexical) stress, contrastive stress and syntactic structure.

The acoustic correlates of the emotional aspects of intonation in conversational speech were first measured by Lieberman (1967). A number of distinct vocal signal changes were identified with the expression of emotion: raised or lowered average fundamental frequency (F_0) when angry; extreme emotion resulted in wider F_0; lowered and narrowed F_0 range; and breaking sentence into many breath-groups or extended breath-groups. Other paralinguistic vocal devices identified include: filled and unfilled pauses; sound pressure level (loudness) raised or lowered; and diminished pitch perturbations of speakers' F_0.

Examination of F_0 and sentence level pitch perturbations has also been carried out in aphasic speakers (Scherer, 1985). Intonational aspects of aphasic speech have been classified as distinct parts of the major aphasic syndromes: Broca's aphasic speech is characterized as being dysfluent, and/or intonationally flat, and Wernicke's aphasic speech is characterized as being generally fluent or even hyperprosodic (Goodglass & Kaplan, 1972; 1983). In studies by Danly and colleagues (Danly, Cooper, & Shapiro, 1983; Danly & Shapiro, 1982), they demonstrated that the impairment of intonation in aphasics, as measured by F_0 contours, was linked in part to the specific syntactic function of the utterance. The speech of Broca's aphasics gives the impression of being dysprosodic due to the slow rate and numerous interruptions in the production of utterances. However, the aphasics were found to exhibit natural terminal falling F_0 and F_0 declination in short, simple sentences, and were impaired relative to the syntactic complexity of longer sentences (Danly & Shapiro, 1982). Moreover, the Broca's aphasic speakers appeared to compensate for their halting speech by typically ending phrases with a sizable continuation rise in F_0. This aids the listener in demarcating syntactic processing units. This continuation rise is often used as a normal conversational discourse device.

Reviewing the evidence

In the 1980's, there was a growing shift of interest from the well-explored linguistic functions of the left hemisphere to the less well-defined functions of the right hemisphere (e.g., Brownell, Michelow, Powelson, & Gardner, 1983; Zaidel, 1976). At that time, Borod, Koff, [Perlman] Lorch and Nicholas commenced a large-scale project investigating the neurological organization of emotional expression. In this project, groups of right brain-damaged patients (RBDs) and left brain-damaged patients (LBDs) with focal cortical lesions, and neurologically healthy control subjects (NCs) were tested on an extensive range of experimental tasks which focused on the various components which comprise human emotional cognition. The initial hypotheses were constructed to test the assumption that right brain damage would lead to impairment on tasks of emotional expression and perception. The results were thus analyzed for differences between the RBDs as compared with the LBDs and NCs (Bloom, Borod, Obler, & Koff, 1990; Borod, Koff, Lorch, & Nicholas, 1985; Borod, Koff, Lorch, & Nicholas, 1986a & b; Borod, Koff, Lorch, Nicholas, & Welkowitz, 1988; Borod, Lorch, Koff, & Nicholas, 1987; Kent, Borod, Koff, Welkowitz, & Alpert, 1988).

In this section, the data from that project will be re-evaluated specifically with regard to the LBD patients' performance. This review is focused on two issues: 1) What intact abilities of emotional expression are evident in the LBDs' performance? and 2) Does their performance differ in any way from NCs? This re-examination may indicate some features of aphasic performance which draw on spared extra-linguistic abilities to compensate in communication.

Experiment I

Subjects

All subjects were drawn from the Neurology Service of the Boston Veterans Administration Medical Center. Subjects were 12 males with unilateral right hemisphere lesions (RBDs) and 15 males with unilateral left hemisphere lesions (LBDs), all resulting from a cerebral vascular accident (CVA), and having an absence of history of psychiatric disorder, psychotropic drug treatment or secondary neurological disorder. Sixteen males without neurological impairment served as controls (NC). All subjects were right-handed by self (or family) report and did not differ on formal assessment of lateral dominance (Coren, Porac, & Duncan, 1979). The subjects did not differ significantly on the demographic variables of age, education and occupational status. Subjects had a mean age of 58 years (S.D. = 8.1 yrs), a mean of 13 years of education (S.D. = 2.8), and a mean occupation level (Hollingshead & Redlich, 1958) of 3.9 (S.D. = 1.5), generally classified as "middle-class white collar worker."

Table 1 displays subject details of age, months post onset, lesion site, aphasia type and scores on the Weschler Adult Intelligence Scale (WAIS) Performance IQ (Weschler, 1958).

ID	Age	Side of lesion	MPO	WAIS PIQ	Lesion site	ID	Age	Side of lesion	MPO	WAIS PIQ	Lesion site
1	51	Right	1	96	P, S	1	56	Left	22	100	F
2	58	Right	58	–	P, S	2	58	Left	79	–	F, S
3	53	Right	7	96	P, T	3	63	Left	59	105	F, S
4	61	Right	80	–	P, T, S	4	44	Left	8	72	F, S
5	52	Right	3	–	P, O, S	5	58	Left	2	106	P, T
6	62	Right	1	–	P	6	63	Left	28	101	P, T
7	71	Right	78	98	F, T, S	7	71	Left	9	94	P, T, S
8	35	Right	3	67	F, P, T, S	8	52	Left	22	95	P, T, S
9	50	Right	1	–	F, P, T, S	9	63	Left	6	122	P, T, S
10	55	Right	2	–	F, P, T, S	10	42	Left	3	104	P, T, O
11	60	Right	2	90	F, P, T, S	11	54	Left	3	72	F, P
12	52	Right	24	–	F, P, O, S	12	54	Left	88	–	F, P, T
						13	54	Left	20	66	F, T, S
						14	55	Left	49	100	F, P., T, S
						15	49	Left	38	89	F, P, T, S

ID = patient identification number
MPO = months post-onset of illness
WAIS PIQ = Weschler Adult Intelligence Scale Performance IQ
F = Frontal
P = Parietal
T = Temporal
O = Occipital
S = Subcortical

Table 1. Characteristics of patients

Method

Spontaneous production of emotion was assessed during the viewing of slides designed to elicit expressions of positive and negative emotions (Buck, 1978). There were 16 slides depicting a variety of emotionally laden scenes—e.g., a baby picking flowers, a beautiful sunset, a couple embracing, a surgical procedure, a young victim of starvation, and a photographic double exposure. Subjects were seated facing a one-way mirror, behind which a video-camera recorded their responses. Each slide was presented for 20 seconds; after 6 seconds, the subject was requested to describe his feelings and reactions to the slide.

Analysis 1—Spontaneous emotional facial expressions

Two judges viewed the videotapes and rated each response for Responsivity—i.e., did a facial response occur, (0 = no, 1 = yes), Appropriateness of the facial expression response (0 = inappropriate, 1 = appropriate) and Intensity of the response (7-point scale, 1 = minimal to 7 = maximal intensity). Mean rating scores were used because inter-rater reliability for a sub-sample of scores was high (Pearson's r = +.74). The mean scores for emotional facial responses to the slides viewed were analyzed by group for each of the 3 ratings described above. The results are displayed in Table 2. (The results are averaged across spontaneous and posed facial expressions.)

Measure	Score range	Means and standard deviations		
		RBD	LBD	NC
Responsivity	(0-1)	0.59 (0.19)	0.74 (0.14)	0.72 (0.14)
Appropriateness	(0-1)	0.83 (0.14)	0.87 (0.09)	0.93 (0.06)
Intensity	(1-7)	3.38 (0.60)	3.62 (0.65)	3.19 (0.56)

Table 2. Mean scores for emotional facial responses

The LBDs were slightly more responsive than NCs, and both groups were more responsive than the RBDs. For the measure of Intensity of responses to the slides, the LBDs produced more intense responses than either the NCs and the RBDs. To examine the relationship among the different response variables, the Spearman rank-order procedure was used to correlate scores for Responsivity (Resp.), Appropriateness (Approp.) and Intensity (Intens.). As can be seen in Table 3, the only significant correlation was between Responsivity and Intensity for the LBD group. This can be interpreted as indicating that for the LBD group, when an emotional response occurred, there was a significantly greater chance that it would be more intensely expressed than for the RBD or NC group.

Group	Resp. vs. approp.	Resp. vs. intens.	Intens. vs. approp.
RBD	0.02	0.67	0.37
LBD	0.14	0.84*	-0.14
NC	-0.44	0.31	-0.02

* p<0.001

Table 3.   Spearman rank-order correlation rhos among parameters of facial emotional expression by subject group

Analysis 2 — Channels of communication

The contribution of the three channels of communication–face, intonation and verbal production–to each subject's response in viewing the 16 slides was rated for "utilization or degree of involvement" on a 4-point scale (0 = noncontributory, 1 = minimal, 2 = moderate, 3 = maximal involvement).

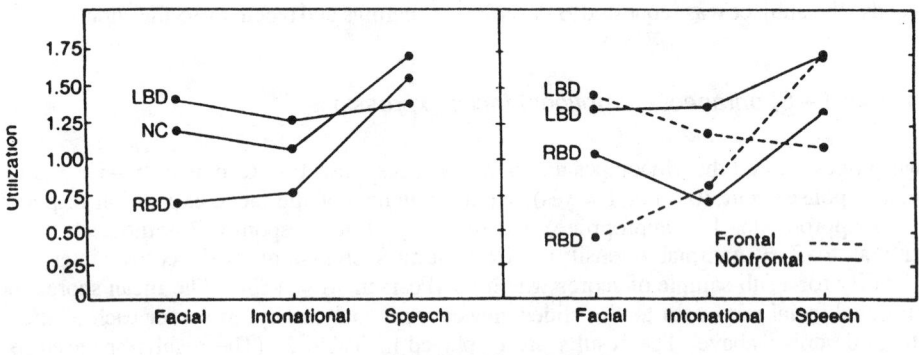

Figure 1.   Communication of emotion as a function of group and channel. LBD indicates left brain-damaged patients; RBD, right brain-damaged patients; and NC, normal controls.

LBDs' use of facial and intonation channels was higher than it was for NCs and significantly higher than for RBDs; see Figure 1. When analyzed for frontal (F) versus nonfrontal (NF) (i.e., posterior) lesions, FLBDs, NFLBDs and NFRBDs used the facial expression channel significantly more frequently than FRBDs.

Clinicians have long observed that extralinguistic channels of communication can supplement or substitute for impaired linguistic expression in spontaneous production. It is not surprising that those with left frontal damage used the speech channel less frequently than the other groups, given their nonfluent aphasia, but it is notable that those with left hemisphere lesions used the intonation channel more than those with right hemisphere lesions regardless of intrahemispheric location of lesion anteriorly or posteriorly.

Analysis 3 — Propositional and nonpropositional contributions to speech communication

The relative contribution of propositional and nonpropositional speech in the subjects' verbal responses to the emotionally ladened slides was also analyzed. A rating of 1 was given to verbal responses which were predominantly propositional with little nonpropositional contribution. A rating of 2 was given to verbal responses with a balance of the two, as seen in modal speech. A rating of 3 was given to verbal responses which were deemed to be relatively more nonpropositional in its composition, with reliance on intonation contours. The LBDs ($\bar{x}$ = 2.14) relied more on nonpropositional speech than the NCs ($\bar{x}$ = 1.92), whereas the RBDs ($\bar{x}$ = 1.64) relied predominantly on speech without a nonpropositional contribution.

Analysis 4—Verbalization of emotion

The linguistic content of the verbal responses to the emotionally-ladened slides was also analyzed. It became apparent that although the participants were all asked specifically to discuss their feelings about the slide being viewed, many, in fact, offered descriptions of the pictorial aspects of the scene represented instead.

Two ratings of linguistic (semantic) content were made on a 4-point scale (0 = not present, 1 = minimally present, 2 = moderately present, 3 = maximally present) for the emotional content (Feelings) and (inappropriate) descriptions of pictorial content (Form). When these responses were analyzed with regard to inter- and intra-hemispheric lesion site groupings, it was found that the aphasic subjects with left nonfrontal/posterior lesions—4 Wernicke's and 2 Conductions—(and the subjects with right frontal lesions) produced the greatest number of these inappropriate pictorial rather than emotional verbal responses.

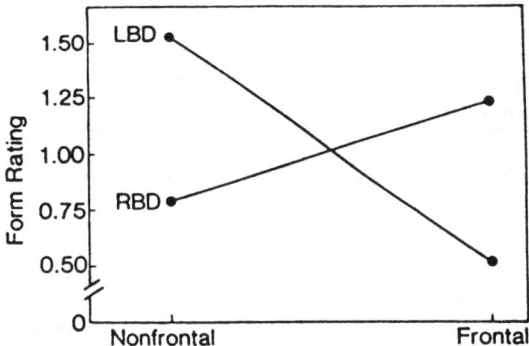

Figure 2.   Use of "form" in describing feelings as a function of lesion side and location. LBD indicates left brain-damaged patients and RBD, right brain-damaged patients.

Experiment II

The finding of greater intensity in facial emotional expression reported above prompted the analysis of the intensity of emotion expressed in the verbal responses. Bloom, Borod, Obler and Koff (1990) carried out a detailed analysis of the verbal responses in a subset of the subjects from the original study. Three LBDs, 3 RBDs and 3 NCs were randomly selected from the 43 subjects used in the original study. Responses to 4 of the original 16 slides viewed were analyzed—2 slides designed to elicit positive emotions and 2 slides designed to elicit negative emotions. The 36 transcripts of verbal reactions to the slides (4 responses x 9 subjects) were submitted to discourse analysis. A filtering procedure was applied to control for the artifactual differences caused by agrammatism present in the LBDs. All free grammatical morphemes (function words) which tend to be omitted in agrammatic aphasics were removed from all of the subjects' transcripts, leaving only substantives (content words) for analysis.

Analysis 1—Category accuracy and valence accuracy

Five naive raters judged how accurately subjects used words to express emotions in the written transcripts of their responses to viewing the slides. Responses were rated for emotional category accuracy and valence accuracy.

Table 4 displays the percentage correct means for category accuracy and valence accuracy by subject group. As can be seen, the LBDs were less accurate that the NCs in communicating in their discourse the category of emotion but were slightly better, particularly at linguistically expressing negative emotion, than the NCs when only positive or negative valence expression was measured.

	Slide type	NC	LBD	RBD
Category	Positive	86.7	46.7	29.2
	Negative	80.8	53.3	50.8
Valence	Positive	93.3	93.3	61.7
	Negative	80.8	93.3	80.8

Table 4. Category accuracy and valence accuracy by subject group

In order to account for the successful communication of emotion in aphasic discourse token/type ratios were calculated. As expected, LBD token/type ratio was higher (2.0) than for the RBD (1.3) and NC (1.5) groups. This reflects an effective pragmatic device in using repetition for emphasis and in expressing intensity for the LBDs.

Analysis 2—Intensity

Four raters characterized the words in the 36 discourse transcripts for intensity of emotion expressed, using a 5-point Likert scale (1 = least, 5 = most).

Slide Type	NC	LBD	RBD
Positive	2.20	2.48	2.16
Negative	2.35	2.42	2.22

Table 5. Mean ratings for intensity by subject group

The LBDs were found to be more intense in linguistic communication of emotion than NCs (and RBDs) despite their aphasia.

Experiment III

In a follow-up study with new subjects, Bloom, Borod, Obler and Gerstman (1993) elicited three types of narratives from pictures: 1) emotional content–a girl whose dog is hit by a car, 2) visual spatial content–moving a box by climbing on books piled on a chair, and 3) procedural/neutral content–how to fry an egg, from a new group of LBD (n = 12), RBD (n = 9), and NC (n = 12) right-handed adults.

The stories produced by the subjects to these 3 pictures were rated by 2 trained judges as "appropriate" or "inappropriate" on 7 pragmatic features adapted from Grice (1975) and Prutting and Kirchner (1987): topic maintenance, conciseness, specificity (i.e., lack of ambiguity), lexical selection, revision and repair, relevancy and quantity of information.

When the narratives were assessed for pragmatic features, the LBDs were found to be significantly impaired relative to the NCs for specificity, revision and repair, lexical selection, and quantity of information but not significantly different from the NCs on conciseness, topic maintenance, and relevancy. While their performance was (as expected) significantly more impaired than the NCs overall with respect to the mean number of pragmatic features (n = 7) rated as appropriate, the LBDs were found to be significantly less impaired on the emotional narrative ($\bar{x}$ = 5.42) than the visual spatial ($\bar{x}$ = 4.08) (p < .05) and procedural ($\bar{x}$ = 4.17) (p < .10) narratives. This held true when the data were examined by individual subject rather than pooled group mean scores. The number of individuals in each group that performed better than or equally in the emotional (E) compared to the nonemotional (NE) condition is displayed in Table 6.

	E > NE	E = NE	E < NE
NC	4	6	2
LBD	9	2	1
RBD	0	1	8

Table 6. Number of individuals in each subject group by performance condition

This improved performance was true for the emotional narrative production in both the LBDs with anterior and those with posterior lesions. Both the anterior and the posterior subgroup of LBDs performed better on the emotional versus nonemotional conditions for mean pragmatic features produced appropriately (anterior LBD emotional and nonemotional mean scores, 6.34 vs. 5.33; posterior LBD emotional and nonemotional mean scores, 5.00 vs. 3.69). Thus, there was a facilitating effect of emotional context on pragmatic performance for the LBDs (with the reverse found for the RBDs).

Emotionally ladened material has been found to provide facilitatory contexts for aphasic persons in a variety of different modalities and tasks: auditory comprehension (Boller, Bole, Vrtunski, Patterson, & Kim, 1979; Reuterskiold, 1991); oral reading and writing (Landis, Graves, & Goodglass, 1982); and repetition (Ramsberger, 1996).

Experiment IV

Facilitation of facial movement by emotional context

Jackson (1878) and Geschwind (1975) stress the role of context for the apraxic patient, who may not be able to carry out a specific movement to verbal command or even to imitation, but will perform spontaneously in a more naturalistic setting. Apraxia is considered a disorder of voluntary movement which is under left hemisphere control (Kimura, 1982). The question addressed in this study was whether emotional cueing would improve the performance of apraxic patients by providing a more "automatic" or affective context. Apraxic patients can carry out spontaneous emotional movements with the same sets of muscles used in praxic skills (Nathan, 1947).

> "The apraxic...is able to smile, lick his lips, express disgust in a normal manner when he *feels* that the occasion demands; but he is unable to act these gestures...when he *thinks* the occasion demands...He remains well adapted to his environment...he is able to play cards, laugh at a joke, eat and drink in the manner customary in his society" (Nathan, 1947, p. 475).

Since the right hemisphere is implicated in the expression of emotion, the possibility that emotional cues could facilitate posed praxic responses in patients with LBD was hypothesized. Such a finding would be interpreted as an instance of compensation for a primary behavioral deficit of the dominant hemisphere by a secondary system.

The original set of patients with LBD and RBD, detailed in Table 1, and NCs were videotaped while being examined on tasks of bucco-facial praxis in both non-emotional and emotional contexts. Six movements were selected for having a neutral (non-emotional) form and an emotional analog. Three movements involved the upper face, and three involved the lower face. This was done in order to control for the effect of (central) facial paralysis of the lower face. The six commands used in the facial movement elicitation task were:

Non-emotional/neutral	Emotional context analog
1) close one eye	1) close one eye like a wink
2) lift your eyebrows	2) lift your eyebrows like you're surprised
3) lower your eyebrows	3) lower your eyebrows like a frown
4) put out your tongue	4) stick out your tongue like you're making a face at me
5) raise the corners of your mouth	5) raise the corners of your mouth like a smile
6) pucker your lips	6) pucker your lips like a kiss

All of the neutral items were presented first; emotionally cued items followed in a second block. This allowed for the establishment of a baseline measure of apraxic performance for each subject to which the degree of improvement due to emotional facilitation could be compared. It was considered unlikely that any improvement in the Emotional condition, presented second, could be attributed to a rehearsal effect. Apraxic patients do not tend to improve with repeated trials (Mateer, 1978).

Two judges rated the execution and accuracy of each videotaped movement with respect to quality. As would be expected, the LBDs demonstrated significant impairment in bucco-facial praxis as indicated by a poor performance on the neutral tasks relative to the RBDs and NCs. However, the LBDs' motor performance was more strongly facilitated by provision of the additional emotional cue (which was more pronounced in those movements involving the lower face) than the other two groups ($F(2,40) = 2.98$; $p = .06$) as displayed in Table 7 below.

	NE	E
NC	2.80	2.90
LBD	2.19	2.58
RBD	2.70	2.91

Table 7. Execution mean ratings for each group by condition

The analogy can be drawn between the facilitation effect of emotional context achieved in this study for LBDs with bucco-facial apraxia and therapeutic techniques which draw on other aspects of cognitive behavior attributed to the right hemisphere used to facilitate speech production in aphasic persons, such as Melodic Intonation Therapy (Sparks, Helm, & Albert, 1974),Visual Communication Therapy (Gardner, Zurif, Berry, & Baker, 1976), and Visual Action Therapy (Helm-Estabrooks, Fitzpatrick, & Barresi, 1982).

Experiment V

Voluntary emotional expression

Posed emotional facial expressions were elicited from both verbal and visual input for 8 emotions (3 positive and 5 negative) with the LBD, RBD, and NC subjects described in Table 1. Performances were rated for intensity of expression, category accuracy and valence accuracy. On ratings of intensity (1 = minimal intensity to 7 = maximal intensity), LBDs ($\bar{x} = 4.14$) posed positive emotions with significantly more intensity than either NCs ($\bar{x} = 3.69$) or RBDs ($\bar{x} = 3.54$). Post-hoc tests to examine valence differences within each subject group demonstrated that positive emotions were posed significantly more intensely than negative emotions by both the LBDs ($\bar{x}s = 4.14$ vs. 3.63) and NCs ($\bar{x}s = 3.69$ vs. 3.35) while there were no differences for the RBDs.

Discussion

In summary, the 5 experiments investigating emotional, linguistic and pragmatic aspects of communication reviewed here demonstrate the wide range of emotional communicative behaviors which are available to aphasic LBDs in speech and facial expression. This large-scale project included a number of other investigations of posed and spontaneous production of emotional facial expressions and the perception of emotional facial expressions reported in detail in Borod et al. (1986a). LBDs were found to be as accurate as NCs for both posed and spontaneous facial expression and face perception, receiving a similar overall mean accuracy score–LBDs ($\bar{x} = 0.69$) and NCs ($\bar{x} = 0.72$), as compared with the impaired performance of the RBDs ($\bar{x} = 0.51$), with no significant differences between conditions or valences.

Evidence has been reported that both fluent and non-fluent aphasics produce more gestures than controls in a free conversational setting (Blonder, Burns, Bowers, Moore, & Heilman, 1995; Feyereisen & Seron, 1982). Feyereisen and Seron (1982) suggest that this indicates that "...social competence could be intact in aphasia." The data presented here from the project by Borod, Koff, Lorch and Nicholas are consistent with that conclusion with respect to emotional communication. Our aphasic subjects were found to be equal to or more highly rated than NCs and RBDs on measures of emotional communication: 1) responsivity and intensity of spontaneous emotional facial expression; 2) the use of both face and intonational aspects of voice as channels of expression; 3) the reliance on nonpropositional speech; 4) expressing the valence (positive/negative) of their emotions verbally; 5) intensity in the linguistic communication of their emotions; 6) the use of discourse in emotional contexts; 7) facilitation of emotional context on pragmatic aspects of discourse; 8) facilitation of emotional context on the quality of bucco-facial praxic movements; and 9) facilitation of emotional context on the intensity of posed facial expressions.

For Smith and Kemp-Wheeler (1996), the question of why we need emotions is considered one of the "unsolved mysteries of the mind." In their discussion, they examine the way emotion may change the way we think. The present review has attempted to explore the way emotion may change the way we communicate, and how aphasic speakers may express emotion more fully than other aspects of language, with speech and voice thus contributing to their communicative success.

Acknowledgments—The first author wishes to acknowledge the invaluable input of Dr. Renata Whurr. This work was supported by NIH Grant No. NS06209 to the Aphasia Research Center of Boston University Medical Center, by NIH Biomedical Research Support Grant No. 1-S07RR07186 to Wellesley College, by NIH Grant Nos. MH42172 and MH44889 to Queens College of the City University of New York, and by the Wellcome Trust. Figures 1 and 2 were reproduced by permission of the American Medical Association.

References

Alajouanine, T., & Lhermitte, F. (1964). Aphasia and physiology of speech. In D. Rioch and E. Weinstein (Eds.), *Disorders of Communication*. Baltimore: Williams and Wilkins Co.

Benson, D. F. (1973). Psychiatric aspects of aphasia. *British Journal of Psychiatry, 123*, 555-566.

Blonder, L. X., Burns, A. F., Bowers, D., Moore, R. W., & Heilman, K. M. (1993). Right hemisphere facial expressivity during natural conversation. *Brain and Cognition, 21*, 44-56.

Blonder, L. X., Burns, A. F., Bowers, D., Moore, R. W., & Heilman, K. M. (1995). Spontaneous gestures following right hemisphere infarct. *Neuropsychologia, 33*, 203-213.

Bloom, R. L., Borod, J. C., Obler, L. K., & Koff, E. (1990). A preliminary characterization of lexical emotional expression in right and left brain-damaged patients. *International Journal of Neuroscience, 55*, 71-80.

Bloom, R. L., Borod, J. C., Obler, L. K., & Gerstman, L. (1993). Suppression and facilitation of pragmatic performance: Effects of emotional content on discourse following right and left brain damage. *Journal of Speech and Hearing Research, 36*, 1227-1235.

Boller, F., Cole, M., Vrtunski, B., Patterson, M., & Kim, Y. (1979). Paralinguistic aspects of auditory comprehension in aphasia. *Brain and Language, 7*, 164-174.

Borod, J. (1992). Interhemispheric and intrahemispheric control of emotion: A focus on unilateral brain damage. *Journal of Consulting and Clinical Psychology, 60*, 339-348

Borod, J. (1993). Emotion and the brain—Anatomy and theory: An introduction to the special section. *Neuropsychology, 7*, 427-432.

Borod, J. C., Caron, H., & Koff, E. (1981). Asymmetries in positive and negative facial expressions. *Neuropsychologia, 19*, 819-824.

Borod, J. C., & Koff, E. (1984). Asymmetries of affective facial expression. In N. Fox and R. Davidson (Eds.), *The psychobiology of affective development*. Hillsdale, NJ: Lawrence Erlbaum Associates.

Borod, J. C., Koff, E., & Caron, H. (1983). Right hemisphere specialization for the expression and appreciation of emotion: A focus on the face. In E. Perecman (Ed.), *Cognitive processing in the right hemisphere*. (pp. 83-110). New York: Academic Press.

Borod, J. C., Koff, E., Lorch, M. P., & Nicholas, M. (1985). Channels of emotional expression in patients with unilateral brain damage. *Archives of Neurology, 42*, 342-348.

Borod, J. C., Koff, E., Lorch, M. P., & Nicholas, M. (1986a). The expression and perception of facial emotion in brain-damaged patients. *Neuropsychologia, 24*, 169-180.

Borod, J. C., Koff, E., Lorch, M. P., & Nicholas, M. (1986b). Deficits in facial expression and movement as a function of brain damage. In J.-L. Nespoulous, P. Perron, and A. R. Lecours (Eds.), *The biological foundations of gestures*. Hillsdale, NJ: Lawrence Erlbaum Associates.

Borod, J. C., Lorch, M. P., Koff, E., & Nicholas, M. (1987). Effect of emotional context on bucco-facial apraxia. *Journal of Clinical and Experimental Neuropsychology, 9*, 147-153.

Borod, J. C., Lorch, M. P., Koff, E., Nicholas, M., & Welkowitz, J. (1988). Emotional and non-emotional facial behaviors in patients with unilateral brain damage. *Journal of Neurology, Neurosurgery and Psychiatry, 51*, 826-832.

Borod, J. C., Koff, E., & White, B. (1983). Facial asymmetry in spontaneous and posed expressions of emotion. *Brain and Cognition, 2*, 165-175.

Borod, J. C., Santschi-Haywood, C., & Koff, E. (1997). Neuropsychological aspects of facial asymmetry during emotional expression: A review of the normal adult literature. *Neuropsychological Review. 7*, 41-59.

Bowers, D., Bauer, R., & Heilman, K. (1993). The nonverbal affect lexicon: Theoretical perspectives from neuropsychological studies of affect perception. *Neuropsychology, 7*, 433-444.

Brownell, H., Michelow, D., Powelson, J., & Gardner, H. (1983). Surprise but not coherence: Sensitivity to verbal humor in right-hemisphere patients. *Brain and Language, 18*, 20-27.

Bryden, M. P., & Ley, R. G. (1983). Right-hemispheric involvement in the perception and expression of emotion in normal humans. In K. M. Heilman and P. Satz (Eds.), *Neuropsychology of Human Emotions* (pp. 6-44). New York: Guilford Press.

Buck, R. (1978). The slide-viewing technique for measuring non-verbal sending accuracy: A guide for replication. *Catalog of Selected Documents in Psychology, 8,* 63.

Buck, R. (1984). *The Communication of Emotion.* New York: Guilford Press.

Buck, R., & Duffy, R. J. (1980). Nonverbal communication of affect in brain damaged subjects. *Cortex, 16,* 351-362.

Caplan, D., & Hildebrandt, N. (1988). *Disorders of syntactic comprehension.* Cambridge, MA: MIT Press.

Chomsky, N. (1965). *Aspects of the Theory of Syntax.* Cambridge, MA: MIT Press.

Coren, S., Porac, C., & Duncan, P. (1979). A behaviorally validated self-report inventory to assess four types of lateral preferences. *Journal of Clinical Neuropsychology, 1,* 55-64.

Damasio, A. R. (1989). Time-locked multiregional retroactivation: A systems level proposal for the neural substrates of recall and recognition. *Cognition, 33,* 25-62.

Danly, M., & Shapiro, B. (1982). Speech prosody in Broca's aphasia. *Brain and Language, 16,* 171-190.

Danly, M., Cooper, W., & Shapiro, B. (1983). Fundamental frequency, language processing and linguistic structure in Wernicke's aphasia. *Brain and Language, 19,* 1-24.

Davidson, R. J. (1984). Affect, cognition, and hemispheric specialization. In C. E. Izard, J. Kagan, and R. Zajonc (Eds.), *Emotions, Cognition and Behavior.* Cambridge: Cambridge University Press.

Downhill, J. E., & Robinson, R. G. (1994) Longitudinal assessment of depression and cognitive impairment following stroke. *Journal of Nervous and Mental Diseases, 182,* 425-431.

Dopson, W., Beckwith, D., Tucker, D. M., & Bullard-Bates, P. (1984). Asymmetry of facial expression in spontaneous emotion. *Cortex, 20,* 243-252.

Feyereisen, P., & Seron, X. (1982). Nonverbal communication and aphasia: A review. *Brain and Language, 16,* 191-212, 213-236.

Fodor, J. (1983). *The modularity of mind.* Cambridge, MA: MIT Press.

Gardner, H., Brownell, H., Wapner, W., & Michelow, D. (1983). Missing the point: The role of the right hemisphere in the processing of complex linguistic materials. In E. Perecman (Ed.), *Cognitive processing in the right hemisphere* (pp. 169-191). New York: Academic Press.

Gardner, H., Zurif, E., Berry, T., & Baker, E. (1976). Visual communication in aphasia. *Neuropsychologia, 14,* 275-292.

Geschwind, N. (1965). Disconnexion syndromes in animals and man. *Brain, 88,* 237-294, 585-644.

Geschwind, N. (1975). The apraxias: neural mechanisms of disorders of learned movement. *American Scientist, 63,* 188-195.

Geschwind, N. (1973). The brain and language. In G. A. Miller (Ed.), *Communication, language and meaning* (pp. 61-72). New York: Basic Books.

Goldstein, K. (1948). *Language and language disturbances.* New York: Grune and Stratton.

Goodglass, H., & Kaplan, E. (1972). *The assessment of aphasia and related disorders.* Philadelphia: Lea and Febiger.

Grice, H. (1975). Logic and Conversation. In P. Cole and J. Morgan (Eds.), *Studies in syntax and semantics: Speech acts.* New York: Academic Press.

Harrington, A. (1987). *Medicine, mind and the double brain.* Princeton, NJ: Princeton University Press.

Head, H. 1926. *Aphasia and kindred disorders of speech.* Cambridge: Cambridge University Press.

Heilman, K. M., Bowers, D., & Valenstein, E. (1985). Emotional disorders associated with neurological diseases. In K. M. Heilman and E. Valenstein (Eds.), *Clinical neuropsychology* (pp. 377-402). Oxford: Oxford University Press.

Helm-Estabrooks, N., Fitzpatrick, P., & Barresi, B. (1982). Visual action therapy for global aphasics. *Journal of Speech and Hearing Disorders, 44,* 385-389.

Hollingshead, A. B., & Redlich, F. C. (1958). *Social Class and Mental Illness.* New York: Wiley.

Jackson, J. H . (1868). On the physiology of language. *The Medical Times and Gazette, 2,* 275-276.

Jackson, J. H. (1878). In J. Taylor (Ed.) (1932), *Selected writings of Jackson.* London: Hodder and Stoughton.

Kent, J., Borod, J. C., Koff, E., Welkowitz, J., & Alpert, M. (1988). Posed facial emotional expression in brain-damaged patients. *International Journal of Neuroscience, 43,* 81-87.

Kimura, D. (1982). Left-hemisphere control of oral and brachial movements and their relation to communication. *Philosophical Transactions of the Royal Society, London, (B) 298,* 135-149.

Landis, T., Graves, R., & Goodglass, H. (1982). Aphasic reading and writing: Possible evidence for right hemisphere participation. *Cortex, 18,* 105-112.

Lebrun, Y. (1994) Selections from the work of Pierre Marie. In P. Eling (Ed.), *Reader in the history of aphasia.* Amsterdam: John Benjamins.

Lesser, R., & Milroy, L. (1993). *Linguistics and Aphasia: Psycholinguistic and pragmatic aspects of intervention.* Harlow: Longmans.

Lieberman, P. (1967). *Intonation, perception and language.* Cambridge, MA: MIT Press.

Lorch, M. P. (1995). Some neurolinguistic evidence regarding variation in interlanguage use. In L. Eubank, L. Selinker, and M. Sharwood Smith (Eds.), *The current state of interlanguage: Studies in honor of William E. Rutherford.* Amsterdam: John Benjamins.

Marie, P. (1906) La troisième circonvolution frontale gauche ne joue aucun rôle spécial dans la fonction du langage. *Semaine Médicale 26,* 241-247. [Translated in Lebrun, 1994.]

Mateer, C. (1978). Impairments of nonverbal oral movements after left hemisphere damage: A followup analysis of errors. *Brain and Language, 6,* 334-341.

Nathan, P. W. (1947). Facial apraxia and apraxic dysarthria. *Brain, 70,* 449-478.

Nelson, L. D., Cicchetti, D., Satz, P., Sowa, M., & Mitrushina, M. Emotional sequelae of stroke: A longitudinal perspective. *Journal of Clinical and Experimental Neuropsychology, 16,* 796-806.

Prutting, C., & Kirchner, D. (1987). A clinical appraisal of the pragmatic aspects of language. *Journal of Speech and Hearing Disorders, 52,* 105-119.

Ramsberger, G. (1996). Repetition of emotional and nonemotional words in aphasia. *Journal of Medical Speech-Language Pathology, 4,* 1-12.

Reuterskiold, C. (1991). The effects of emotionality on auditory comprehension in aphasia. *Cortex, 27,* 595-604.

Robinson, R. G., & Szetela, B. (1981). Mood changes following left hemispheric brain injury. *Annals of Neurology, 9,* 447-453.

Ross, E. (1985). The modulation of affect and nonverbal communication by the right hemisphere. In M.-M. Mesulam (Ed.), *Principles of Behavioral Neurology* (pp. 239-257). F. A. Davis: Philadelphia.

Ross, E., & Mesulam, M.-M. (1979). Dominant language functions of the right hemisphere? Prosody and emotional gesturing. *Archives of Neurology, 36,* 144-148.

Sackeim, H. A., Greenberg, M. S., Weiman, A. L., Gur, R. C., Hungerbuhler, J. P., & Geschwind, N. (1982). Hemispheric asymmetry in the expression of positive and negative emotions: Neurological evidence. *Archives of Neurology, 39,* 210-218.

Sarles, H. B. (1977). *Language and human nature.* Minneapolis: Minnesota Press.

Scherer, K. (1985). Methods of research on vocal communication: Paradigms and parameters. In K. Scherer and P. Ekman (Eds.), *Handbook of methods in nonverbal behavior research* (pp. 136-198). New York: Cambridge University Press.

Smith, P. T., & Kemp-Wheeler, S. M. (1996). Why do we need emotions? In V. Bruce (Ed.), *Unsolved mysteries of the mind.* Hove: Taylor and Francis.

Sparks, R., Helms, N., & Albert, M. (1974). Aphasia rehabilitation resulting from melodic intonation therapy. *Cortex, 10,* 303-306.

Weinstein, E. A., & Kahn, R. L. (1955). *Denial of illness: Symbolic and physiological aspects.* Springfield, IL: Charles C. Thomas.

Weschler, D. (1958). *The Measurement and appraisal of adult intelligence.* Baltimore: Williams and Watkins.

Zaidel, E. (1976). Auditory vocabulary of the right hemisphere following hemidecortication. *Cortex, 12,* 191-211.

**Pergamon**

J. Neurolinguistics, Vol. 11, Nos 1–2, p. 119–136, 1998
© 1998 Published by Elsevier Science Ltd. All rights reserved
Printed in Great Britain
0911-6044/98 $19.00 + 0.00

PII: S0911-6044(98)00009-8

Processing of lexical ambiguity in patients with traumatic brain injury

Karen L. Chobor* and Avraham Schweiger†

*Department of Neurology, New York University Medical Center;
†Department of Speech and Hearing Sciences, City University of New York Graduate Center

Abstract—The goal of this study was to determine the type of cognitive functions required for the interpretation of lexical ambiguity. Brain-injured and normal subjects performed lexical decision (reaction time) and matching tasks, using three categories of ambiguous words: homonymy, polysemy and metaphor. Patients showed much slower reaction times on metaphor than normals, while accuracy rates for matching were comparable, suggesting relative intactness of the semantic interpretations of these words despite a retrieval deficit.

Introduction

Most words possess some indeterminacy in their meanings (Burgess & Simpson, 1988), so ambiguity may be a characteristic that pervades natural language processing (Simpson, Burgess, & Peterson, 1987; Swinney, 1982). In English, it has been estimated that over 50% of words have more than one meaning (Ziff, 1967).

A significant amount of research has been done in the field of lexical ambiguity comprehension in both normal and brain-injured subjects, using the distinct categories of homonymy, polysemy, and metaphor. Results of these investigations have been interpreted according to models of selective or multiple access, the former referring to the access of a single meaning of an ambiguous term over other meanings according to word frequency and/or context effects, and the latter referring to the access of all meanings of an ambiguous term. In practice, it is not always clear into which category a given ambiguous word falls. The reason is that both etymological and psychological factors are relevant, and these may sometimes contradict each other. For example, words that are readily perceived as having quite distinct senses, such as port (harbor) and port (alcoholic drink) turn out to have a (tortuous but meaningful) link in a common Latin source (Durkin & Manning, 1989).

The majority of lexical ambiguity studies in normals have been interpreted according to the multiple, or exhaustive, access model. In studies using brain-injured subjects, interpretation has centered on localizing the neurological substrate for such linguistic processing rather than determining the necessary cognitive mechanisms. Right hemisphere function has often been assumed to be necessary for successful processing of lexical ambiguity (Gardner et al., 1983; Brownell et al., 1984).

A comprehensive review of the research in lexical ambiguity using both normal and brain-injured populations does not reveal such clear-cut interpretations. There are, in fact, a substantial number of studies that speak to selective access in favor of multiple access in normals. Further, results of studies using brain-injured subjects reveal much evidence suggesting cognitive difficulties traditionally assigned to the frontal lobes as responsible for the handling of multiple meanings. Such cognitive abilities include the conscious manipulation of dual meaning and the ability to shift from one category to another, functions otherwise subsumed under the category of abstraction.

The purpose of this study was to clarify the cognitive functions required for the successful interpretation of lexical ambiguity. Brain-injured and normal subjects served as the study populations.

Review of the literature

Lexical ambiguity is a phenomenon whereby a single word may have more than one meaning. The two most common forms of lexical ambiguity are homonymy and polysemy, followed by metaphor. Homonymy is defined as one word having two or more distinct meanings, while both polysemy and metaphor refer to an instance whereby one word may have two or more related meanings. A finer distinction between the latter two reveals that metaphor is seen as instances of elliptical simile, where "like" or "as" is implied for the purpose of using one thing to stand for another. Thus, polysemy is an aspect of metaphor while remaining a distinct phenomenon.

Beardsley (1962) states that a term may have a central meaning (its ordinary designation) and a marginal meaning (its connotation). The standard designation of wolf, for example, might include "mammal", "four legged", and "canine", whereas the marginal meaning would include "fierce", "voracious", "clever", etc. Meanings other than the central, or dominant, meaning are those which constitute polysemy and metaphor. The two divergent meanings of homonymy are, of course, both central meanings, such as the meanings for ball ("toy" or "a formal party for dancing").

Lexical ambiguity and brain function

Right hemisphere function and lexical ambiguity

The right hemisphere has traditionally been considered to be responsible for the comprehension of meanings secondary to the central meaning of a word (Gardner et al., 1983; Brownell et al., 1984). This includes alternative meanings (Chiarello, 1988) and the appreciation and integration of relationships in verbal discourse and narrative materials (Brownell, Potter, & Michelow, 1984) as well as the understanding of jokes (Gardner, 1994).

Patients with right hemisphere brain damage exhibit a striking amount of difficulty in handling complex linguistic material. They exhibit clear difficulties in abstracting morals and in acquiring a sense of the overall gestalt of linguistic entities; they seem unable to appreciate the relations among key points of a story or joke. They may exhibit difficulties in interpreting phrasal metaphor or using verbal context to make linguistic judgments. Confronted with complex linguistic entities, these patients exhibit clear and recurring difficulties relating to the abilities to conceptualize the unit as a whole, to appreciate its purpose and form, and to integrate specific elements appropriately within these forms. Many of these patients seem insensitive to the context in which these linguistic entities are produced and utilized (Burns et al., 1983). The insensitivity to non literal meaning has also been observed at the single word level (Gardner et al., 1983). The right hemisphere also makes some contribution to language comprehension and reading processes (Krashen, 1976; Searleman, 1977; and Gazzaniga, 1983).

Though there is overlap between right hemisphere functions and tasks of lexical ambiguity, it is clear from a summary of the work in this area that dichotomizing the left and right hemisphere will not suffice in explaining brain-damaged patients' difficulty with such material. In analyzing the cognitive processes that are necessary for the interpretation of ambiguous words, particularly the "abstract" demands in the accessing of the multiple meanings of a lexical item, it appears that these may be more localized to the frontal lobes. Thus, the anterior/posterior dichotomy of brain function may be of equal importance or of greater importance than (left/right-) hemispheric asymmetry when investigating lexical ambiguity.

Frontal lobe function and lexical ambiguity

Traditionally, damage to the frontal lobes has been associated with impairments in abstract thinking (Goldstein, 1948). It has been argued that abstraction and metaphoric thinking depend on a common underlying mechanism of part-whole relationship (Brown, 1997).

This study investigated the relationship between the comprehension of lexically ambiguous words and the ability to apprehend an item as belonging to more than one category, the ability to simultaneously hold more than one category (or concept) in one's mind, and the ability to shift from one category (or concept) to another. Such conceptual or categorical thinking skills, first described by Goldstein (1948) as an "abstract attitude", have traditionally been assigned to the frontal lobe (Luria, 1966). In concrete behavior, the category cannot be accessed from the instance, e.g., the color "red" is not abstracted independent of a red apple. If the patient can accomplish this task, he may not arrive at two or more categories from an instance, e.g., that an apple is a member of the category of shape (round), or color, and of food. The impairment of abstraction can be relatively selective, as in defects of color naming or sorting, or it can be generalized and affect a great many perceptual tasks. In the latter cases, the deficiency is the basis for impairments on tests such as the Wisconsin Card Sorting Test (Heaton, 1981), in which the patient is required to sort objects along several dimensions. Given an instance of a category, e.g., shown a round red object and asked to group it with similar objects, the patient cannot derive the target category from the member items, nor can the patient sort along several dimensions (color, shape, etc.).

Lezak (1995) describes conceptual difficulties, or problems with abstraction, as involving at least (1) an intact system for organizing perceptions even though specific perceptual modalities may be impaired; (2) a well-stocked and readily available store of remembered learned material; (3) the integrity of the cortical and subcortical interconnections and interaction patterns that underlie thought; and (4) the capacity to process two or more mental events at a time.

It is proposed that the basis for difficulty with lexical ambiguity involves problems with mental shift considered a cognitive (conceptual) function associated with the frontal lobe rather than the reduced utility of the right hemisphere with linguistic items. On this view, brain damaged patients who have difficulty (disproportionate to performance on other standardized measures) on tests which require the ability to appreciate two concepts at a single time, to shift from one interpretation of a word or concept so as to emphasize a different one, would be expected to have difficulty handling lexical ambiguity.

In sum, it is proposed that the same mental process is responsible for (a) appreciating lexical ambiguity and for (b) mental shift tasks. The prediction is that a subject who performs poorly on (a) will also perform poorly on (b). It is also predicted that subjects who perform well on (a) will perform well on (b). While intact mental shift is necessary for lexical ambiguity tasks, it is a general cognitive function that is not only applied to such tasks.

Effects of brain injury on the comprehension of multiple meanings

It is commonly known by clinicians that brain damaged patients without aphasia have difficulty with polysemous words. For example, given a target word such as bank, and asked to point to words such as money, river, etc., that go with the target word, they will often select only one meaning. Other cases, asked the color of an orange, may say *yellow* or *red*.

A similar phenomenon occurs with verbal nouns, e.g., "what do you shovel snow with?" Such patients are unable to deal with more than one meaning or interpretation at a time. There is an inability to retrieve the concept, perhaps due to blocking or persistence of the initial interpretation out of which the ambiguous item develops.

There have been relatively few studies that speak to the question of lexical ambiguity in the brain-injured population, and findings among those that do exist have been inconsistent. It is likely that this arises from (1) difficulty in drawing distinctions between homonymous, polysemous, and metaphoric words which results in a lack of uniformity within and across studies of the selection of stimuli, and (2) differences in the manner of presentation of a task and the required response mode. These issues are illustrated below.

Winner and Gardner (1977) presented tasks of metaphoric competence, i.e., matching metaphoric phrases to pictures followed by verbal description, to groups of normal and brain damaged subjects. For the right hemisphere brain damaged (RBD) patients, 43% of the initial responses were metaphoric ones (40% of the time). However, when asked to describe metaphors, these patients did quite well. The left-hemisphere brain damaged (LBD) patients selected the metaphoric picture in their initial choice 58% of the time, with literal pictures chosen 18% of the time. These patients had difficulty in describing metaphors. This double dissociation of performances on the two metaphor tasks clarifies the contributions made by each hemisphere to linguistic and aesthetic functioning. More importantly, however, competent performance of the RBDs on the verbal description condition invalidates the common assertion that they are insensitive to metaphor.

Brownell et al. (1990) examined metaphoric sensitivity to single polysemous words, whereby subjects were asked to judge semantic relatedness along a set of common adjectives: warm, loving, cold and hateful. The results of their study suggest a qualitative difference in the manner in which left- and right- brain-damaged patients process certain aspects of word meaning. The authors point out the striking finding that relative to the LBDs, the RBDs do not appreciate metaphoric meaning fully even at the single word level. This is inconsistent with the study by Winner and Gardner (1977) since in the latter study, the RBDs could use language to interpret metaphor.

Tompkins (1990) presented an auditory lexical decision task to LBDs and RBDs, in which they heard a priming phoneme string and then a target phoneme string, either a real word or a non-word. Subjects were asked to indicate whether the target string was a word or not; accuracy and time data were collected. Results of this study showed that brain damaged subjects' auditory lexical decisions for ambiguous target words can be facilitated by primes related to the metaphoric or literal meanings of those words. Brain damaged subjects' performance in conditions conducive to automatic processing resembled that of controls in all ways, except for absolute speed. Of note, RBDs did not have difficulty accessing metaphoric aspects of word meaning, perhaps because access proceeded relatively automatically in this task, and perhaps because the anterior-posterior distinction was not made.

Using word triad and dyad tasks, Gagnon et al. (1994) did not find a clear disparity between LBD and RBD patients' ability to handle metaphor. The RBDs performed similarly to the LBDs for the metaphoric and neutral triad task, but significantly worse than the controls on the metaphoric triads. The LBD group always performed significantly lower than the controls in the dyad task. However, the RBDs performed similarly to the controls and much better than the LBDs when they had to perceive the presence of a primary meaning and an alternative neutral meaning in the dyad task. The RBD group performed like the LBD aphasic group and were significantly worse than the control group when they had to perceive the presence of an alternative semantic metaphoric meaning in a pair of words.

Milberg et al. (1987) studied the processing of lexical ambiguities in Wernicke's and Broca's aphasic patients on performance of a lexical decision task. They were to decide whether the third word of an auditorily presented triplet series of words was "real" or not. The first and third words of each triplet were related to one, both, or neither meaning of the second word which was semantically ambiguous. The performance pattern of the Wernicke's aphasics was similar to that of normals; they showed selective access to different meanings of the ambiguous words, as demonstrated by the fact that the context provided by the first word affected semantic facilitation on the third word. In contrast, Broca's aphasics showed no semantic facilitation in the priming condition. These results are consistent with previous findings, suggesting that semantic representations may be largely spared in Wernicke's aphasics. The authors state that the complete failure of Broca's aphasics to demonstrate semantic facilitation in any condition in the experiment suggests that they may have a deficit either in the underlying representation of words or in accessing this information via automatic processing routines. The latter is consistent with the idea that impaired processing of lexical ambiguity requires intact frontal function.

In contrast to these findings, Katz (1988) found that on an auditory lexical decision task with stimulus pairs containing ambiguous (semantically-related) or unambiguous (unrelated) words as primes, Broca's aphasics produced a pattern of results similar to normal subjects; namely, faster reaction times for target words preceded by semantically related than unrelated words (i.e., semantic priming). These results do not support Milberg et al. (1987) but rather suggest that the subjects in the Milberg et al. (1987) study had difficulty with the word triplet paradigm use.

It is clear from the studies reviewed here that there is no consensus on the neurological or cognitive localization of lexical ambiguity. However, there is ample evidence that questions the contribution of the right hemisphere as necessary for this type of language processing.

While there is controversy concerning most aspects of lexical ambiguity and its neural representation, some points can be made from the literature. First, all known meanings of a word are available to normal subjects (Simpson, 1984). Context, if present, then constrains interpretation (Onifer & Swinney, 1981). There is controversy over how long context effects take, whether meaning frequency or dominance affects the order and strength of meaning selection and whether meaning selection occurs automatically or under the control of the subject. However, once a meaning has been selected, additional but non-relevant meanings of the ambiguous word are suppressed or eliminated within a third of a second so as not to interfere with the appropriate interpretation (Swinney, 1979).

While neuropsychological studies are contradictory, damage to the right hemisphere does not produce a consistent relative sparing. There is, however, evidence in the clinical literature that the frontal lobes play an important role in word meaning interpretation or in the underlying ability to scan multiple interpretations serially or simultaneously.

Research questions

The purpose of the present investigation is to examine the possible contributing factors for the observations documented above, particularly the inconsistency of brain-damaged patients' abilities to access alternate word meanings. This includes the lack of any clear-cut disparity in the abilities of left- and right-hemisphere damaged patients to perform such tasks and the uncertainty of any difference in the automatic and conscious processing of ambiguous words. The following specific questions will be addressed:

(A) Do patients with abstraction problems (as measured by neuropsychological tests) show reduced priming for ambiguous words and their multiple meanings, relative to unrelated words, despite their abstraction problems? To answer this question, subjects will participate in a lexical decision task that will measure speed of response using target ambiguous words paired with their multiple meanings, along with unrelated words and non-words.

(B) Are problems with abstraction (as measured by neuropsychological tests) related to decreased ability to perceive lexical ambiguity of the three types tested here, i.e., homonymy, polysemy, and metaphor? To answer this question, experimental subjects will be administered a matching task whereby they will be required to match target ambiguous words with their multiple meanings, presented along with unrelated word choices.

(C) Comparing brain-injured and normal subjects, are there differences in performance on tasks of lexical ambiguity of the three types tested here, i.e., homonymy, polysemy and metaphor (with specific attention to metaphor, given the greater abstraction demands on this word category)?

Methodology

Subjects

Two groups of subjects were selected for this study. One of these groups consisted of 16 traumatically brain-injured (TBI) patients selected from an area outpatient rehabilitation center according to the criteria listed below. The other group consisted of 22 normal subjects drawn from the same rehabilitation setting (personnel and visitors), also according to the criteria listed below. Ages ranged from 19 to 45 years. Monolingualism was determined by having subjects rate their general experience with languages other than English according to a Likert scale (see Appendix A). Those scoring 2 or below (range: 0-4) were included in the study. All subjects were right handed; handedness was determined by A. J. Harris' *Test of Lateral Dominance* (1958). Education level was determined by asking the subject to indicate the highest degree attained by the time of testing, and only those subjects with a minimum of a high school education were included in the study. All subjects had normal or corrected vision and audition. The TBI subjects consisted of patients who had sustained a single traumatic brain injury over six months prior to testing, with no other history of neurological or psychiatric disorders. The normal subjects had no history of neurological or psychiatric disorders.

In order to ensure adequate comprehension of the experimental procedure, subjects were required to achieve 100% yes/no reliability for six personally relevant or general information questions from the *Western Aphasia Battery* (Kertesz, 1982). Subjects were required to perform at a minimum of 90% accuracy on a 40-item practice run on a computer-generated reaction time task to ensure the ability to perform Task One. They were also required to demonstrate at least 90% accuracy in matching 4 choices to a target synonym arrayed in the manner in which Task Two was presented so as to indicate that visual/reading ability was sufficient to process printed stimuli physically similar to those of the task.

The age range for the experimental group was 19-45 with a mean of 36.8 years. Twelve of the 16 patients in this group graduated from college, while the remainder had a high school education. Eight of the 16 were male and 8 were female. Duration post-onset ranged from 8 to 48 months.

Test procedures

Each TBI subject was required to perform Tasks One and Two (see below) in addition to undergoing a neuropsychological evaluation described below, while each normal control was required to perform only Tasks One and Two, in addition to receiving the *Raven's Progressive Matrices* (Raven, 1960) and the *Shipley Institute of Living Scale* (Zachary, 1991), both of which can be converted into IQ scores.

Half of the subjects underwent Task One followed by Task Two; the other half underwent the reverse. Further, stimuli for both Task One and Task Two were randomized. The following ensured attention to the task: (a) subjects were visually and auditorily alerted to Task One by a large cross presented in the center of the screen, along with an audible tone; (b) only those subjects with accuracy scores of 80% or above on Task One were included; (c) subjects were given 2 ten-minute breaks at equal intervals during Task One; and (d) only those experimental subjects who were able to sustain 40 minutes of therapy were included in the study.

Task One (for both groups of subjects)

Task One consisted of a lexical decision task. Subjects were asked to indicate whether a letter string was a word or not by pressing a key on the computer's keyboard designated "yes" or "no". Each subject was seated in front of a Macintosh quadra605 computer with a 14 inch colored monitor which presented stimuli at regular intervals, described below. Stimuli were presented in black on white background via the software package, Mac Probe, which collects reaction time and accuracy data.

The task was introduced to subjects with the following instructions:
"You will hear a sound which will alert you to pay attention to the computer screen. At the same time, you will see a cross on the screen which is also intended to hold your attention. Then you will see two different words on the computer screen, shown one after the other. The first will be shown for only a short while, and may or may not be related to the second word. Do not pay attention to the first word. When the second word comes on the screen, press the "yes" key if you think it is a word in English and the "no" key if you do not think it is a word in English. Press the chosen key as quickly as possible."

These instructions were followed by:
(a) An audible "beep" lasting 500 msec which served as a warning signal intended to direct the subject's attention to the computer screen.
(b) A "priming" word (e.g., "beam") was presented for 100 msec.
(c) The target word, that is, a word related in meaning to the prime (e.g., "light" or "plank" for "beam") was presented for 300 msec.

Interstimulus intervals (ISIs) were 500 msec.

There were 60 lexically ambiguous primes, each paired with 1 non-word and 2 related words, which served as targets. Additionally, there were 60 non-ambiguous primes, each paired with 1 non word and 2 real words (chosen from the list of related words which were paired with the ambiguous words above), which also served as targets. Non-words were constructed by combining one syllable from each of two primes into pronounceable letter strings. Words and non words were randomized for presentation.

This priming task was chosen to address the automatic levels of processing lexically ambiguous stimuli. The rationale for its use was to explore the pervasiveness of abstraction deficits on handling such material, i.e., levels of processing that are presumed to precede more conscious comprehension such as is required by a matching task. Thus, if patients with abstraction difficulties were found to have longer reaction times for lexically ambiguous material, then this would serve as evidence for impairment beyond the level usually tapped by studies of this topic.

Task Two (for both groups of subjects)

Task Two consisted of a word matching task that required subjects to determine whether 4 choices matched a target word in meaning. Stimuli consisted of a single target word printed at the top center of a 5 x 8 index card. Four other words were printed underneath this target word, spaced at regular intervals.

This task was introduced with the following instructions: "You will see a single target word followed by 4 other words. Circle those that are related to the target word in meaning. There may be more than one correct choice."

This matching task was designed to address conscious processing of lexically ambiguous material, and so, to provide converging evidence to data from the priming (automatic processing) task described above. In other words, the combination of the priming task with a matching task aimed to make the expected study findings more robust than if one or the other task were used.

Materials

For this study, the stimuli used to investigate the understanding of alternate meanings of words were of two types, as follows: (a) a priming word followed by target words of the following types: either of two of that (real word) prime's multiple meanings, unrelated words, or non-words, presented on a computer screen, and (b) a single target word or non-word followed by 4 choices, including two or three of that target's multiple meanings and unrelated words. Stimuli for part (b) were presented on 5 x 8 inch index cards with word sprinted in black lettering.

The stimuli for Task One and Task Two consist of the following types of words: polysemous words, homonymous words, and metaphoric words (see Appendices A, B and C); non-words; and unrelated words. There are a total of 20 primes for each of the first three categories (homonymy, polysemy, metaphor)—totaling 60 ambiguous words. Additionally, there were 120 nonwords and 60 unrelated words which served as targets. Note that for the priming task, the words listed in Appendices A, B, and C served as the primes. For example, the target word "chicken" primed the target word "coward".

Most of the polysemous and homonymous words were selected according to a list provided by Durkin and Manning (1989), compiled from a variety of sources, mainly form previous norm-gathering studies (e.g., Gilhooly & Logie, 1989; Gorfein, Viviani, & Leddo, 1982; Nickerson & Cartwright, 1984), with the majority of words chosen from the Gilhooly and Logie (1980) list as this list contains the most extensive normative data about the frequency, imagery, age of acquisition, familiarity, and concreteness of each ambiguous word. Word choices for these were matched for part of speech, length, and frequency.

The metaphoric word list was generated by the investigator, along with two assistants (see acknowledgments) by using the concept of elliptical simile, described earlier. Further, two normal subjects (matched for educational level and age) were asked to choose 20 metaphoric words form the total list of 60 lexically ambiguous words, given instructions to "find those words in the "choices" column that can be used to stand for, or in place of, their corresponding targets." There was high interrater reliability: one rater correctly identified 19 items and the other, 17 items. One rater made 1 false positive response and the other, 0, while one rater made 1 omission and the other rater made 3 omissions.

Word lists were created according to the following rules for selection: All of the polysemous and homonymous words had at least one noun meaning and verb meaning each. For Task Two, four words in each of the polysemous and homonymous word lists had three choices that were potentially correct (i.e., matching meaning); the remainder had only two potentially correct choices. For those that had a verb meaning as a correct choice, at least one foil was also a verb. Though many words in the total list had more than two meanings, viable synonyms were considered only those which were assigned a minimum relatedness score of 12.17 and a maximum relatedness core of 4.0 as assigned by Durkin and Manning (1989), where available. Eighty-five percent of the words used had this information available.

As the features (i.e., frequency, imagery, age of acquisition, familiarity, and concreteness) considered for matching homonymous and polysemous words were not available for the metaphoric words, the latter category was treated as a separate category. Six of the metaphoric words and verb meanings. The metaphoric words as a group had two synonyms that were considered potentially correct choices for task two.

Results

Task One: Reaction time task

Prior to analyses, trials on which reaction time measures fell farther than 2 standard deviations from the group centroid (by computing Mahalanobis d-squared for all raw scores within each subject) were eliminated, as these were considered to represent lapses of attention (a common practice in studies using reaction times). Using the two standard deviation guide, no more than 2% of the total number of trials for any one subject were eliminated from the analysis.

The first step in the analysis of these data was to calculate the difference between the reaction times for the prime-target pairs, e.g., "chicken - coward" (metaphoric pairing) and "chicken - bird" (literal pairing). For the metaphoric word category, the order of subtraction was metaphoric minus literal values. As the two related homonymous and polysemous targets were matched for frequency, etc., the order of subtraction was random. The next step in the analysis was to obtain averages of the above calculations by category (homonymy, polysemy, and metaphor) for each subject, followed by averaged across subjects. These figures were used in all of the following analyses with the exception of the final analysis of variance (ANOVA).

Initial simple correlations of all reaction time scores showed an unexpected correlation between homonymy and metaphor. Such a correlation between variable violated the assumption of ANOVA with repeated measures and artificially inflates the chance of finding spurious significance. Therefore, the initial method of analysis was a Multivariate Analysis of Covariance (with age as a covariate), or the Hotelling's T-Squared, which takes into account the dependence between variable when evaluating differences between two groups. This is a stringent measure of the difference between groups (patients and normals) across the reaction time variable (homonymy, polysemy, and metaphor). The overall multivariate results were evaluated using Wilk's Lambda (df = 3,24) which resulted in a value of 0.761 with p = .024.

Since the previous analysis demonstrated differences between the groups, the next question of interest concerned which of the variables they differed on. Results of t-tests for each of the three variables (homonymy, polysemy, and metaphor) are depicted in Table 1.

Dependent variable	t	df	2-tailed P
Homonym	-.606	36	.547
Polysemy	.462	36	.646
Metaphor	2.905	36	.006

Table 1.   T-Test values for independent samples comparing patients and normals on the three category types. These values correspond to Graph 1.

There was a significant difference between the groups for metaphors, with an apparent difference between mean difference for metaphors and the other two variables in the patient group, i.e., it suggests a significant slowing in responding to metaphoric prime-target pairs when compared to polysemous and homonymous pairs in patients but not in normals. In fact, when compared to the unrelated prime-target pairs, there was no priming of metaphoric prime-target pairs when compared to polysemous and homonymous pairs in patients but not in normals. In fact, when compared to the unrelated prime-target pairs, there was no priming of metaphoric prime-target pairs. In order to statistically examine this relationship between the means, an ANOVA with planned comparison was performed (as it was hypothesized that responses to metaphoric pairs would be slower than the other two word categories): the mean differences for the metaphor category was compared with the combined means for the homonymy and polysemy categories since no difference was found between the latter two. The result of this analysis confirms the significance of the difference between metaphor and the other two variables {F(df = 1,36) = 4.16, p < .05}. Actual reaction times (means and S.D.) for patients and normals are shown in Table 2.

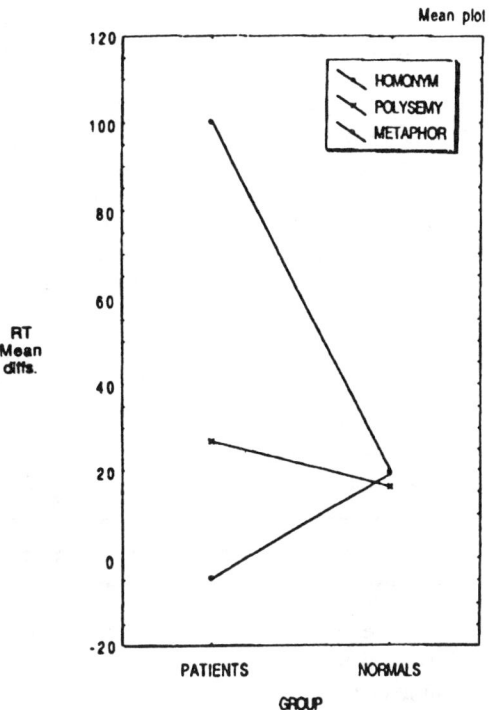

Graph 1.   Mean differences in reaction times for three word types. For each prime, reaction times for two related targets were computed and averaged across trials and across subject groups. This depicts a statistically significant difference between metaphor and the other two variables, i.e., it shows significant slowing in responding to metaphoric prime-target pairs when compared to polysemous and homonymous pairs in patients but not in normals.

	Patient mean	Patient S.D.
H	721	269
P	746	281
M1	782	266
M	720	271
	Normal mean	Normal S.D.
H	597	111
P	610	86.8
M1	639	134
M	605	125

Table 2.   Reaction Time performance (Mean and S.D.) for patients and normals on the following conditions: homonymous pairs (H), polysemous pairs (P), metaphoric pairs (M1) and literal (M) word pairs.

The next question for Task One concerned the relationship between performance on the neuropsychological battery and the reaction time data for the three variables (word categories). A canonical correlation showed that they are significantly correlated with the following outcome: R = .987; chi-squared (df = 27) = 43.69, p = .022. A more specific question concerning the relationship between tests of abstraction and reaction time performance was investigated using Backward Stepwise Multiple Regression Analysis. In this analysis, reaction time differences for metaphors was used as the dependent variable with scores on the neuropsychological battery as predictors. The following tests were noted to contribute most to the prediction of reaction time data in the regression equation: Wisconsin Card Sort Test, Trails B, Verbal I.Q., Word Fluency (FAS), Boston Naming Test, and Raven's. The results of this latter analysis with the standardized beta weights are presented in Table 3.

Variable	Standardized beta values
VIQ	-1.836
BNT	1.151
Trails-B	.658
FAS	.506
WCST-C	.483
Raven's	.377

Table 3.   Results of the multiple regression analysis for neuropsychological tests and the "metaphor effect" (F = 3.887, df = 6,8, p > .05, Multiple R = .863). The metaphoric-prime target pairs minus the same prime-literal meaning pairs. This shows the significance of the combined performance on tests of abstraction in predicting reaction time differences for metaphoric and literal word pairs, i.e., performance on BNT contributed the most weight to reaction time performance.

In order to determine the presence of overall priming, an ANOVA was performed and the results showed a main effect for relatedness of prime for the patients (F = 4.92, p = .032), but no interaction, thus demonstrating the priming effect for patients but no difference for normals. As expected, the patients performed significantly slower than the normals with means as follows: TBI group: 764.04 and Normals: 612.72.

Task Two: Matching task

In contrast to the reaction time findings, ANOVA showed no main effect for group for the matching task (F = .93, p > .05). On the other hand, significant differences were noted among categories for both groups (F = 5.63, p = .01), also unlike the reaction time task. Note that even though the differences are statistically significant, they appear to be relatively small. There was a statistically significant difference among the three categories for performance on the matching task, with performance on homonymous words more accurate than polysemous and metaphoric words, and no significant difference between metaphor and polysemy (See Tables 4 and 5). However, this may be a spurious finding due to unknown variables, since this pattern was not revealed in the reaction time data, and since the actual differences are, indeed, small.

	df	MS effect	F	p-level
Group	1,36	88.81	1.88	0.18
Word Category	2,72	29.09	5.63	0.01
Interaction	2.72	04.82	0.93	0.40

Table 4. Results of the analysis of variance for matching performance

	Homonymy	Polysemy	Metaphor
Patients	93.9	92.1	93.2
Normals	95.6	94.2	94.2

Table 5. Means of percent correct for patients and normals, by categories

Discussion

The goal of this study was to determine the type of cognitive functions required for the successful interpretation of lexical ambiguity. The primary contention held that patients exhibiting problems with abstraction would demonstrate impairments in the sensitivity to alternate meanings of words and that this would be evident on both the lexical decision (reaction time) task and the matching task. Further, this was expected for the three categories of ambiguous words used: homonymy, polysemy, and especially for metaphor, given the greater demands on abstraction for this word type.

The experimental procedures were developed to evaluate the difference between brain-injured and normal subjects on these tasks, and were designed to allow analysis of any differential effect of word type (i.e., homonymy, polysemy, and metaphor) on lexical access and on a comprehension task.

There is sufficient evidence that questions the degree of right hemisphere involvement in lexical ambiguity interpretation (Winner & Gardner, 1977; Gagnon et al., 1994) and other evidence that points to frontal lobe involvement in these types of tasks (Milberg et al., 1987). The results of this study support the latter, as is evidenced by the fact that tests of abstraction contributed much more weight than tests thought of as affecting right hemisphere function to predicting performance on lexical ambiguity tasks.

Task One

WAIS-R Verbal I.Q. scores contributed most heavily to predicting reaction time performance for metaphors, i.e., the higher the Verbal I.Q. score, the smaller the difference between the appreciation of concrete and metaphoric meanings of words in this category, indicating approximately equal facility of access of word meaning. The other tests found to correlate in this direction were WCST category subtest, Trails B, Word Fluency (FAS), BNT, and Raven's, with Ravens showing the lowest weight of this group of tests, which, save for the BNT are widely considered to be tests of abstraction. Note, however, that the standardized weight of the Raven's, while relatively small, was still in the same direction. It is notable that while the Raven's is a test of abstraction, it also requires intact visuo-spatial processing, and may thus also be considered a test of right hemisphere function.

Reaction time performance for polysemy was most related to performance on WMS-R Verbal subtests and Trails B, and to a lesser degree with the WCST Perseveration subtest, Stroop, and Raven's. For homonymy, reaction time performance was most related to Word Fluency, Verbal I.Q., and BNT, and to a lesser degree, WCST, Trails B and Stroop.

There was the expected finding that patients performed with longer response latencies on the reaction time task than did normals. Another expected finding was slower reaction times for words related to the target in meaning than for words unrelated to the target in meaning, for the patient group. The most striking finding on the reaction time task was that the patients evidenced significantly slower reaction times on metaphor than on polysemy and homonymy both within their own group and as compared to the normals.

Task Two

On the matching task, which used the same ambiguous words as the lexical decision task, patients performed at accuracy rates which were comparable to the normals. While this finding is surprising, it is important to note that the matching task involved untimed multiple choice. The finding that patients' performance was comparable to normals' performance suggests relative intactness of the semantic representations of these words in spite of retrieval deficits. It is noteworthy that patients and normals showed higher accuracy for responding to homonyms than to polysemous or metaphoric words on the matching task, despite the fact that the homonymous and polysemous words were matched closely in terms of frequency, etc. This suggests relative ease in discerning discrete multiple meanings such as for homonymous words as compared to a relative deficit in fully appreciating the marginal meanings of the polysemous and metaphoric word stimuli. Recall that homonyms consist of only central meanings, while polysemous and metaphoric words consist of both central and marginal meanings, which, within categories, are not mutually exclusive.

General issues

This study has revealed numerous methodological difficulties confronted in the field of lexical ambiguity. For example, divisions between polysemy and metaphor are somewhat blurred and often arbitrary. Indeed, Brownell et al. (1990) investigated brain damaged patients' appreciation of alternative word meanings using a word list which consisted of "metaphoric polysemous" words. In considering the above findings in light of the widely held definitions of metaphor, it would be prudent for researchers to clearly define the categories of lexically ambiguous words and to control for their individual parameters.

In the classical literature on this topic which dates back to Aristotle, metaphor is seen as an instance of novel poetic language in which words are not used in their normal everyday sense (Lakoff, 1989). In contrast to language as the central realm of metaphor, Ortony (1993) considers metaphor to be more in the domain of thought, or the mapping of conceptual domains. Thus, the focus of metaphor is in the way we conceptualize one mental domain in terms of another, i.e., cross-domain mapping. For example, metaphor can be seen in terms of mapping from a source domain (tabloid) to a target domain (rag). Here, there is a conceptual correspondence between one domain of objects (newspapers) and a very different domain of objects (cloth). Such cross-domain mapping requires cognitive operations which are necessary for testes of abstraction, namely, set-shifting (WCST,

Trails B), flexibility of thought (Word Fluency), and conceptualization (Verbal I.Q.)—all of which were strongly related to performance on reaction time for metaphor. Generally, then, the same cognitive functions are necessary for metaphor and tests of abstraction.

To date, there is great variability in the methods used in studies on this topic, including subject selection, task requirements, and response mode. Aside from the obvious difficulties with issues such as these, they raise special concern with regard to the intention of this study. While the primary goal was to isolate the cognitive functions responsible for lexical ambiguity interpretation, an extension of this would involve identifying its neurological substrate.

Thus, it was presumed that patients who exhibited difficulties in abstraction would also evidence frontal lobe impairment, and thus, difficulty on the study tasks. While most authors claim to use patient populations with discrete right or left lesions, there may exist an inherent bias toward selecting patients with anterior lesions, as a common exclusion criterion for right hemisphere patients is visual neglect—a symptom associated with right posterior damage. Further, most studies do not use the same lexically ambiguous stimuli over a variety of tasks, as this one did.

A further methodological difference in the literature on this topic concerns the use of sentences vs. words as stimuli. Single words were chosen for this study as this would assist in eliminating the possible influences of impaired attention or memory in the TBI subjects. Additionally, differences exist among priming studies which use sentences. For example, Onifer and Swinney (1981) showed differences in priming ability when word choices were presented immediately following the occurrence of the ambiguity in a sentence, contrary to when these word choices were presented 1.5 seconds following occurrence of the ambiguity.

Additionally, metaphoric target phrases require significantly longer processing times than literal target phrases if reading time measures are taken at the end of the target phrase rather than at the end of the target sentences in which the phrases are embedded.

Given that there were dissimilar findings for both tasks in this study, this calls into question the validity of findings of studies which use one or the other type of task. Indeed, the matching data in this study revealed the relative facility of access even in the brain-injured population, while the reaction time data showed this population to be quite below normal levels of performance.

As there is some question as to the issue of how the number of meanings for lexically ambiguous words affects their comprehension, a sampling of performance of both patients and normals on the matching task was taken in order to see if accuracy was affected by this parameter. Of the sixty target items, eight had three potentially correct matches, while the remainder had only two. A preliminary analysis of these data indicated that patients and normals performed at approximately equal accuracy rates (48% and 51%, respectively) for those items with three potentially correct matches, with much higher accuracy for those items with only two (92% and 94%, respectively). Though this study was not set up to address the question of selective vs. exhaustive access, this preliminary analysis would point in the direction of supporting selective access, whereby the access of two meanings on an ambiguous word may suppress the third. Certainly, this issue should be addressed in a more rigorous manner, as it raises important questions regarding the possible contributing factors to disambiguation.

Thus, in addition to providing support for the cognitive localization of lexical ambiguity to function (such as abstraction) related to the frontal lobe, this study has highlighted some of the concerns related to other investigations of this topic. The literature is replete with studies on lexical ambiguity, particularly with normals, and it is evident that findings using the brain-injured population would add new perspective on the topic, when such studies are stringently controlled.

Conclusions

The most common way of investigating sensitivity to lexical ambiguity is through the use of tasks which require conscious processing, such as the matching of words or phrases. The present study differed in that it used both a matching task and a priming (reaction time) task; the latter was included in an attempt to contribute to the robustness of the expected findings, i.e., worse performance for patients with abstraction difficulties. However, a discrepancy in performance on these tasks was found, with brain-injured subjects performing as well as normals on the matching task, though performing with much slower reaction times on the priming task. This is important in highlighting the stage of cognitive processing (automatic vs. conscious) at which disambiguation takes place, and also points to the necessity of careful design in studies on this topic.

Given the large discrepancies found in other studies of the topic of lexical ambiguity using brain-injured subjects, serious consideration was given to other possible confounding variables in the design of this particular study. These include subject selection, task requirements, response mode, and selection of stimuli.

While this study sets out to isolate the cognitive processes involved in lexical ambiguity comprehension, it did not rely on a patient population with clear-cut lesion sites. Instead, it used the procedure of evaluation to determine purported lesion site, and provide findings regarding right hemisphere and frontal involvement in lexical ambiguity. The findings provided support for the cognitive mechanisms underlying the comprehension of lexical ambiguity, all of which are traditionally assigned to the frontal lobes. These include the ability to simultaneously apprehend one or more item and to shift from one item (or mental set) to another. A natural progression in the study of this topic would be to attempt to localize the neurological underpinnings for these cognitive mechanisms, a task made simple by the use of a population of subjects with focal brain damage, i.e., left, right, anterior, or posterior.

References

Beardsley, M. (1962). The metaphorical twist. *Philosophy and Phenomenological Research, 22,* 293-307.
Brown, J. W. (1997). Process and creation. In A. Andersson and N.-E. Sahlin (Eds.), *The complexity of creativity* (pp. 35-50). Dordrecht: Kluwer.
Brownell, H. H., Potter, H. H., Michelow, D., & Gardner, H. (1984). Sensitivity to lexical denotation and connotation in brain damaged patients: A double dissociation? *Brain and Language, 100,* 717-729.
Brownell, H. H., Simpson, T. L., Bihrle, A. M., Potter, H. H., & Gardner, H. (1990). Appreciation of metaphoric alternative word meanings by left and right brain damaged patients. *Neuropsychologia, 28,* 375-383.
Chiarello, C. (1988). Semantic priming in the intact brain: Separate roles for the right and left hemispheres? In C. Chiarello (Ed.), *Right hemisphere contributions to lexical semantics* (pp. 59-69). New York: Springer-Verlag.
Durkin, L., & Manning, J. (1989). Polysemy and the subjective lexicon: Semantic relatedness and the salience of intraword sentences. *Journal of Psycholinguistic Research 18,* 577-612.
Gagnon, L., Goulet, P., & Joanette, Y. (1994). Semantic processing of ambiguous words after a right hemisphere lesion. Poster presented at the International Neuropsychological Society. Angers, France.
Gardner, H., Brownell, H., Wapner, W., & Michelow, D. (1983). Missing the point: The role of the right hemisphere in the processing of complex linguistic materials. In E. Perecman (Ed.), *Cognitive processes in the right hemisphere* (pp. 169-191). New York: Academic Press.
Gardner, H. (1994). The stories of the right hemisphere. In W. Spaulding (Ed.), *Forty-first Nebraska symposium on motivation 1992-1993.* Lincoln, NE: University of Nebraska Press.

Gazzaniga, M. S. (1983). Right hemisphere language following brain bisection: A 20-year perspective. *American Psychologist, 38,* 525-537.

Gilhooly, K. J., & Logie, R. H. (1980). Meaning-dependent ratings of imagery, age of acquisition, familiarity, and concreteness for 387 ambiguous words. *Behavior Research Methods and Implementation, 2,* 428-450.

Goldstein, K. (1948). *Language and language disturbances.* New York: Grune and Stratton.

Harris, A. J. (1958). Harris Tests of Lateral Dominance. *Manual of directions for administration and interpretation* (3rd ed.). New York: The Psychological Corporation.

Heaton, R. K. (1981). *Wisconsin Card Sorting Test.* Odessa, FL: Psychological Assessment Resources, Inc.

Kertesz, A. (1982). *Western Aphasia Battery.* Ontario: The Psychological Corporation.

Krashen, S. D. (1976). Cerebral asymmetry. In H. Whitaker and H. A. Whitaker (Eds.), *Studies in Neurolinguistics* (Vol. 2, 157-191). New York: Academic Press.

Lakoff, G. (1993). The contemporary theory of metaphor. In A. Ortony (Ed.), *Metaphor and thought.* Cambridge University Press: New York.

Lezak, M. D. (1995). *Neuropsychological assessment* (second ed.). New York: Oxford University Press.

Luria, A. R. (1966). *Higher cortical functions in man.* New York: Basic Books.

Milberg, W., Blumstein, S. E., & Dworetzky, B. (1987). Processing of lexical ambiguities in aphasia. *Brain and Language, 31,* 138-150.

Ortony, A., (1993). *Metaphor and thought* (2nd ed.). Cambridge: Cambridge University Press.

Onifer, W., & Swinney, D. A. (1981). Accessing lexical ambiguities during sentence comprehension: Effects of frequency of meaning and contextual bias. *Memory and Cognition, 9,* 225-236.

Raven, J. C. (1960). *Guide to the Standard Progressive Matrices.* London: H. K. Lewis.

Searleman, A. (1977). A review of right hemisphere linguistic capabilities. *Psychological Bulletin, 84,* 503-528.

Simpson, G. B. (1984). Lexical ambiguity and its role in models of word recognition. *Psychological Bulletin, 96,* 316-340.

Simpson, G. B., Burgess, G. B., & Peterson, R. R. (1987). Human comprehension processes and the indeterminacy of meaning. *Cognitive Systems, 2,* 213-232.

Swinney, D. A. (1979). Lexical access during sentence comprehension: (Re)consideration of context effects. *Journal of Verbal Learning and Verbal Behavior, 18,* 645-659.

Swinney, D. A. (1982). The structure and time course of information interation during speech comprehension: Lexical segmentation, access, and interpretation. In J. Mehler, E. C. T. Walker, and M. Garrett (Eds.), *Perspectives on mental representation: Experimental and theoretical studies of cognitive processes and capacities.* Hillsdale, NJ: Lawrence Erlbaum.

Winner, E., & Gardner, H. (1977). The comprehension of metaphor in brain-damaged patients. *Brain, 100,* 719-727.

Zachary, R. (1991). *Shipley Institute of Living Scale.* Manuel, CA: Western Psychological Services.

Ziff, P. (1967). Some comments on Mr. Harmon's confabulations. *Foundations of Language, 3,* 403-408.

Appendix A

Metaphoric words	Primes (for RT task)	Targets (for RT task)	Targets (matching task)	Choices (matching task)
BUG	INSECT	SNAKE	PESTER	RESERVE
BLUE	COLOR	SMELL	SAD	HAPPY
PIG	SLOB	FRUIT	COMEDIAN	ANIMAL
SHARK	BIRD	THIEF	SAILOR	FISH
RAG	CLOTH	PILLOW	TABLOID	CHECK
NUTS	VEGETABLES	CRAZY	RELAXED	SEEDS
DOUGH	BREAD	MEAT	TOWEL	MONEY
PEANUTS	SNACK	SMALL	WHEAT	LARGE
LEMON	FRUIT	FENCE	MINERALS	CAR
DOG	PURSUE	ANIMAL	PLANT	KNOW
BUTTERFLIES	INSECTS	FURNITURE	FUNNY	NERVOUS
SHADES	CURTAINS	SUNGLASSES	EAT	WATCH
EGG	INDUCE	FOOD	FURNITURE	GESTURE
FRESH	OPEN	NEW	FRAYED	BRAT
CHICKEN	BIRD	REPTILE	COWARD	LAZY
WHEELS	BUTTONS	HOUSE	TIRES	CAR
BOOT	CHAIN	KICK	GLUTTON	SHOE
TUBE	STEREO	TELEVISION	CYLINDER	STEM
BANANAS	CRAZY	SNEAKY	FRUIT	WACKY
GREEN	COLOR	TEXTURE	OLD	NEW

Appendix B

Homonymous words	Primes (for RT task)	Targets (for RT task)	Targets (Matching task)	Choices (Matching task)
ROCK	MINERAL	GAME	SWEEP	SWAY
SCALE	FISH	LEAF	PAINT	BALANCE
PLANT	VEAL	FLOWER	BUILDING	CASTLE
FILE	WATER	TOOL	CABINET	BUSH
BAT	STOOL	VEGETABLE	CLUB	ANIMAL
BILL	TAIL	DOLLAR	SHOE	BIRD
STABLE	STEADY	KITCHEN	HORSE	FAST
POST	FLOAT	MAIL	AFTER	WOOD
BANK	FREEZER	VALLEY	MONEY	RIVER
DECK	ROLL	SHIP	CARDS	DOOR
BOLT	SCREW	SPOON	SWIM	RUN
BALL	DANCE	PARADE	CLOTHING	TOY
COUNT	ADDITION	NOBILITY	SING	NURSE
SEASON	TIME	BURN	PEACE	FLAVOR
DATE	TIME	FRUIT	CALENDAR	POEM
PUNCH	HIT	DRINK	READ	ROSE
SHIP	BOAT	TRACTOR	THROW	SEND
COACH	BUS	TEACH	HELICOPTER	PAINT
PORT	HARBOR	WINE	CITRUS	SUITCASE
PALM	HAND	CABIN	KNIFE	TREE

Appendix C

Polysemous words	Primes (for RT task)	Targets (for RT task)	Targets (matching task)	Choices (matching task)
HIDE	CONCEAL	SKIN	VENEER	RUB
PART	SECTION	STAPLE	ROLE	COIL
MOUTH	PEAK	FACE	RIVER	ROOF
LAP	SQUEEZE	PAGES	KNEES	LICK
CHOP	CLEAN	SKIRT	CUT	MEAT
STEP	WALK	DRIVE	RIBBON	STAIR
BOX	TYPE	CONTAINER	STEP	FLIGHT
LIGHT	CABOOSE	STEM	IGNITE	LAMP
SINK	RECEPTACLE	SHED	SUBMERGE	IRON
TRIP	FOUNTAIN	FALL	JOURNEY	SIT
NOTE	MESSAGE	SEED	MUSIC	PLAY
CHARM	QUALITY	FLATTER	JEWELRY	DRESS
LETTER	ALPHABET	STAMP	SPOKE	OPERA
PLUG	STOPPER	PUMP	ELECTRIC	HOSE
STUMP	TREE	LEG	PUZZLE	JOKE
BEAN	LADDER	SMILE	PLANK	LIGHT
BLOCK	CUBE	WOOD	STOP	DIVE
SHEET	PAPER	UMBRELLA	BRAID	BED
FORK	ROAD	CLOUD	COMB	UTENSIL
MARCH	JUMP	PROTEST	MONTH	WALK

J. Neurolinguistics, Vol. 11, Nos 1–2, p. 137–152, 1998
© 1998 Published by Elsevier Science Ltd. All rights reserved
Printed in Great Britain
0911-6044/98 $19.00 + 0.00

Pergamon

PII: S0911-6044(98)00010-4

Shifting the burden to the interlocutor: Compensation for pragmatic deficits in signers with Parkinson's disease

Judy A. Kegl and Howard Poizner

Center for Molecular and Behavioral Neuroscience, Rutgers, The State University of New Jersey

Abstract—This paper takes a qualitative look at the behavior of interlocutors in the context of conversations with deaf signers of American Sign Language (ASL) who have Parkinson's disease (PD), a neurodegenerative disease affecting sensori-motor planning and execution, attentional systems, and memory. Deficits in these areas, while non-linguistic, impact pragmatic aspects of conversational interaction. Three signers with PD at varying levels of severity (mild, moderate-to-severe, and severe) are compared and contrasted with attention given to the effects of PD on their use of conversational regulators and turn-taking devices. The behavior of persons engaging in conversation with signers who have PD is shown to corroborate in tangible ways what we have identified as the attentional and sensori-motor deficits of PD. Interlocutors' complex repositioning to maximize visual acuity, increased use of shift regulators, hyperattentiveness to sign initiation, as well as increased reiteration and expansion of information all serve to amplify and reinforce the contributions of signers with PD to ASL conversations. Over the course of the disease, unimpaired interlocutors gradually "become" the attentional system of signers with PD. Increased tapping and attention-getting devices used to explicitly re-orient attention compensate for attentional deficits and maintain the participation of signers with PD in multi-participant conversations.

Introduction

Pragmatics can be thought of as language in context, both the real world context and discourse context. Real world context focuses on using shared knowledge to inform interpretation. Discourse context focuses upon chunks of communication beyond single sentences or phrases that cohere in an interdependent fashion. These chunks can be comprised of independent sentences produced by a single speaker that are strung together in a narrative or comprised of utterances interwoven by several participants in the context of a conversation.

Analysis of language in context must take into account not only grammatical competence on the part of producers of language output but also the viability of the message as received by the interlocutor. Sometimes the behavior of the non-neurologically impaired interlocutor can be the best indicator of the nature of a deficit that impacts conversations they are participating in.

This paper takes a qualitative look at the behavior of the interlocutor in the context of conversations with Deaf signers of American Sign Language who have Parkinson's disease, a neurodegenerative disease involving motor and attention deficits that impact pragmatic aspects of conversational interaction. We begin by reviewing aspects of the deficits of unilaterally cortically-lesioned signers as a basis of comparison.

Signers with cortical lesions

Damage to posterior portions of left cerebral cortex

The sentence has been claimed by linguists such as Noam Chomsky to have a privileged linguistic status. Examination of a left temporo-parietal lesioned signer of ASL (as compared with his neurologically intact identical twin) revealed a dissociation of linguistic versus affective facial expression focused specifically at the sentence level (Kegl & Poizner, 1997). The aphasic signer exhibited preserved abilities to use word-level adverbial facial expressions, discourse level markers of topic, and affective facial expression, while exhibiting specific problems with clause-level grammatical facial markers that are associated with complementizer position (such as wh-question markers, markers of conditional expressions, etc.) as well as syntactically moved topics. This deficit in the use of clause level grammatical facial expressions coexists with reduced production of syntactically possible word orders and construction types in ASL. This interdependence makes sense since permutations in word order are frequently signaled by differences in the occurrence and spread of facial expressions. In addition, this signer tends neither to consistently inflect signs with subject and object agreement nor to consistently use other aspects of spatial grammar.

Compensations

Despite the syntactic deficits identified in this left-lesioned signer, interlocutors use a variety of strategies, often unconscious, to maximize communicative success. For example, in signing to him, his wife and brother signed slower and used a much higher proportion of ASL sentences using canonical SVO ordering. Sentences produced by this left-lesioned signer were often repeated in an expanded and inflected form by the interlocutors and correct interpretation was verified.

Damage to posterior portions of right cerebral cortex

Sentence level processes in both signed and spoken languages seem particularly susceptible to damage that occurs in the Perisylvian regions of the left hemisphere, while discourse level disruptions occur more frequently with right hemisphere damage. Neurological evidence for a distinction between sentence and discourse was found when we examined differential impairment of parallel grammatical processes that occur at both the level of syntax and discourse in a right temporo-parietal lesioned hearing signer of ASL (Kegl & Poizner, 1991; Poizner & Kegl; 1992). This right-lesioned signer was able to establish and consistently maintain both person and locative agreement within the confines of the sentence, but failed to do so between sentences strung together in a discourse.

Agreement is a process that links two grammatical entities. In person agreement, noun phrase arguments are linked to morphological markers of subject and object on the verb via association with the same location in the signing space. This process signals which noun phrase functions as the subject and which functions as the object. In locative agreement, oblique arguments are similarly linked to their respective verbs, registering the locative or directional relations they bear in a given clause. From sentence to sentence within a discourse, the person agreement and locative agreement of noun phrases with the same preestablished locations in the signing space, serve as a cohesive device that maintains reference throughout a narrative or conversation.

This same right-lesioned signer exhibited problems in the use of discourse level markers of role shift (breaking of eyegaze with interlocutor, shift of the body or head orientation toward the referent whose role is being taken on, facial caricature, etc.), while consistently marking role shift at the sentence level via a change of third person to first person grammatical agreement (Loew, Kegl, & Poizner, 1997) and while consistently using eye gaze and head tilt for grammatical agreement. In transitive sentences in ASL, eyegaze to a referent marks object agreement, while head tilt toward a referent is a marker of subject agreement (see Bahan, 1996).

Compensation

Except for some problems with the linguistic encoding of direct visuo-spatial mapping (using verbs of motion and location and classifiers to describe real world spatial relations) the communication of this right-lesioned signer was highly grammatical and intelligible at the level of the sentence. Beyond the sentence, the inconsistencies of coreference from sentence to sentence are compensated for in two ways. The signer herself compensates for inconsistent discourse coreference by redundantly re-establishing the referential indices of referential noun phrases in each sentence. So, in essence, each sentence becomes its own encapsulated discourse. The interlocutor compensates, almost unwittingly, by reconstructing these redundant sentence sized narratives into a larger cohesive text. In the case of role play, the interlocutor typically uses shared knowledge to reconstruct insufficiently signaled cases of role play. Another cue is to attribute a role shift to situations in which maintaining the same first person referent would be illogical. In any case, the burden of reconstructing either coreference through a narrative or instances of role shift are placed upon the interlocutor rather than the narrator.

Summary

Cortical lesions tend to affect circumscribed components of linguistic (or other cognitive) capacities, leaving other aspects intact. The compensations made on the part of the interlocutor tend to occur at the level of error correction in the course of interpretation, and are frequently so automatic that the errors and stylistic aberrations on the part of the right-lesioned signer actually go unnoticed.

Signers with Parkinson's disease

In contrast with cortical lesions, subcortical degeneration of the sort observed in Parkinson's disease (PD) affects sensori-motor planning and execution, attentional systems, and memory. Yet, deficits in these areas, while non-linguistic in nature, can directly impact language production and comprehension.

Parkinson's disease

Patients with Parkinson's disease are thought to decompose high level motor plans, being unable to automatically execute multiple motor tasks either sequentially or simultaneously (Marsden, 1982; Marsden & Obeso, 1994; Harrington & Haaland, 1991; Weiss et al., 1997). They seem unable to program movement parameters for the next motor act (Marsden, 1987; Benecke et al., 1987), or to scale movement amplitude (force) to fit task requirements (Margolin & Wing, 1983; Hallett & Khosbin, 1980).

Manifestations of Parkinson's disease in signers of ASL

Manifestations of PD tend to be characterized by across-the-board effects on sign articulation, attention, and memory rather than by circumscribed linguistic errors.

Flattening of distinctions

Typically, language tends to maintain a delicate balance between ease of articulation and ease of perception. Overall, signers coping with the effects of Parkinson's disease adjust this balance to favor ease of articulation over ease of perception, putting the burden of perception on their addressees. In other publications (Kegl & Poizner, 1993; Brentari & Poizner, 1994; Brentari, Kegl, & Poizner, 1995; Loew, Kegl, & Poizner, 1995) we have discussed the nature of the ease-of-articulation changes that characterize PD. The examples discussed below can be found in those publications.

Accommodations favoring ease of articulation include an overall laxing of handshape, movement, orientation and place of articulation distinctions such that each handshape is only minimally distinguished from the others. With handshape, this can involve failure to fully open or close the fist; failure to maximally spread, extend, or contact the fingers; or failure to suppress the involvement of non-selected fingers in a given sign articulation.

Movements are laxed by being only minimally realized. For example, a translatory movement that extends from slightly in front of the signer to near the addressee can be reduced to orientation from signer to addressee with the most minimally perceptible movement from one toward the other. Hand internal movements are laxed by minimizing the separation between or bending and closure of the fingers. Whole arm movements are reduced to movement at fewer or even to a single joint. For example, shoulder and elbow movements may be distalized to the wrist or even fingers.

Orientation is often shifted to an unmarked position. For example, palm upward orientation of the basehand in a sign like NEW, or READ, may be shifted to a sidewards orientation, forcing the entire articulation of these two-handed signs to follow suit.

Place of articulation is often laxed by failure to make contact with the location on the body with which the sign is associated. In addition, the shortening of sign movements can lead to a shortening of the distance between, and therefore a change in the locus of, sign articulations. While raising can occur as a result of a reduced transition between a sign made at the face and a following sign made at chest level (typically only in signers with mild PD), signers with PD tend to produce signs lower than usual, possibly as a result of lessened resistance to gravity in an attempt to reduce effort. In fact, there is an across-the-board shrinking and lowering of the signing space over the course of the disease.

Facial expressions are also compromised. Affective facial expressions are replaced by an almost masklike expression. Grammatical facial expressions, responsible for carrying the bulk of syntactic information in ASL are slower to disappear.

One accommodation is that maximum articulation of a given grammatical facial gesture is not attained. The eyebrows are only slightly raised; the brow only slightly furrowed. In fact, facial expressions can also become distalized. For example, the raised eyebrows of a yes/no-question are often realized instead by a slight raising of the upper eyelid.

Another accommodation involves changes in the way facial expressions are mapped onto the signed sentence. Aarons, Bahan, Kegl, and Neidle (1992) note that grammatical facial expressions are associated with a given position in a sentence and spread from that position over a circumscribed syntactic domain (known as a c-command domain). As a result, the facial expression builds to or wanes from a maximal articulation associated with that position (Bahan, 1996).

In signers with PD, this mapping of an independently modulated facial expression to a distinct position on an independently signed manual sequence of signs is the kind of sensory motor act of planning and coordination that poses problems. Instead, the facial expression, reduced but unmodulated for intensity is coupled with the manual string and spread evenly over the clause. (See Kegl, Cohen, & Poizner (in press) for a description of this alternate form of non-manual mapping to a manual sequence of signs and Brentari, Kegl & Poizner (1995) for parallel examples of the coupling of articulatory components solely within manual articulations of signing as produced by signers with PD.)

Short of the most severe cases, it should be noted that the laxing and flattening of distinctions noted in all PD signers are not a mandatory result of a compromised motor system. Signers with PD are able to produce good exemplars of signs in isolation. The ease of articulation accommodations mentioned here occur in extended sign sequences such as signing full sentences or chunks of discourse. Thus, requirement for production of extended sign sequences make it much more likely that a signer with PD will show the reductions and laxing discussed above.

Attentional problems

In terms of attention, signers with PD tend to have a hard time disengaging attention from one person to another, so that in conversations involving more than two participants, signers with PD find it difficult to shift attention from one interlocutor relinquishing a turn to the next interlocutor who has recently taken the floor. In a signed language conversation where visual contact is crucial to receiving the message, failure to shift one's attention to the next contributor to the conversation can leave one lost. Furthermore, failure to give the appropriate addressee response that signals willingness to yield the turn, may leave the signer initiating a turn unsure of what to do.

Summary

We are working from the premise that the linguistic competence of Parkinsonian signers is left intact, and that they are, in fact, using aspects of this competence to systematically lax and reduce signs, making them easier to produce. However, the modifications made in Parkinsonian signing combined with attentional problems can interfere greatly with the pragmatic aspects of language use, particularly participation in conversations. Flattening of distinctions in sign articulation makes the Parkinsonian signer more difficult to understand, problems with initiating signed utterances and attentional problems make it difficult for the signer with PD to maintain participation in the conversation at all. Moreover, their failure to give the appropriate addressee responses in timely fashion may even lead to their being seen as disruptive of the conversational flow.

Regulators and turn-taking in ASL conversations

Before turning to our study of the impact of Parkinson's disease on ASL conversations, we need to present some basic background on the nature of conversational regulators and turn-taking devices in ASL. Basically, in any conversation, participants take on one of two roles: 1. the participant who at any given moment claims the speaking turn (Duncan, 1972) and 2. the participants that at any given moment do not claim the speaking turn (Kendon, 1967).

The foundational article in this area was written by Baker (1977), who used a classification system proposed by Weiner and Devoe (1974) to analyze a corpus of conversational data elicited from two dyads of deaf signers. Baker videotaped two conversations and analyzed the initial portions for regulator behaviors. These excerpts were short, about 58-82 signs. The forms Baker (1977:218-219) found for initiation, continuation, and shift regulators are summarized below. These regulators form the basis for our study of Parkinsonian signers, although we will not limit ourselves to dyadic conversation.

Initiation regulators

According to Baker's very careful analysis, a signer initiates a turn by raising or extending the hands from a full-rest (arms at sides, relaxed), half-rest (waist-height, palms facing signer), or quarter-rest (resting one hand against the body above the waist) position, followed by some attention-getting gesture (pointing, touching, waving the hand), and an optional lean forward toward the addressee. Statements begin with a breaking of eyegaze once the addressee's attention is confirmed, while questions maintain eyegaze with addressee. Addressees signal their permission for the signer to initiate a turn by making eye contact[1] with the signer and by assuming a rest position and not signing.

Continuation regulators

The signer currently in control of a conversational turn signals that they will maintain the turn after a bounded information chunk has been completed or after a brief pause by breaking eyegaze with the addressees, optionally increasing one's signing speed, and by not returning to a rest position. Other methods of holding the floor include the use of filled pauses that indicate thinking or pondering such as looking up, using a quizzical expression, shaking the index finger or palm, or by some postural shift. Another option is holding the final configuration and position of the last sign articulated throughout the pause.

---

[1] While maintaining eye contact may seem necessary for the mundane purpose of visually processing the signed message, it is also an important feature of the addressee's response. One bit of evidence for this comes from conversational interactions between the right-lesioned signer mentioned earlier and her interlocutors. This subject tends to monitor her interlocutor's signing and visually track (using smooth pursuit, which necessitates breaking eyegaze with her addressee) those signs that move into the lower quadrant of her left visual field. This behavior is immediately noticed by her interlocutors and is extremely disruptive. Despite knowing that this behavior is a consequence of her lesion, it is difficult for her interlocutors not to register this behavior as rudely violating conversational protocol. This insuppressible reaction to a pragmatic violation contrasts sharply with the almost unconscious compensations made by these same interlocutors in response to the right-lesioned signer's failure to mark role shifts or to maintain coreference across a discourse.

The addressee signals continued attention and willingness to remain the observer by maintaining eyegaze and offering various subtle back-channel cues such as head-nodding, smiling, postural shifts, facial activity, uncertainty, or confusion. Addressees optionally point toward the signer after each proposition or repeat the ends of sentences. As might be expected, this continuation regulator feedback is common and more pronounced in the feedback that sign language interpreters give to deaf speakers as they process their signed communication for subsequent voice interpretation.

Shift regulators

A signer signals willingness to yield the turn to another participant in the conversation by reestablishing eyegaze with the addressee; by an optional decrease in signing speed near the end of the turn, by optionally calling for an addressee response via a "what do you think?" gesture; by pointing to the addressee; by holding or raising the last sign while making a questioning facial expression; or by returning the hands to a rest position.

The addressee signals a turn-claim during another's turn by optionally increasing and intensifying head nodding and pointing at the prior claimant; by switching palm orientation from toward oneself to palm upward with the heel of the hand higher than the fingertips, by moving out of rest position, or via a number of options enumerated in the initiating and continuation regulators sections above. To initiate a turn by interrupting the current claimant, the addressee repeats their first few signs until the current claimant makes eyegaze and yields the floor or until the current claimant suppresses the addressee's turn claim.

Summary

As can be seen, the mechanics of conversational control are complex, multi-leveled and require constant monitoring of the interlocutor(s). Claiming a turn becomes even more complex as we move from the conversational dyad to larger groups of participants. With increasing demands on the number of participants to be monitored and on the number of competing claims to fend off, we would expect the Parkinsonian signer to become increasingly disadvantaged in the conversational context.

Turn-taking in signers with Parkinson's disease

We examined conversations involving three ASL-signing[2] Deaf men in their seventies, each of whom had been previously diagnosed with Parkinson's disease, one mild, one moderate to severe, and the other severe. Data were drawn from a two-hour first intake interview, where background information concerning cause of deafness, schooling, employment and family history were collected. Each interview incorporated many opportunities for open-ended conversation as well. In the course of these interviews each subject engaged in both dyadic (two-way) conversations and conversations with more than two participants. In these two types of situations, we noted their behavior with respect to

---

[2] All three individuals were prelingually deaf, culturally integrated into the Deaf community, and considered ASL as their primary and preferred form of communication.

monitoring the participation of others and in claiming and yielding conversational turns. In particular we noted instances of physically touching the subject to get attention, eye or head orientations toward an interlocutor (JL, JK, VJ and JC are all testers serving as interlocutors; all but the author JK are Deaf), cases where the subject failed to observe a turn claim by another participant, signals of turn claiming and turn yielding on the part of the subject, the use of back channel signals like head nods, copying the ends of sentences, and gestures of understanding or disbelief, and finally cases where the subject continued to sign simultaneously with another participant in the conversation (indicating competition for the floor or repetition by the interlocutor of the subject's signing as a means of verifying meaning.) While our analysis is based upon behaviors in the interviews as a whole, in each of the sections below, we present coding results from a few minutes of conversation from each of the three signers with PD to give a flavor of the differences between their performances.

Mild Parkinsonian (LN)

Table 1. presents a few minutes of conversational turn-taking by mild Parkinsonian subject LN. Roughly each row corresponds to a turn.[3] indicating that in dyadic conversation, where his efforts can be focused upon claiming and yielding turns, his performance falls within the normal range. He alternated turns with ease and consistently offered back-channel signals of attentiveness and understanding whenever yielding the floor to his interlocutor. He frequently initiated turn claims and at one point (see the 2.5 second overlap in signing), he actually competed with his interlocutor for the turn.

When we move from dyadic to more than two conversational participants, the monitoring of cues from multiple participants puts added attentional demands on LN. Despite fairly natural turn-taking behavior, a stark reduction in his attempts to claim a turn suggest that he may be overburdened by the added attentional demands. In fact, in the one turn he finally negotiates, he asks his wife to make his point for him. LN did miss two turn shifts between tester JL and his wife, but these constituted a side conversation rather than formal turn taking in the course of the main conversation.

In the broader context of LN's conversational interaction, we were able to observe instances of all the initiation, continuation and shift regulators noted by Baker (1977).

Among these regulators, were simultaneous turns which occur when a conversant claims a turn and another fails to yield. These occurred with LN, the longest overlap lasting a full five seconds. Furthermore, simultaneous sign also occurred in the context of back channel behavior. In contrast with Baker's reports that these overlapping stretches tend to be much longer in ASL (average length 1.5 seconds with the longest stretch she documented lasting 4.3 seconds (Baker, 1977: 216)) than in spoken language conversations (mean duration less than .5 second (Jaffe & Feldstein, 1970)), factoring out simultaneous turns, LN's simultaneous signing was considerably shorter and less frequent than observed in standard ASL conversations. One of LN's preferred forms of feedback is questioning. For example, when the tester describes the make up of her family he responds with the equivalent of "Oh, so you have one child?" These tend to get coded as turn initiations, but if they were coded with back channel signals, they would up the length of simultaneous signing during back channeling considerably.

---

[3] The only exception to this is when the time code at the beginning of a row appears in parentheses (see Table 2.). This indicates a shift of eyegaze or body orientation that, while significant for grammatical agreement or some other function that resembles the gesturing involved in turn taking, does not constitute a conversational turn.

Dyad 02:40	Attention touch	Orient.	To	Missed shift	Turn claim	Turn yield	Back channel	Simultaneous signing	With
2:41			JL			1	1		
2:58			--		1				
2:59			--			1	1		
3:03			--		1			.5	JL
3:04			--			1	1		
3:11			--		1				
3:12			--			1	1		
3:13			--		1		1		
3:21			--		1			2.5	JL
3:23			--			1	1		
3:32			--		1				
3:35			--			1	1		
3:36			--		1				
3:40			--			1			
Multiple 00:00	Attention touch	Orient.	To	Missed shift	Turn claim	Turn yield	Back channel	Simultaneous signing	With
00:02	1	1	JL			1	1		
00:27		1	JK			1	1		
1:24		1	wife						
1:25		1	JK				1		
2:09			wife*	1					
2:14			JL*	1					
2:18		1	wife*						
2:19		1	JL*			1			
2:24			wife*	1			1	5.0	JL
2:30		1	JK			1	1		

Table 1. Conversational turntaking in mild Parkinsonian subject LN

Moderate to severe Parkinsonian (JH)

In contrast with LN, whose symptoms of PD are milder and more confined to the planning, execution, and sequencing of motor plans than to attentional problems, JH's conversational performance strikes one as more disrupted across the board. Still we see throughout his conversations evidence of his mastery of the rules for turn-taking in ASL. Yet his turntaking is still pragmatically off and jarring to the interlocutor.

Table 2. presents the results of coding several minutes of JH's conversational interactions in both a dyad and a larger group. In the dyadic conversation, the lack of back channel signaling on JH's part is notable. He strongly signals a turn yield with very fixed eyegaze on his addressee, but he offers little in the way of nodding, copying of final signs in an utterance or other signals that he is following along with the thread of the conversation. Yet, his contributions to the conversation are timely and appropriate, as well as numerous, demonstrating that he is indeed following just fine. He just fails to give much feedback. When he does it is typically a very reduced headnod or a smile and little more.

Most of JH's conversations, even those with more than two participants, tend to become dyadic because once he focuses upon one interlocutor, he has trouble disengaging and shifting to another one. There is an effect of where relative to JH a given interlocutor is positioned. In the conversational stretches coded in Table 2., VJ is on JH's left, while JK is across the room basically right of center. This attentional difficulty is a form of neglect and in JH's case seems to be more strongly manifested with individuals situated to his left.

Dyad	Attention touch	Orient.	To	Missed shift	Turn claim	Turn yield	Back channel	Simult. signing	With
00:00		1	VJ		1				
00:03			--			1			
00:07		avert	--		1				
00:10		1	VJ			1			
00:12	1	avert			1				
00:14	1	avert			1				
00:15		1	VJ			1	smile		
00:18		avert			1				
00:20	1	1	VJ			1			
00:25		1	JK			1			
00:26		1	VJ			1			
00:30		avert			1				
00:32		1	VJ			1			
00:35			--		1				
00:36			--			1			
01:09			--		1				
Multiple	Attention touch	Orient.	To	Missed shift	Turn claim	Turn yield	Back channel	Simult. signing	With
07:18			JK			1	1		
07:20			--		1				
07:30			--			1	1		
07:32		avert	--		1				
07:57		1	JK		1				
08:04			--		1				
08:08			--			1			
08:09			--		1				
08:14			--			1	1		
08:23			--		1				
08:32			--			1			
08:39					1				
08:44			--		1				
08:49	1		--						
08:52	1	1	VJ			1			
08:56		avert			1			.2	VJ
08:57		avert			1				
09:06		1	VJ			1	1		
09:42			--		attempt				
09:47			--		1				
09:57		1	JK			1			
09:58			--		attempt				
09:59	1	1	VJ			1			
10:23		1	JK			1	1		
11:07			--		1				
(11:19)		1	index(VJ)		1				
(11:20)		1	JK			1	1		
11:36			--		1				
11:40			--			1			
11:44	1	1	VJ		1				

Table 2. Conversational turntaking in moderate-to-severe Parkinsonian subject JH

In observing JH's conversations, we found that there was a higher than usual occurrence of physical touches on the part of interlocutors to shift his attention to them as they took the floor, however, most of these touchings came from VJ, the interlocutor on JH's left. She used them to disengage attention from JK as well as to claim a turn from JH himself. When JH claimed a turn from VJ, he averted his gaze from her (as is typical in turn initiation), but once averted, he did not easily disengage or monitor VJ for signals that she wanted to re-initiate a turn.

In addition, JH exhibited very little simultaneous signing and was frequently unsuccessful in claiming a turn away from another interlocutor. He would partially raise his hands from a full-rest position, but not forcefully enough to take the floor.

In contrast with LN, much of the communication addressed to JH was in the form of questions that required answers. This means that he was frequently drawn into claiming a turn, rather than left to initiate a turn on his own.

Severe Parkinsonian (JW)

JW constitutes one of our most interesting cases. Being in the late stages of PD, JW has a very masklike facial expression and his signing is completely distalized to the wrist and fingers. In fact, in addition to realizing most of the movement of signs with his fingers alone, even using the fingers as articulators, he restricts their movement to the most minimal of bending or separating possible. His signing gives the impression of a person very sporadically and very slowly bending the fingers while keeping them in an outstretched plane. The major movement change in the hand is at the metacarpal joints where the fingers meet the palm of the hand.

Table 3. presents a glimpse of JW's conversational interaction. Basically, he arranges himself in a single postural configuration where his head is tilted to the side to allow him to look at an interlocutor and he remains in that position throughout the conversation. We don't see any of the characteristic changes in head orientation or eyegaze that signal claiming a turn. However, it is also the case that JW rarely needs to negotiate for a turn. In general his interlocutors speak to him in questions and wait sometimes for extremely long periods of time for the slightest of answers.

Like JH, JW has difficulty disengaging attention. For example, in the coded conversation among multiple conversants in Table 3., JW is unable to disengage attention from a picture of himself in an army uniform that is serving to illustrate his story about faking being hearing in order to join the army, which he did. Over the course of 8-9 turns by his wife and the tester, with repeated touching to gain his attention, they were unable to draw his attention away from the picture. They eventually combined repeated touching of JW with moving their hands right onto the picture to capture his gaze. It is also notable that both the tester and his wife positioned themselves very distinctly on JW's direct right. In fact, they were in a line with him (his wife almost behind him) rather then facing him, which allowed them to physically place their hands/signs between JW and the picture he was fixated upon.

A turn initiation by JW was generally not signaled by anything more than beginning to sign or fingerspell a response. Signals of turn claiming, turn yielding and back channel information were totally absent from his signing, but interlocutors were highly attentive to the slightest indications that he was beginning to participate. Simultaneous signing on JW's part was nonexistent. In fact the lag time for a response could be several seconds. However, his interlocutors would frequently sign simultaneously with JW copying and clarifying what he was signing, decoding for themselves and each other the minimal, yet linguistically decodable, articulation of his signs.

Dyad	Attention touch	Orient.	To	Missed shift	Turn claim	Turn yield	Back channel	Simult. signing	With
00:00			JC			no change			
00:04			--		1			.5	JC
00:08			--			no change			
00:14			--	no claim					
00:21			--		1			.5	JC
00:28			--			no change			
00:39		eyes up left			?				
00:40		avert eyes			?				
00:41		1	JC			no change			
00:51	1		--		no claim				
00:56			--			no change			
01:04			--		1			.5	JC
01:08		1	filmer			1			
01:12		1	JC			1			
01:13		1	filmer			1			
01:15		1	JC			1			
01:16		avert			1				
01:33		1	JC			1			
Multiple	Attention touch	Orient.	To	Missed shift	Turn claim	Turn yield	Back channel	Simult. signing	With
17:45	1		pix	wife		no response			
17:46			--	JC		no response			
17:47	touch pix		--	JC/wife		no response			
17:53			--	JC		no response			
17:55			--	wife		no response			
17:56			--	JC		no response			
17:59		1	JC			1			
18:02			pix	JC		no response			
18:06	1		--	wife	1				
18:12		1	JC			1			
18:13	1	1	hand	wife		no response			
18:14		1	JC			1			
18:17			--		1				
18:23		1	--			1			
18:26	touch pix	1	pix		1				
18:38	touch pix		pix	wife		no response			
18:40			--	JC/wife		no response			
18:44	1			wife		no response			
18:51	1	1	JC			1			
18:55	1	1	JC			1			
19:03	1		pix	wife		no response			
19:05	1	1	pix	wife		no response			
19:10		1	JC			1			

Table 3. Conversational turntaking in severe Parkinsonian subject JW

Conversational compensation by signing interlocutors

A perusal of Baker's summary of conversational regulators and turn-taking in ASL reveals a complex and orchestrated interweaving of manual and non-manual signals on the part of both the speaker and addressee(s). All participants in the conversation are constantly giving and monitoring eyegaze as well as gestural cues that confirm their status in the conversational process. Participating in a conversation is clearly a multi-tasking operation —something individuals with Parkinson's disease are very bad at. The more components of conversational monitoring were required, the more compromised conversation became— even for mildly affected LN.

We need to recognize, however, that conversation is not always a competition for control of the floor. Sometimes other interlocutors expressly seek to include the Parkinsonian participant in a conversation. For example, the three signers with PD studied here were extremely interesting personalities. LN was very active in the Deaf community with lots of stories to tell. JH was a pop artist, a competitive diver and a professional wrestler. He was doing motorcycle acrobatics like standing on his head on the handlebars of a moving motorcycle controlled only by himself at age 65. JW was actually inducted into the army and almost sent overseas in World War II when it was finally discovered that he could not hear and had been fooling everyone. The testers were mesmerized by the stories and anecdotes these subjects shared with them. In such cases, since the burden of making conversational interaction work can no longer be shared equally, the lion's share of responsibility for conversational success falls to the participants without PD. And, they eagerly take it on.

In the remainder of this paper, we focus upon the burden placed on the interlocutor when conversing with a Parkinsonian signer, the compensatory strategies recruited to maximize comprehension and maintain the flow of information, and what these compensations reveal about the nature of the deficits associated with PD. We examine how interlocutors compensate for three characteristic deficits associated with Parkinsonian signing: 1. the flattening of distinctions in sign articulation; 2. problems initiating signed utterances; and 3. problems related to disengaging attention.

Adjusting the target of foveal vision

Sacrificing perceptual saliency exacts a steep price in terms of the demands it places on the addressee. In interactions with LN and JH, we saw interlocutors simply pay closer attention to their signs and reiterate sentences for clarification. But, with JW the processes of easing articulation were taken to their maximal extent. A hearing speaker at this stage would probably be inaudible and unintelligible and certainly would be beyond the technology we have for capturing and analyzing speech. But, JW was still understandable– albeit with great effort.

Shifting of attention to multiple interlocutors was no longer a possibility for JW, although we do see one instance of his shifting of gaze to monitor the person behind the camera for possible input to the conversation. Instead, one of the interlocutors (or each in turn) takes on the task of receiving and decoding the communication from JW and transmitting it to the others. We saw this clearly in the interaction between tester JC and JW's wife. But given the maximal reduction of JW's signs, simple monitoring was not enough. The "designated decoder" unconsciously adjusted his position to maximize the perceptibility of JW's signing.

In ASL, signs are distributed in the signing space which ranges from the top of the head to a semi-circle about waist high, extending a bent elbow's length in front of the signer such that signs with more marked handshapes and movements occur near the chin and neck in the area of the interlocutor's foveal vision (Siple, 1978). Fixating at this point also allows facial expressions (crucial to the grammar of ASL) to fall in the area of highest visual acuity. Typically, the addressee focuses at about the signer's chin and uses peripheral vision, rather than adjusting fixation, to process signs (mandatorily signs with unmarked movements and handshapes) that occur in the periphery.

But Parkinsonian signing is typically lowered and reduced such that all signs are in essence marked. Furthermore, facial expression is severely attenuated. While facial masking is characteristic of PD, we see the preservation of minimal eyegaze and grammatical expressions maintained as long as possible.

But JW's interlocutors converged on a solution. Over the course of the conversation, JW's tester can be observed to shift his position gradually downward getting JW's waist height hands in the area of foveal vision, but at the expense of facial information. Eventually, we see the full solution. By the end of the conversation, the tester has unknowingly maneuvered himself such that he is kneeling on the floor with his head tilting and looking up at JW's face through JW's hands. The tester moved to the one vantage point where JW's hands and face could both be processed in the interlocutor's area of foveal vision. The tester reestablished the balance between ease of articulation and ease of perception by adjusting his position.

Increasing the use of shift regulators

Although less so in the case of LN, all the testers maximized the participation of signers with PD in conversations by increasing the use of shift regulators. Their communication became primarily questions and restatements. Testers also elaborated on statements by the signers with PD, fleshing out the story and offering additional opportunities for the signers with PD to chime in with additional information. More and more turns on the part of the tester and family members became questions demanding a response, and that response was patiently waited for, although sometimes prodded out with various means of contacting or starting the signer along his way.

Becoming the PD signer's attentional system

Finally, in multi-participant conversation, the interlocutors as a group, became the PD signer's attention system, systematically tapping him or working together to draw his attention where needed. In cases like that of JH, where the attention system is asymmetrically affected, we see this asymmetry reflected in the behavior of the interlocutors. VJ (on JH's left), for example, begins to consistently tap JH before every conversational turn, where the other participants continue to use visual cues to reorient him. JW's interlocutors literally touch him and interrupt his line of sight to disengage his attention from one target and reorient it to another.

Conclusion

In contrast with the individualized nature of sentence production or interpretation, the pragmatic aspects of conversation are a mutually negotiated operation. When one participant is not fully up to the task, the situation demands compensation from the remaining members of the conversation. The rules of pragmatics are such that errors cannot just be ignored or corrected covertly in the interlocutor's interpretation. When a signer with PD fails to signal that a turn is being yielded, the person claiming the turn is actually thwarted from continuing. Something about this group endeavor leads pragmatic violations to be more jarring and more salient than a simple slip of the tongue or agreement error that might be encountered from a signer with aphasia.

Observing the behavior of persons engaging in conversation with signers who have Parkinson's disease corroborates in very tangible ways what we know to be the attentional as well as sensori-motor planning and execution deficits associated with its progression. Their complex repositionings to maximize perception of the reduced and laxed signing demonstrate clearly the consequences of the PD signer's prioritization of ease of articulation. Their increased use of shift regulators, patient attentiveness to the PD signers' every move, and taking over in many cases responsibility for clarifying or amplifying the PD signers' minimally signaled turntaking claims registers sensitivity to their deficits in quickly planning and initiating signed motor acts. And finally, we see single and multiple interlocutors actually become (via their tapping and attention getting devices) the attentional system of conversational participants with PD, registering the impact that neglect plays in Parkinson's disease. All of these accommodations occur in the absence of any knowledge of the nature of the disease, but rather are the consequence of repairing breakdowns in conversational turntaking to keep the process going.

As is the case with their linguistic capacities, we do not see any pragmatic deficit per se in signers with PD. Rather what we see are extralinguistic factors of sensori-motor planning and execution as well as attention that impede the ability of these rather eager and interesting conversationalists to successfully employ the turntaking devices they know to hold their own in conversations, especially those with more than two participants. The elaborate compensations made by interlocutors in conversations with signers who have PD highlights for us the important role pragmatics plays in our communicative lives.

Acknowledgements—The authors are grateful to numerous individuals who participated in the collection and transcription of these data including Joanne Lauser, Jimmy Challis, Vicki Joy Sullivan, Patricia Trowbridge, and Toni Fuller. We are also grateful to the participants in the Aphasia Committee meeting (IALP) in Montreal (July, 1997) for insightful questions and constructive comments that contributed greatly to our thoughts on this subject. Portions of this study were partially funded by National Science Foundation grants #SBR-9410562 and #IRI-9528985 to Boston University and Rutgers University, National Science Foundation grants #BNS-9000407 and #SBR-9513762 and grant #DC01656 from the National Institute on Deafness and Other Communication Disorders of the National Institutes of Health to Rutgers University. Correspondence should be addressed to Judy Kegl, Center for Molecular and Behavioral Neuroscience, Rutgers University, 197 University Avenue, Newark, NJ 07102.

References

Aarons, D., Bahan, B., Kegl, J., & Neidle, C. (1992). Clausal structure and a tier for grammatical marking in American Sign Language. *Nordic Journal of Linguistics, 15,* 103-142.

Bahan, B. (1996). *Non-manual realization of agreement in American Sign Language.* Doctoral Dissertation, Boston University.

Benecke, R., Rothwell, J. C., Dick, J. P. R., Day, B. L., & Marsden, C. D. (1987). Disturbance of sequential movements in patients with Parkinson's Disease. *Brain, 110,* 361-379.

Brentari, D., & Poizner, H. (1994). A phonological analysis of a Deaf Parkinsonian signer. *Language and Cognitive Processes, 9,* 69-100.

Brentari, D., Poizner, H., & Kegl, J. (1995). Aphasia and Parkinsonian signing: Differences in phonological disruption, *Brain and Language, 48,* 69-105.

Duncan, S. (1972). Some signals and rules for taking turns in conversations. *Journal of Personality and Social Psychology, 23,* 283-292.

Hallett, M., & Khoshbin, S. (1980). A physiological mechanism of Bradykinesia. *Brain, 103,* 301-314.

Harrington, D. L., & Haaland, K. Y. (1991). Sequencing in Parkinson's Disease. *Brain, 114,* 99-115.

Jaffe, J., & Feldstein, S. (1970). *Rhythms of dialogue.* New York: Academic Press.

Kegl, J., Cohen, H., & Poizner, H. (In press). Articulatory consequences of Parkinson's Disease: Perspectives from two modalities. *Brain and Cognition.*

Kegl, J., & Poizner, H. (1991). The interplay between linguistic and spatial processing in a right-lesioned signer. *Journal of Clinical and Experimental Neuropsychology, 13,* 38-39.

Kegl, J., & Poizner, H. (1993). Preservation of syntactic distinctions in a moderate to severe Parkinsonian signer. Presented at the Sixth Annual CUNY Sentence Processing Conference. University of Massachusetts at Amherst, March 18-20.

Kendon, A. (1967). Some functions of gaze direction in social interaction. *Acta Psychologica, 26,* 22-63.

Loew, R., Kegl, J., & Poizner, H. (1995). Flattening of distinctions in a Parkinsonian signer. *Aphasiology, 9,* 381-396.

Loew, R., Kegl, J., & Poizner, H. (1997). Fractionation of the components of roleplay in a right-lesioned signer. *Aphasiology, 11,* 263-281.

Margolin, D. I., & Wing, A. M. (1983). Agraphia and micrographia: Clinical manifestations of motor programming and performance disorders. *Acta Psychologica, 54,* 263-283.

Marsden, C. D. (1982). The mysterious motor function of the basal ganglia. *Neurology, 32,* 514-539.

Marsden, C. D. (1987). What do the basal ganglia tell premotor areas? In *Motor areas of the cerebral cortex* (Ciba Foundation Symposium 132). Chichester: Wiley.

Marsden, C. D., & Obeso, J. A. (1994). The functions of the basal ganglia and the paradox of stereotaxic surgery in Parkinson's Disease. *Brain, 117,* 877-897.

Poizner, H., & Kegl., J. (1992). The neural basis of language and motor behavior: Perspectives from American Sign Language. *Aphasiology, 6,* 219-256.

Poizner, H., & Kegl, J. (1993). Neural disorders of the linguistic use of space and movement. In P. Tallal, A. M. Galaburda, R. Llinas, and C. von Euler (Eds.), *Temporal information processing in the nervous system: Special reference to dyslexia and dysphasia* (vol. 682, pp. 192-213). Annals of the New York Academy of Sciences. New York: The New York Academy of Sciences.

Siple, P. (1978). Visual constraints for sign language communication. *Sign Language Studies, 19,* 95-110.

Weiner, M., & Devoe, S. (1974). Regulators, channels, and communication disruption, Research Proposal, Clark University.

Weiss, P., Stelmach, G. E., & Hefter, H. (1997). Programming of a movement sequence in Parkinson's Disease. *Brain, 120,* 91-102.

Pergamon

J. Neurolinguistics, Vol. 11, Nos 1–2, p. 153–177, 1998
© 1998 Published by Elsevier Science Ltd. All rights reserved
Printed in Great Britain
0911-6044/98 $19.00 + 0.00

PII: S0911-6044(98)00011-6

Pragmatics in frontal lobe dementia and primary progressive aphasia

Joseph B. Orange, *Andrew Kertesz[†] and Jennifer Peacock[*]*

[*]School of Communication Sciences and Disorders, University of Western Ontario;
[†]Department of Clinical Neurological Sciences, Lawson Research Institute, St. Joseph's Health Centre

Abstract—The purpose of this study was to document the pragmatic performance of subjects with frontal lobe dementia (FLD), non-fluent primary progressive aphasia (PPA), and fluent PPA using topic-directed conversation-based interviews. Unique profiles of pragmatic performance emerged for the three subject groups using objective measures of discourse, pragmatics, and scores from the *Profile of Communicative Appropriateness (PCA)* (Penn, 1985). The differences in performance provide a clearer understanding of the selective influences of frontal and frontotemporal pathology on the pragmatics of patients with FLD and PPA.

Introduction

Frontal lobe dementia (FLD) and primary progressive aphasia (PPA) are distinct, yet inter-related clinical syndromes, that are characterized by progressive declines in selected spheres of cognitive and language abilities. In the case of FLD, disturbances in cognition, behaviour, and emotion are prominent early features (Brun, Mann, Englund, Neary, et al., 1994). These disturbances include, but are not limited to, memory deficits, poor judgement and problem solving abilities, lack of insight, inflexibility of thought and behaviour, aggressiveness, impulsivity, denial of problems, changes in personality, decreased initiative, social withdrawal, flattened affect, disinhibition, and emotional lability (Brun, Mann, Englund, Neary et al. 1994; Gustafson, 1993; Snowden, Neary, & Mann, 1996). FLD has been associated with motor neuron diseases (Caselli, Windebank, Petersen, Komori, Parisi, Okasaki et al., 1993; Neary, Snowden, Mann, Northen, Goulding, & Mcdermott, 1990; Strong, Grace, Orange, & Leeper, 1996).

First described by Mesulam (1982), PPA is a clinical syndrome characterized by progressive deterioration in language with relative preservation of cognitive abilities such as memory and attention, personality, and insightfulness within the first few years following onset of language problems (Karbe, Kertesz, & Polk, 1993; Weintraub, Rubin, & Mesulam, 1990). Problems with cognition may emerge only after several years of disease progression, sometimes upwards of five to seven years following the onset of the language disturbances (Green, Morris, Sandson, McKeel, & Miller, 1990; Kesler, Artzy, Yaretzky, & Kott, 1995). One of the unifying features of FLD and PPA is the presence of language disturbances. Descriptions of the language characteristics of patients with FLD and patients with PPA have appeared in the literature over the past decade.

General patterns of the speech and language characteristics of patients with FLD have been described in small groups of patients. Dysarthric-like speech output and dysprodic disturbances are apparent, with relatively preserved language until the late stage of the disease where echolalia, perseveration, and mutism occur (Neary, 1990). Reduced spoken output, or logopenia, is one of the prominent features in patients with FLD, and is reflected in lower verbal fluency scores and reduced conversational initiation (Barber, Snowden, & Craufurd, 1995; Gustafson, 1993; Neary, 1990; Neary, Snowden, Mann, Northen, Goulding, & Mcdermott, 1990). Increased talkativeness and circumlocution also have been described in a select of group of patients with FLD (Gustafson, 1993; Neary, 1990). Patients in the early stage of FLD may show reduced and aspontaneous verbal output, few elaborations of information, and stereotyped and repetitive utterances (Gustafson, 1993; Neary et al., 1990). Others may confabulate and exhibit anomia in the form of semantic paraphasic errors, although these features have been observed less frequently (Johanson & Hagberg, 1989; Neary et al., 1990). With disease progression, language may be digressive and socially inappropriate, reflecting the advancing disinhibition and losses in social awareness (Brun, Mann, Englund, Neary et al., 1994).

The language of patients with PPA varies considerably, but primarily includes features observed in non-fluent type aphasias (Karbe, Kertesz, & Polk, 1993; Kempler, Metter, Riege, Jackson, Benson, & Hanson, 1990; Kertesz, Hudson, Mackenzie, & Munoz, 1994). The language of the non-fluent subgroup of patients PPA is characterized by telegraphic utterances, phonemic paraphasias, reduced spoken output, limited phrase length, reduced grammatical complexity, and relatively preserved auditory and reading comprehension (Duffy & Petersen, 1992; Karbe, Kertesz, & Polk, 1993; Kertesz, Hudson, Mackenzie, & Munoz, 1994; Thompson, Ballard, Tait, Weintraub, & Mesulam, 1997; Weintraub, Rubin, & Mesulam, 1990). Anomia may be present and prominent in some cases (Mesulam, 1982; Weintraub et al., 1990), but is usually less severe than the limited spoken output (Duffy & Petersen, 1992). Apraxia of speech also may be evident in patients with more advanced progression (Karbe, Kertesz, & Polk, 1993).

Recent evidence suggests that two other subgroups of PPA patients also exist. One group exhibits fluent-like language and impaired comprehension (Thompson et al., 1997), while the other exhibits dysfluency and impaired comprehension (Snowden, Neary, Mann, Goulding, & Testa, 1992). Thompson et al. (1997) examined the grammatical complexity of four subjects with non-fluent-like PPA from discourse samples collected over a 3-7 year period. They found both non-fluent and fluent profiles of language in their sample. The fluent-like PPA subgroup primarily show word-retrieval difficulties and comprehension problems with relatively preserved morphosyntactic structures. Snowden and her colleagues (1992) noted that the third subgroup of non-fluent PPA patients show changes in personality and social interaction skills similar to alterations observed in patients with bilateral frontal lobe deterioration.

The recent studies of the language of patients with FLD and PPA have described a wide range of linguistic abilities but few have examined their pragmatic performance in spontaneous spoken tasks. The frontal and anterior frontotemporal cortical sites of neuropathology in FLD and PPA (and related subgroups) suggest that disturbances in pragmatics may be prominent. It is clear that pre-frontal symptoms such as disinhibition, distractibility, impersistence, attention problems, among others, may contribute to disruptions in pragmatics and discourse. Horner (1985) noted over a decade ago that analysis of the pragmatics of patients with progressive language deterioration would likely identify the presence of repetition of ideas, stereotyped phrases, self-corrections and revisions, off-topic utterances, and egocentric comments. The purpose of this study was to document the pragmatic performance of subjects with FLD, non-fluent PPA, and fluent PPA. Our goal was to describe unique profiles which would distinguish the three groups. It was anticipated that any observed differences among the FLD and the two PPA subgroups would advance our understanding of the theoretical perspectives of pragmatic performance in the presence or absence of cognitive impairments, help clarify the influence of progressive frontal and frontotemporal pathology on pragmatic abilities, and provide clinical intervention options for patients, family, and professional caregivers.

Method

Subjects

Fourteen subjects participated in the study. All were native English speakers, right handed, and had a minimum of Grade 8 education. There were 7 females and 7 males. Ages ranged from 41 to 81 (see Table 1). None had a history of psychiatric, medical, or neurological conditions prior to their evaluations and diagnosis by a neurologist (AK). Comprehensive medical, neurological, neuroimaging, psychiatric, neuropsychological, and language assessments revealed the presence of FLD and PPA congruent with currently accepted clinical diagnostic criteria (Brun, Mann, Englund, Neary, et al., 1994; Mesulam, 1982). Extensive histories were obtained from patients or family members to document the presenting complaints and to compile an historical profile of problems. Three subjects were diagnosed with FLD, five subjects were diagnosed with non-fluent PPA, and six with fluent PPA by a neurologist (AK).

Cognitive and language testing

Neuropsychological testing was conducted to measure performance across a number of cognitive systems (e.g., memory, attention, nonverbal intelligence, etc.). A core battery of neuropsychological tests was administered to the majority of subjects while selected tests and subtests were given to several of the subjects. The core battery of neuropsychological tests included the *Mattis Dementia Rating Scale* (Mattis, 1976), *Wechsler Adult Intelligence Scale - Revised* (Wechsler, 1981), *Raven's Coloured Progressive Matrices* (Raven, 1947), *Wechsler Memory Scale - Revised* (Wechsler, 1987), and the *Rey-Osterrieth Complex Figure* (Osterrieth, 1944; Rey, 1941). The other frequently administered tests included all of or selected subtests from the *California Verbal Learning Test* (Delis, Kromer, Kaplan, & Obler, 1987), *Benton Visual Retention Test* (Benton, 1974), *Facial Recognition Test* (Benton & Van Allen, 1972), and the *Hooper Visual Organization Test* (Hooper, 1958). FLD subjects' performances on the neuropsychological battery of tests and subtests showed significant problems with memory, nonverbal intelligence, and frontal lobe-based processes including judgement, insightfulness, and inhibition. The non-fluent and fluent PPA subjects' scores on the neuropsychological tests generally ranged from low-normal to low-average and did not indicate the presence of obvious cognitive impairments.

The *Western Aphasia Battery* (*WAB*, Kertesz, 1982) was administered to all but one of the subjects (i.e., FLD subject CPY). The most prominent disturbance among all subjects on the *WAB* was their anomia. *WAB* Aphasia Quotient scores for all subjects who completed it, except one (Subject CP), were below the cut-off criterion of normal language performance and ranged from 7.2 to 91.4 (see Table 1).

| Features | Subject Groups | | | | | | | | | | | | | |
| | FLD (n = 3) | | | Non-fluent PPA (n = 5) | | | | | Fluent PPA (n = 6) | | | | | |
	EB	CPY	CP	RH	VH	FS	AS	BV	VB	JH	IM	RR	KS	JS
Sex	M	F	F	F	F	F	M	F	F	M	M	M	M	M
Age (years)	65	41	60	78	77	55	65	65	75	81	76	77	69	58
Education (years)	12	19	14	8	N/A	17	10	11	22	N/A	15	12	19	16
Handedness	R	R	R	R	R	R	R	R	R	R	R	R	R	R
WABAQ	80.2	NA	97.2	38.0	16.0	7.2	16.7	77.7	82.6	81.8	73.9	65.8	90.0	91.4
MDRS	100	NA	114	untestable	untestable	untestable	66	94	109	82	83	89	129	N/A

Note:
N/A = Not available
WAB AQ = Western Aphasia Battery Aphasia Quotient (Kertesz, 1982); total possible score is 100
MDRS = Mattis Dementia Rating Scale (Mattis, 1976); total possible score is 144; subjects untestable because of severe language problems

Table 1. Subject demographic information

Procedure

To obtain an extended sample of spontaneous language, all subjects were invited to participate in a topic-directed interview as part of their comprehensive neuropsychological and language assessment. A single, trained examiner acted as the partner in the topic-directed interviews for all fourteen subjects. Topic-directed interviews have been used previously for subjects with dementia, and have been shown to be useful in generating extended samples of spoken discourse suitable for analyses of discourse and pragmatic features (Garcia & Joanette, 1994; Illes, 1989; Mentis, Whittaker, & Gramigna, 1995; Ripich & Terrell, 1988).

Topic-directed interviews were conducted in a quiet, distraction-free room. Interviews were video recorded. The pre-selected topics were introduced by the examiner either during the Spontaneous Speech section of the *WAB* or separately during the assessment using the open-ended request, "Tell me about_____". The following five topics were used: (a) your family, (b) your health right now, (c) what you do each day, (d) where you were born and raised, and (e) the jobs you had or the work that you did. The topics were introduced in the same order for all subjects. The topics are similar to those used by Ripich and Terrell (1988) in their study of the discourse of subjects with Alzheimer's disease, and those used by Illes (1989) in her study of the discourse of subjects with various forms of cortical and subcortical dementias.

During the interview, the examiner was limited to using a single, simple prompt for extending talk on a topic, and in using only back-channelling responses that indicated her comprehension or agreement (e.g., "Yes", "Umhum", facial expressions and head nods, etc.). When subjects made a concluding statement, stopped speaking for fifteen seconds *and* where there were no indications of word searching behaviours (such as upward/downward eye gazing, prolongations of sounds, or fillers such as "Uhm", "Uh", "Like", "Just a minute", etc.), or otherwise signalled the temporary completion of the monologue (used downward intonation, made eye contact and paused for turn shift, gestured for turn shift, etc.), the interviewer made the single prompt "Tell me more about _____.". If subjects indicated (verbally or nonverbally) that there was nothing more to be said on that topic, the interviewer introduced the next topic until all five had been discussed. If subjects shifted topic during their extended monologue, the interviewer did not interrupt but waited until subjects had concluded their remarks. No time limits were imposed on any topic or the interview task.

All samples were orthographically transcribed and segmented into turns and utterances according to the criteria established in the *Shewan Spontaneous Language Analysis* system (Shewan, 1988). The segmentation of utterances was based on morphosyntactic, semantic, and prosodic criteria.

Utterances in the samples were coded for several discourse and pragmatic features. These are listed in Tables 2 and 3. Operational definitions for the discourse and conversational topic management measures are presented in Appendix A. The measures were selected based on findings from previous work on the discourse and pragmatic performances of subjects with various forms of dementia and aphasia which showed their effectiveness in differentiating the performance of subjects (Illes, 1989; Nicholas & Brookshire, 1993); Ripich & Terrell, 1988; Tomoeda & Bayles, 1993; Ulatowska, Allard, Reyes, Ford, & Chapman, 1992.

In addition to identifying the discourse and pragmatic measures outlined in Appendix A, the transcript of each subject was scored using the definitions and guidelines of the *Profile of Communicative Appropriateness (PCA)* (Penn, 1985; Penn, 1988). The items comprising the *PCA* are outlined in Appendix B. All items from the *PCA*, with the exception of Control of Direct Speech in the category Sociolinguistic Sensitivity, were scored by a trained rater (author JP). This single item was not scored and was eliminated from the analyses of *PCA* ratings as its definition was non-specific and could not be coded uniformly across subjects. The *PCA* was selected as an outcome measure because it has been used successfully in previous studies to document the pragmatic performances of individuals with various forms of dementia (Penn, Sonnenberg, & Schnaier, 1988), and for profiling pragmatics in individuals with aphasia (Penn, 1988).

Agreement studies

To assess inter-rater agreement for the scoring and the rating of the outcome measures used in this study, point-by-point percent agreement scores were calculated. Two trained raters re-scored the entire transcripts from two randomly selected subjects (i.e., 14% of the data base). Inter-rater agreement for utterance segmentation was 95.5%. The mean percent agreement for coding the discourse and pragmatic outcome measures was 94%. Agreement scores for only three outcome measures were below 80%. These were categories (a) number of could not evaluate utterances/topic (75%), (b) percent of could not evaluate utterances/topic (75%) (this category occurred only for the non-fluent PPA subjects), and (c) percent of utterances in side sequences - external (60%). The most likely reason for the low agreement on these measures was their infrequent occurrence in the transcripts of all subjects. The mean percent inter-rater agreement for *PCA* ratings was 91.5%.

Data analysis

The small number of subjects who participated in this preliminary study warranted against using parametric and non-parametric statistical analyses for group comparisons. We undertook a cautious approach in interpreting absolute differences in scores among the groups. For the purposes of data analysis, percentages, rates, and time-based scores were calculated. Time-based measures were calculated using the speaking time of the samples (i.e., total time of the sample minus all inter- and intra-pauses longer than 5 seconds). Means and standard deviations were calculated for the topic management and *PCA* outcome measures. These values were used as a basis for descriptive comparisons.

Results

Data derived from the samples obtained in the topic directed interviews for the FLD subjects, the non-fluent PPA subjects, and the fluent PPA subjects are presented in Tables 2 through 6. Individual data points, summary means, and standard deviations are shown.

			Subjects																		
Discourse measures	FLD (n = 3)				Non-fluent PPA (n = 5)						Fluent PPA (n = 6)						
	EB	CPY	CP	x̄ (SD)	RH	VH	FS	AS	BV	x̄ (SD)	VB	JH	IM	RR	KS	JS	x̄ (SD)
# words	61	813	90	321.3 (426)	108	115	22	18	211	65.8 (52.9)	741	537	176	1451	572	124	600.2 (480.7)
MLU	5.6	14.7	4.7	8.3 (5.5)	5.6	5	1.2	1.3	5	3.62 (2.2)	9.4	9.2	4.5	7.9	10.3	5.4	7.8 (2.3)
# utterances	11	61	21	31 (26.5)	21	26	18	16	42	24.6 (10.4)	92	58	40	176	78	23	77.8 (54.2)
% complete utt.	100	100	95	98.3 (2.6)	81.0	96.2	88.9	100	78.6	88.9 (9.3)	91.3	93.1	62.5	83.5	98.7	100	88.2 (13.9)
% incomp. utt.	0	0	5	1.7 (2.9)	19.0	3.8	11.1	0	21.4	11.1 (9.3)	8.7	6.9	37.5	16.5	1.3	0	11.8 (13.9)
% stereotype	0	4.9	0	1.6 (2.8)	19.0	46.2	44.4	6.3	9.5	25.1 (19.1)	2.2	5.2	17.5	25	1.1	0	8.5 (10.3)
% overt	0	0	0	0 (0)	0	0	0	0	11.9	2.4 (5.3)	4.3	1.7	2.5	2.8	1.7	0	2.2 (1.4)
#self correct/utt.	0.1	0	0	0 (0.1)	0.2	0	0.2	0	0.2	0.1 (0.1)	0.4	0	0.2	0.3	0.1	0.5	0.3 (0.2)
% pronoun	10.3	14.8	13.3	12.8 (2.3)	4.9	18.3	0	0	18	8.2 (9.3)	19.3	18.6	17	21.8	16.8	14.5	18 (2.5)
% pwa	0.1	0.8	0	0.3 (0.4)	29.2	66.7	0	0	10.5	21.3 (28.1)	4.2	6	6.7	27.8	1	0	7.6 (10.2)
% demonstrative	20	11.6	5	12.2 (7.5)	4.2	66.7	0	0	26.7	19.5 (28.6)	20.2	25	3.3	16.5	13.5	11.1	14.9 (7.5)
# words/min.	38.9	157.9	41.3	79.4 (68.0)	52.9	53	16.3	13.5	65.9	40.3 (23.8)	82.2	120.1	51	96.3	66.8	40.8	76.2 (29.5)
# comp. utt./min.	6.4	11.8	9.2	9.1 (2.7)	6.4	10.2	9.6	4.5	7.5	7.6 (2.3)	8.7	11.6	7.3	9.2	7.1	6.9	8.5 (1.8)
# incomp.utt./min.	0	0	0.5	0.2 (0.3)	2.5	0.9	2.2	0.8	5.3	2.3 (1.8)	1.4	1.3	3.8	2.3	0.6	0.3	1.6 (1.3)
# reps./utterance	0.1	0	0	0 (0.1)	0.6	0.6	0.5	0.6	0.5	0.6 (0.1)	0.5	0.2	0.5	0.5	0.2	0.9	0.5 (0.3)

Table 2.   Summary of discourse measures from topic directed interview

Topic measures	FLD Subjects (n = 3)			
	EB	CPY	CP	x̄ (SD)
Total # utterances	11	61	21	31.0 (26.5)
Total # topic units	5	5	10	6.7 (2.9)
# utterances / topic	2.2	12.2	2.1	5.5 (5.8)
# on-topic utterances / topic	2.2	12.2	2.1	5.5 (5.8)
# off-topic utterances / topic	1.8	0	0	0.6 (1.0)
# could not evaluate utterances / topic	0	0	0	0 (0)
% on-topic utterances	82	100	100	94 (10.4)
% off-topic utterances	18	0	0	6.0 (10.4)
% utterances in side sequences	9.1	1.6	4.8	5.2 (3.8)
% utterances in side sequences - metastatements	0	1.6	4.8	2.1 (2.4)
% utterances in side sequences - external	9.1	0	0	3.0 (5.3)
% perseverative utterances	0	0	0	0 (0)
% intrusive utterances	0	0	0	0 (0)

Table 3. Measures of topic analysis for FLD subjects

Topic measures	Non-fluent PPA subjects (n = 5)					
	RH	VH	FS	AS	BV	x (SD)
Total # utterances	21	26	18	16	42	24.6 (10.4)
Total # topic units	10	11	5	6	10	8.4 (2.7)
# utterances / topic	2.1	2.4	3.6	2.7	4.2	2.7 (0.7)
# on-topic utterances/topic	0.7	0.8	1.2	0.4	3.8	1.4 (1.3)
# off-topic utterances/topic	0.8	1	1.6	0.1	0.2	0.7 (0.6)
# could not evaluate utterances/topic	0.6	0.5	0.8	0.5	0.2	0.5 (0.2)
% on-topic utterances	33	34.6	33	37.5	90.5	45.7 (25.1)
% off-topic utterances	38	42.3	44	12.5	4.8	28.3 (18.3)
% utterances in side sequences	9.5	0	5.6	25	40.5	16.1 (16.5)
% utterances in side sequences - metastatements	0	0	5.6	12.5	38.1	11.2 (15.9)
% utterances in side sequences - external	9.5	0	0	12.5	2.4	4.9 (5.8)
% perseverative utterances	0	3.8	44	0	0	9.7 (19.3)
% intrusive utterances	4.8	19.2	0	0	0	4.8 (8.3)
% could not evaluate utterances/topic	29	23.1	22	50	4.8	25.8 (16.3)

Table 4. Measures of topic analysis for non-fluent PPA subjects

Topic measures	Fluent PPA subjects (n = 6)						
	VB	JH	IM	RR	KS	JS	x (SD)
Total # utterances	92	58	40	176	78	23	77.8 (54.2)
Total # topic units	10	8	12	15	8	6	9.8 (3.3)
# utterances / topic	8.7	8.6	15	7.4	12.8	0	8.8 (5.1)
# on-topic utterances/topic	8.1	6.8	1.5	7.7	9.3	3.8	6.2 (3.0)
# off-topic utterances/topic	1.1	0.5	1.8	4.0	9.3	0	2.8 (3.5)
# could not evaluate utterances/topic	0	0	0	0	0	0	0 (0)
% on-topic utterances	88	93	45	66	0	100	65.3 (37.9)
% off-topic utterances	12	7	55	34	0	0	20.1 (20.52)
% utterances in side sequences	9.2	7.3	3.3	11.7	9.8	3.8	7.5 (3.4)
% utterances in side sequences - metastatements	8.7	6.9	15	6.3	100	0	35.9 (45.1)
% utterances in side sequences - external	0	1.7	0	1.1	0	0	0.47 (0.75)
% perseverative utterances	0	0	5	0	0	0	0.8 (2.04)
% intrusive utterances	0	0	12.5	6.8	0	0	3.2 (5.3)

Table 5. Measures of topic analysis for fluent PPA subjects

PCA Measures	Percentages for subject groups		
	FLD subjects (n = 3)	Nonfluent PPA subjects (n = 5)	Fluent PPA subjects (n = 6)
Response to interlocutor			
Inappropriate	0	42.9	0
Mostly inappropriate	0	21.4	8.3
Subtotal	0	64.3	8.3
Some appropriate	0	7.1	0
Mostly appropriate	33.3	14.3	33.3
Appropriate	66.7	14.3	58.3
Subtotal	100	35.7	91.6
Could not evaluate	20.0	44.0	20.0
Semantic content			
Inappropriate	5.8	38.9	8.3
Mostly inappropriate	5.8	50.0	33.3
Subtotal	11.7	88.9	41.6
Some appropriate	0	11.1	2.8
Mostly appropriate	17.6	0	16.6
Appropriate	70.6	0	38.9
Subtotal	88.2	11.1	58.3
Could not evaluate	5.6	40.0	0
Cohesion			
Inappropriate	0	18.2	0
Mostly inappropriate	0	18.2	10.8
Subtotal	0	36.4	10.8
Some appropriate	0	9.0	5.4
Mostly appropriate	0	27.3	8.1
Appropriate	100	27.3	75.7
Subtotal	100	63.6	89.2
Could not evaluate	9.5	68.6	11.9

Table 6.   Summary of percentage scores on the Profile of Communicative Appropriateness for FLD, Nonfluent PPA, and Fluent PPA subjects

Table 6 (continued)

PCA Measures		FLD Subjects (n = 3)	Nonfluent PPA Subjects (n = 5)	Fluent PPA Subjects (n = 6)
			Percentages for subject groups	
Fluency				
	Inappropriate	0	57.1	17.1
	Mostly inappropriate	0	23.8	19.5
	Subtotal	0	80.9	36.6
	Some appropriate	0	4.7	14.6
	Mostly appropriate	6.3	14.3	21.9
	Appropriate	93.7	0	26.8
	Subtotal	100	19.0	63.3
	Could not evaluate	23.8	40.0	2.4
Sociolinguistic sensitivity				
	Inappropriate	0	50	5.9
	Mostly inappropriate	0	10	8.8
	Subtotal	0	60	14.7
	Some appropriate	27.3	10	11.8
	Mostly appropriate	9.1	20	29.4
	Appropriate	63.6	10	44.1
	Subtotal	100	40	85.3
	Could not evaluate	61.9	50.0	29.2
Non-verbal communication				
	Inappropriate	3.6	4.3	1.7
	Mostly inappropriate	14.3	15.2	1.7
	Subtotal	17.9	19.5	3.4
	Some appropriate	10.7	8.7	11.8
	Mostly appropriate	10.7	10.9	10.2
	Appropriate	60.7	60.9	74.6
	Subtotal	82.1	80.5	96.6
	Could not evaluate	15.2	16.4	10.6

Discourse measures

As a backdrop to the results obtained from the analyses of the pragmatic performance of the FLD, non-fluent, and fluent PPA subjects, several measures of discourse were first calculated. A summary of the findings are provided in Table 2. These measures reflect traditional measures of language performance (e.g., number of words, number of utterances, mean length of utterance (MLU), as well as discourse measures reflecting intra- and inter-sentential relationships of lexical and referential elements (e.g., stereotyped phrases, pronouns without antecedents, and demonstrative pronouns).

A review of the mean scores of the discourse measures for all three groups reveal the following. From the standpoint of overall quantity of production, non-fluent PPA subjects generated the fewest words ($\bar{x} = 65.8$), fewest utterances ($\bar{x} = 24.6$), and the fewest number of words per minute ($\bar{x} = 40.3$). The fluent PPA subjects, on the other hand, generated the highest number of words ($\bar{x} = 600.2$), the highest number of utterances ($\bar{x} = 77.8$), and had a substantially higher rate of words per minute ($\bar{x} = 76.2$). The MLU for the non-fluent PPA subjects ($\bar{x} = 3.62$) was less than one half the values for the FLD subjects ($\bar{x} = 8.3$) and the fluent PPA subjects ($\bar{x} = 7.8$). These findings are in keeping with current views on the reduced spoken output of non-fluent PPA patients and the normal or near normal levels of spoken output among fluent PPA subjects (Thompson et al., 1997). The quantity of the spoken output of the FLD subjects was skewed upwards by the scores of one subject (CPY). High verbal output in FLD subjects has been documented previously (Gustafson, 1993; Neary, 1990). The mean number of words and utterances of the other FLD subjects, however, were below those of the highly verbal fluent PPA group.

The percentage of incomplete utterances by the non-fluent and fluent PPA groups were substantially higher ($\bar{x} = 11.1$ versus $\bar{x} = 11.8$) than the percentage for the FLD group ($\bar{x} = 1.7$). The use of stereotyped phrases, pronouns with antecedents, and demonstrative pronouns was highest among the non-fluent PPA group and second highest among the fluent PPA group. These measures reflect, in part, the word finding difficulty described in subjects with PPA by other investigators, and particularly those with a fluent-like PPA (Weintraub et al., 1990). Scores for the FLD group were lower than those of the PPA subjects, suggesting that they experienced less word finding difficulty in the topic-directed interviews. The following example from non-fluent PPA subject BV illustrates the incomplete utterances and stereotyped utterances characteristic of the discourse samples of the non-fluent PPA group.

Example 1

Examiner:	and your address?
Subject:	*um it's uh R or …*
Subject:	*no wait a minute!*
Subject:	*its um …*
Subject:	that's not right um today.
Subject:	it it's in Clinton.
Examiner:	uhhuh.
Subject:	*and and um it's uh four no …*
Subject:	*oh dear.*
Subject:	*um let's see.*
Subject:	*Clinton and um …*
Subject:	*oh gosh.*
Subject:	*my uh address is um ah …*
Subject:	*oh gosh.*

Topic

Differences in pragmatic performance among the FLD, non-fluent, and fluent PPA subjects were more pronounced on measures of conversational topic management. The summary of scores for the FLD subjects, the non-fluent PPA subjects, and the fluent PPA subjects are presented in Tables 3, 4, and 5, respectively.

The non-fluent PPA subjects produced the fewest number of utterances per topic ($\bar{x}$ = 2.7) compared to the fluent PPA subjects ($\bar{x}$ = 8.8) and the FLD subjects ($\bar{x}$ = 5.5). The non-fluent PPA subjects also produced the fewest number of on-topic utterances per topic ($\bar{x}$ = 1.4) and had the lowest percentage of on-topic utterances ($\bar{x}$ = 45.7). The fluent PPA subjects had only two-thirds of their utterances rated as being on-topic ($\bar{x}$ = 65.3%), while the FLD subjects talked predominantly on-topic ($\bar{x}$ = 94%). The non-fluent PPA subjects also exhibited the highest percentage of off-topic utterances ($\bar{x}$ = 28.3) compared to the fluent PPA subjects ($\bar{x}$ = 20.1) and the FLD subjects ($\bar{x}$ = 6.0). The following examples illustrate the off-topic utterances produced by the non-fluent PPA, fluent PPA, and FLD subjects. In Example 2, non-fluent PPA subject RH produces off-topic utterances that are unrelated to the examiner's question about her name.

Example 2

Examiner:	tell me your full name.
Subject:	R G H (full name spoken).
Subject:	*and I don't very much very very much about uh um um day.*
Subject:	*and my uh my father's dead.*
Subject:	*my father uh you know he was really…*
Subject:	we'd he's awful.

In Example 3, the fluent PPA subject RR talks about unrelated issues following the examiner's questions.

Example 3

Examiner:	when did you notice you had language problems?
Subject:	uh the the things that uh you know in in talking.
Examiner:	in talking?
Subject:	*well when I was in working and and uh I had…*
Subject:	*I get the people talking about…*
Examiner:	uhhuh.
Subject:	*that me…*
Examiner:	yeah.
Subject:	*okay?*
Examiner:	yeah.
Subject:	*now if they don't do the things that uh I think that's not right then I would you know just give them something and say well "wouldn't it be …"*
Subject:	*you know things like that.*

Although of low occurrence, FLD subjects did produce off-topic comments. In Example 4, FLD subject EB responds off-topic to a question by the examiner:

Example 4

Examiner:	what problems have you been having Mr. B.?

Subject:	*uh I had an M I R (spelled each letter) and and a and and I had a C A T scan (spelled)*
Subject:	*and I proper proper name is and the and it's called...*
Examiner:	but those are the tests that you've had.
Examiner:	what is your problem?

Additional detailed analyses were undertaken to examine the percentage of off-topic talk in which the subjects temporarily suspended discussion about the primary topic (i.e., side sequence) (Jefferson, 1972). Side sequences included (a) talk on a topic related to the primary topic (i.e., a subtopic), (b) comments concerning the lexical or emotional difficulty of talking about the primary topic (i.e., side sequence - meta-statement), and (c) comments that addressed matters in the immediate context of the assessment (e.g., video camera, microphone, lighting, etc.) or were otherwise unrelated to the primary topic. The non-fluent PPA subjects had the highest percentage of utterances categorized in side sequences ($\bar{x}$ = 16.1), with fluent PPA subjects (x = 7.5) and FLD subjects ($\bar{x}$ = 5.2) exhibiting less than half the percentage of the non-fluent group. The fluent PPA group, however, had the highest percentage of utterances in side sequences that were meta-statements ($\bar{x}$ = 35.9) compared to the non-fluent PPA subjects ($\bar{x}$ = 11.2) and the FLD subjects ($\bar{x}$ = 3.0). This latter finding is not surprising considering that the fluent PPA group exhibited the greatest word finding difficulty, as measured by the discourse measures and their ratings on the appropriate use of items in the Semantic Content and Fluency categories of the *PCA* (see below). The non-fluent PPA subjects had the highest percentage of side sequence utterances that were unrelated to the primary topic ($\bar{x}$ = 4.8). For all three groups, however, the percentage of utterances categorized as unrelated and external to the primary topic was relatively low.

The following examples illustrate the meta statements made by fluent and non-fluent PPA subjects. In Example 5, the fluent PPA subject VB comments on her word finding difficulty in response to the examiner's question.

Example 5

Examiner:	have you been here before?
Subject:	where?
Examiner:	in this hospital?
Subject:	for mmh something that happen to me?
Examiner:	uhhuh.
Subject:	no.
Subject:	yesterday in with professor or doctor oh ....
Subject:	*damn.*
Subject:	*I am thinking of his name.*
Subject:	*it's coming but uh ...*
Subject:	*when I when I am tense everything goes (gestures with hands).*

In Example 6, non-fluent PPA subject BV comments that she would like to have a relative come in and explain her difficulty.

Example 6

Examiner:	What kinds of problems have you been having?
Subject:	Well I don't know how you know I'm ....
Subject:	*I wish K (name of relative) would be could her to now ...*
Subject:	*to tell yeah.*
Examiner:	I know you'd like to ask her yeah.

Subject: but um um well uh uh I'm I'm in the um in the ...

The following example illustrates the external utterances produced almost exclusively by the non-fluent PPA subjects. In example 7, non-fluent PPA subject RH makes unrelated comments to the examiner's question before continuing on the topic.

Example 7

Examiner: What kind of work did you use to do?
Subject: oh uh it I don't know and uh ...
Subject: he had a store for a while you know and it runs and its and it runs and it gobble up.
Subject: it wasn't in on Owen Owen Sound it was later
Subject: we have to get home to get an an an I I didn't.
Examiner: that's okay.
Examiner: you have a little bit longer you daughter in law knows and so does B.

For the last two measures on conversational topic, non-fluent PPA subjects exhibited the highest percentage of perseverative ($\bar{x} = 9.7$) and intrusive utterances ($\bar{x} = 4.7$). The fluent PPA subjects produced few perseverative and intrusive utterances, while FLD subjects did not produce any at all. The majority of the perseverative utterances produced by the non-fluent PPA subjects were stereotyped phrases including "Like no" and "Okay yeah".

Profile of Communicative Appropriateness (PCA)

As a final measure of the pragmatic performance of the FLD subjects, the non-fluent PPA subjects, and the fluent PPA subjects, ratings were made (i.e., Inappropriate, Mostly inappropriate, Sometimes appropriate, Mostly appropriate, and Appropriate) for items in each of the six domains of pragmatics comprising the *PCA*. Domain items of the *PCA* are listed in Appendix B. Ratings for items within each domain were collapsed for subjects in the FLD, non-fluent, and fluent PPA groups. Percent scores for the ratings were calculated for items that could be evaluated in each domain. These figures are presented in Table 6.

For the pragmatic domain of Response to Interlocutor, the non-fluent PPA subjects' performance on items that could be evaluated, such as requesting information, using clarification requests, and replying to questions, were rated frequently as either inappropriate or mostly inappropriate (64.3%). The fluent PPA subjects (91.6%) and the FLD subjects (100%), on the other hand, were rated as appropriately using responses with their conversational partner the majority of the time.

For the pragmatic domain of Semantic Content, the non-fluent PPA subjects showed the highest ratings of inappropriate or mostly inappropriate use of elicited items related to topic management, word finding, and idea completion (88.9%). These ratings concur with the objective measures of topic management discussed above wherein the non-fluent PPA subjects exhibited the highest percentage of off-topic utterances. The fluent PPA subjects' ratings also showed that they experienced problems with idea sequencing, idea completion, and maintaining or shifting topics (41.6%). The ratings for the FLD subjects showed that they experienced little difficulty with issues related to topic and idea completion (88.2%). Again, the objective measures of topic maintenance discussed previously for the FLD subjects support these ratings.

In the domain of Cohesion, which includes of use of references, tense structure, and other inter- and intra-sentential linguistic elements, the non-fluent PPA subjects showed the poorest ratings for items that could be evaluated (36.4%). It should be noted, however, that a high percentage of items could not be evaluated as either the non-fluent PPA subjects could not produce the items and therefore the items did not appear in the transcript, or because the task was not sufficient in structure, complexity, or length to elicit the items.

The ratings for the non-fluent PPA subjects for items that could be evaluated in the Fluency domain of pragmatics, such as repetitions, revisions, incomplete phrases, and word finding difficulties, were either inappropriate or mostly inappropriate (80.9%). For the fluent PPA subjects, nearly two thirds of the items in the Fluency domain that could be evaluated were used appropriately or somewhat appropriately (63.3%). FLD subjects were rated with few problems for revision and word findings abilities.

For items that could be evaluated in the pragmatic domain of Sociolinguistic Sensitivity, such as the use of stereotypes, indirect speech acts, and acknowledgements, the non-fluent PPA subjects were rated as using the items inappropriately or mostly inappropriately the majority of the time (60%). The fluent PPA subjects and the FLD subjects experienced fewer difficulties, as reflected in their high ratings of appropriate use of these complex elements of pragmatic performance. It should be noted, however, that many of the items comprising this domain could not be evaluated for the FLD and non-fluent PPA subjects for reasons similar to those discussed for items in the Cohesion domain.

Finally, the FLD subjects, the non-fluent PPA subjects, and the fluent PPA groups were rated quite highly in the appropriate use of speech and gestural components of non-verbal communication. Most of the items in the domain could be evaluated for all three subject groups.

Discussion

The data derived from topic-directed interviews and reported herein are the first to address in a systematic manner the pragmatic performances of subjects with FLD, non-fluent PPA, and fluent PPA. The use of three separate sets of measures of discourse and pragmatics show consistent performance among related constituent elements of pragmatics for all three subject groups. Differences in pragmatic performance among the three diagnostic groups, however, reveal three separate, but overlapping profiles of abilities and weaknesses.

One of the important findings of our study is that the two subgroups of PPA subjects can be identified and distinguished on the basis of their pragmatic performance. This supports the assertions made by others, on the basis of analyses of language and neuropsychological data, that there are two, and possibly three, subgroups of PPA patients (Snowden et al., 1992; Thompson et al., 1997). The non-fluent PPA subjects' pragmatic performance, based on the multiple objective and rating measures used in this study, can be characterised best as inappropriate use of topic maintenance skills, production of off-topic comments, poor use of responses to partner's questions and requests, difficulty tracking and using devises of reference, and limited contributions to the semantic development of topics, most likely the result of word finding difficulties and incomplete utterances.

The pragmatic performance profile of the fluent PPA subjects in this study, while similar in some respects to that of the non-fluent group, is distinguished by their more subtle difficulties in topic maintenance such that they are able to develop topics more fully and respond more appropriately to requests by their discourse partner. The percent of their talk that is rated as appropriate to the requests and demands of their partners is greatly superior to that of non-fluent PPA subjects. Moreover, the fluent PPA subjects display an insightfulness and awareness of the linguistic and pragmatic problems that is not apparent in the non-fluent PPA subjects. The fluent PPA subjects' comments about their difficulty communicating reflects an awareness of problems that is not evident among the non-fluent PPA group. It is fully acknowledged that the extent of differences in the pragmatic performances between the non-fluent and fluent PPA groups is subtle and may only be a matter of the degree of the ratings within the same domain of pragmatics. Further detailed analyses of additional components of pragmatics, such as the correct use and interpretation of figurative language or the appropriate use of paralinguistic phenomena (e.g., affective prosody), is warranted to explore more fully differences in the pragmatic performances of the PPA subgroups.

A second important finding is that the pragmatic performances of the FLD subjects in this study are generally well preserved and uniquely different from the PPA subjects. Whereas the PPA subjects exhibited greater difficulty keeping their comments on the topics presented by the examiner, FLD subjects showed greater ability to generate comments that were on-topic and relevant to the theme of discussion. The FLD subjects also were judged as more appropriate in their responses to the requests of the examiner and in their acknowledgements to the responses of the examiner, despite reduced verbal output for two of the three FLD subjects in the discourse sampling task. The pragmatic performance of the FLD subjects in this study can best be described by Grice's (1975) general principles, or maxims, of conversation. That is to say, our FLD subjects' pragmatic performances are appropriate in relevance, manner, and quality, although quantity may be reduced. This finding is at variance with previous descriptions of the language and communicative performance of FLD subjects, such as increased talkativeness, perseveration, irrelevance, reduced conversational initiation, insensitivity, social inappropriateness, and confabulation (Barber et al., 1995; Gustafson, 1993; Neary, 1990; Neary et al., 1990). Previous reports of the language and communication individuals with FLD are generally based on accounts of performance recorded during standardized neuropsychological testing (Miller, Cummings, Villanueva-Meyer, Boone, Mehringer, Lesser, & Mena, 1991; Neary et al., 1990), from interview-based responses of relatives of patients (Barber et al., 1995; Kertesz, Fox, & Davidson, 1997), or from selective tests of language (e.g., verbal fluency) (Pasquier, Lebert, Grymonprez, & Petit, 1995). The features of conversation-elicited stories given by relatives is difficult to capture on formal, standardized tests. The nature of the topic directed interview task used in this study, however, may very well have provided the linguistic and thematic support needed by the FLD subjects to overcome their inertia in the initiation, maintenance, and relevance of their responses. This is a limitation of the topic directed interview. Future studies might explore the limits of the pragmatic performance of FLD subjects under less constraining contexts (spontaneous conversation with a unfamiliar partner). Moreover, studies also need to be undertaken to explore the relationship among general cognitive domains (such as attention and memory), language, frontal lobe-based behaviours (e.g., inhibition, judgement, set-shifting), and multiple measures of pragmatics such as those used in this study. Exploring the nature and strength of the relationships in a group of FLD patients will advance our theoretical understanding of the cognitive and linguistic influences on pragmatics, and how their relative influence changes over the progression of the illnesses.

The final important finding of the study is that there was considerable agreement and overlap of findings among the three approaches used in the study to examine the pragmatic performance of the FLD and PPA subjects. The results from the detailed and very time-consuming transcription, coding, and analyses of discourse and pragmatic features, such as percent pronouns without antecedents, percent stereotyped utterances, and percentage of on- and off-topic utterances, among others, were similar to the rating profiles obtained on the

PCA. The ability of the *PCA* to discriminate the pragmatic performances of two individuals with different types of dementia has been shown (Penn et al., 1988). It may very well be the case that the *PCA* can be used to examine and distinguish the pragmatic performances of subjects suspected of having FLD or PPA rather than undertaking lengthy transcription and micro-analysis procedures. Future studies should examine the potential of the *PCA* in this regard.

Acknowledgements—The authors are grateful for the help of Pat McCabe in collecting the discourse samples. The assistance of Christina Aere in coding the discourse samples also is acknowledged. The support and participation of the subjects and their families in this study are gratefully appreciated and acknowledged.

References

Barber, R., Snowden J., & Craufurd, D. (1995). Frontotemporal dementia and Alzheimer's disease: Retrospective differentiation using information from informants. *Journal of Neurology, Neurosurgery, and Psychiatry, 59*, 61-70.

Benton, A. L. (1974). *Revised Visual Retention Test* (4th ed.). New York: The Psychological Corporation.

Benton, A. L., & Van Allen, M. W. (1972). Prosopagnosia and facial discrimination. *Journal of the Neurological Sciences, 15*, 167-172.

Brun, A., Mann, D. M. A., Englund, B., Neary, D. et al. (1994). Consensus on clinical and neuropathological criteria for frontal lobe dementia. *Journal of Neurology, Neurosurgery, and Psychiatry, 57*, 416-418.

Caselli, R. J., Windebank, A. J., Petersen, R. C., Komori, T., Parisi, Okazaki, H., et al. (1993). Rapidly progressive aphasic dementia and motor neuron disease. *Annals of Neurology, 33* (2), 200-207.

Delis, D. C., Kromer, J. H., Kaplan, E., & Obler, B. A. (1987). *California Verbal Learning Test.* New York: The Psychological Corporation.

Duffy, J. R., & Petersen, R. C. (1992). Primary progressive aphasia. *Aphasiology, 6*, 1-15.

Garcia, L. J., & Joanette, Y. (1994). Conversational topic-shifting analysis in dementia. In R. L. Bloom, L. K. Obler, S. De Santi, & J. S. Ehrlich (Eds.), *Discourse analysis and applications: Studies in adult clinical populations* (pp. 161-183). Hillsdale, NJ: Erlbaum.

Green, J., Morris, J. C., Sandson, J., McKeel, D. W., & Miller, J. W. (1990). Progressive aphasia: A precursor of global dementia? *Neurology, 40*, 423-429.

Grice, H. P. (1975). Logic and conversation. In P. Cole and J. L. Morgan (Eds.), *Syntax and semantics 3: Speech acts* (pp. 41-58). New York: Academic Press.

Gustafson, L. (1993). Clinical picture of frontal lobe degeneration of the non-Alzheimer type. *Dementia, 4*, 143-148.

Hooper, H. E. (1958). *The Hooper Visual Organization Test: Manual.* Beverley Hills, CA: Western Psychological Services.

Horner, J. (1985). Language disorder associated with Alzheimer's dementia, left hemisphere stroke, and progressive illness of uncertain aetiology. In R. H. Brookshire (Ed.), *Clinical Aphasiology Conference Proceedings* (pp. 149-158). Minneapolis, MN: RBK Publishers.

Illes, J. (1989). Neurolinguistic features of spontaneous language production dissociate three forms of neurodegenerative disease: Alzheimer's, Huntington's and Parkinson's. *Brain and Language, 37*, 628-642.

Jefferson, G. (1972). Side sequences. In D. Sudnow (Ed.), *Studies in social interaction* (pp. 294-338). New York: Free Press.

Johanson, A., & Hagberg, B. (1989). Psychometric characteristics in patients with frontal lobe degeneration of non-Alzheimer type. *Archives of Gerontology and Geriatrics, 8*, 128-137.

Karbe, H., Kertesz, A., & Polk, M. (1993). Profiles of language impairment in primary progressive aphasia. *Archives of Neurology, 50*, 193-201.

Keenan, E. O., & Schieffelen, B. B. (1976). Topic as a discourse notion: A study of topic in the conversations of children and adults. In C. N. Li (Ed.), *Subject and topic: A new typology of language* (pp. 335-384). New York: Academic Press.

Kempler, D., Metter, E. J, Riege, W. H., Jackson, C. A., Benson, D. F., & Hanson, W. R., (1990). Slowly progressive aphasia: Three cases with language, memory, CT, and PET data. *Journal of Neurology, Neurosurgery, and Psychiatry, 53*, 987-993.

Kertesz, A. (1982). *Western Aphasia Battery.* New York: Grune & Stratton.

Kertesz, A., Fox, H., & Davidson, W. (1997). Frontal Behavioural Inventory: Diagnostic criteria for frontal lobe dementia. *Canadian Journal of Neurological Sciences, 24*, 29-36.

Kertesz, A., Hudson, L., Mackenzie, I. R. A., & Munoz, D. G. (1994). The pathology and nosology of primary progressive aphasia. *Neurology, 44*, 2065-2072.

Kesler, A., Artzy, T., Yaretzky, A., & Kott, E. (1995). Slowly progressive aphasia, a left temporal variant of probable Pick's disease: 15 years of follow-up. *Israel Journal of Medical Science, 31*, 626-628.

MacWhinney, B. (1991). *The CHILDES project: Tools for analyzing talk*. Hillsdale, NJ: Lawrence Erlbaum Associates.

Mattis, S. (1976). Mental status examination for organic mental syndrome in the elderly patient. In L. Bellak and T. B. Karasu (Eds.), *Geriatric Psychiatry*. NY: Grune & Stratton.

Mentis, M., Whittaker, J., & Gramigna, G. D. (1995). Discourse topic management in senile dementia of the Alzheimer's type. *Journal of Speech and Hearing Research, 35*, 1054-1066.

Mesulam, M.-M. (1982). Slowly progressive aphasia without generalized dementia. *Annals of Neurology, 11*, 592-598.

Miller, B. L., Cummings, J. L., Villanueva-Meyer, J., Boone, K., Mehringer, C. M., Lesser, I. M., & Mena, I. (1991). Frontal lobe degeneration: Clinical, neuropsychological, and SPECT characteristics. *Neurology, 41*, 1374-1382.

Neary, D. (1990). Dementia of the frontal lobe type. *Journal of the American Geriatrics Society, 38*, 71-72.

Neary, D., Snowden, J. , Mann, D. M. A., Northen, B., Goulding, P. J., & McDermott, N. (1990). Frontal lobe dementia and motor neuron disease. *Journal of Neurology, Neurosurgery, and Psychiatry, 53*, 23-32.

Nicholas, L. E., & Brookshire, R. (1993). A system for quantifying the informativeness and efficiency of the connected speech of adults with aphasia. *Journal of Speech and Hearing Research, 36*, 338-350.

Osterrieth, P. A. (1944). Le test de copie d'une figure complexe: Contribution à l'étude de la perception et de la mémoire. *Archives de Psychologie, 30*, 286-356.

Pasquier, F., Lebert, F., Grymonprez, L., & Petit, H. (1995). Verbal fluency of frontal lobe type and dementia of the Alzheimer type. *Journal of Neurology, Neurosurgery, and Psychiatry, 58*, 81-84.

Penn, C. (1985). The Profile of Communicative Appropriateness: A clinical tool for the assessment of pragmatics. *South African Journal of Communication Disorders, 32*, 18-23.

Penn, C. (1988). The profiling of syntax and pragmatics in aphasia. *Clinical Linguistics & Phonetics, 2*, 179-207.

Penn, C., Sonnenberg, B., & Schnaier, Y. (1988). Dementia and communication pathology: Two case examples. *South African Journal of Communication Disorders, 35*, 65-74.

Raven, J. C. (1947). *Coloured Progressive Matrices Sets A, Ab, and B*. London: H. K. Lewis.

Retherford, K. S. (1993). *Guide to analysis of language transcripts*. Eau Claire, WI: Thinking Publications.

Rey, A. (1941). L'examen psychologique dans les cas d'encéphalopathie traumatique. *Archives de Psychologie, 28*, 286-340.

Ripich, D. N., & Terrell, B. (1988). Patterns of discourse cohesion and coherence in Alzheimer's disease. *Journal of Speech and Hearing Disorders, 53*, 8-15.

Shewan, C. M. (1988). The 'Shewan Spontaneous Language Analysis (SSLA)' system for aphasic adults: Description, reliability, and validity. *Journal of Communication Disorders, 21*, 103-138.

Snowden, J. S., Neary, D., & Mann, D. M. A. (1996). *Fronto-temporal lobar degeneration: Fronto-temporal dementia, progressive aphasia, semantic dementia*. New York: Churchill Livingstone.

Snowden, J. S., Neary, D., & Mann, D. M. A., Goulding, P. J., & Testa, H. J. (1992). Progressive language disorder due to lobar atrophy. *Annals of Neurology, 31*, 174-183.

Strong, M. J., Grace, G. M., Orange, J. B., & Leeper, H. A. (1996). Cognition, language, and speech in amyotrophic lateral sclerosis: A review. *Journal of Clinical and Experimental Neuropsychology, 18*, 291-303.

Thompson, C. K., Ballard, K. J., Tait, M. E., Weintraub, S., & Mesulam, M. (1997). Patterns of language decline in non-fluent primary progressive aphasia. *Aphasiology, 11*, 297-321.

Tomoeda, C. K., & Bayles, K. A., (1993). Longitudinal effects of Alzheimer disease on discourse production. *Alzheimer Disease and Associated Disorders, 7*, 223-236.

Ulatowska, H. K., Allard, L., Reyes, B. A., Ford, J., & Chapman, S. (1992). Conversational discourse in aphasia. *Aphasiology, 6*, 325-331.

Wechsler, D. (1981). *Wechsler Adult Intelligence Scale-Revised*. NY: Psychological Corporation.

Wechsler, D. (1987). *Wechsler Memory Scale - Revised*. NY: Psychological Corporation.

Weintraub, S., Rubin, N. P., & Mesulam, M.-M. (1990). Primary progressive aphasia: Longitudinal course, neuropsychological profile, and language features. *Archives of Neurology, 47*, 1329-1335.

Appendix A

Definitions for outcome measures

Discourse measures

Words
Words were identified according to Nicholas and Brookshire's (1993) definitions and guidelines. According to their definition, words are "intelligible in context to someone who knows the ...topic being discussed". They "do not have to be accurate, relevant, or informative relative to the ...topic being discussed (p.348)."

Mean length of utterance
The average number of morphemes per utterance (Retherford,1993, p. 272).

Utterance
An utterance, as defined by Shewan (1988), is a "complete thought, usually expressed in a connected grouping of words, which is separated from other utterances on the basis of content, intonation contour, and/or pausing" (p.124). Utterances may also be expressed a nonverbal actions.

Complete utterance
A complete utterance was defined as an utterance which expresses the speaker's complete, uninterrupted thought.

Incomplete utterance
Incomplete utterances were identified following the definitions and guidelines of the Childes Project (MacWhinney, 1991). According to this definition, incomplete utterances are "incomplete but not interrupted" utterances in which the speaker "trails off" without completing the thought (p. 43).

e.g., *Subject:*   *The doctor here was in touch with him and told him…*
Subject: Oh the B-12 shot was something that I got for a long time.

Stereotype utterance
A stereotype utterance contains an idiosyncratic, non-propositional phrase or phrases, such as *fine thank you, pretty good, okay now*, etc.

e.g., *Subject:*   *I was very good.*
Examiner: Tell me about all the jobs that you've had.
Subject: *I was very good um…*

Overt statements regarding anomia
Overt statements regarding anomia refer to comments made by the speaker which indicate that he/she is experiencing difficulty recalling an intended word.

e.g., Subject: Yesterday with doctor, or professor…
Subject: *I am thinking of his name.*
Subject: *I can't seem to recall it.*

Self correction
Self corrections refer to all instances of "retracing with correction" as defined and described in the Childes Project (MacWhinney, 1991). Self corrections refer to instances where " a speaker starts to say something, stops, repeats the basic phrase, changes the syntax but maintains the same idea" (p.52).

e.g., Subject: It just *it was just* terrible to try and do it.

Pronoun without antecedent
A pronoun without antecedent includes a referent which was not clearly established in the preceding discourse and thus is unknown or ambiguous to the conversational partner.

Demonstrative pronoun
A pronoun that points out that which it modifies (*e.g., this, that, those, these*) (Retherford, 1993, p. 272). The pro-forms *here* and *there*, which designate location, are also included in this category.

e.g., Subject: I think *those* sleeping pills helped me.
e.g., Subject: I've never been *here* before.

Repetitions
Repetitions refer to all instances of "retracing without correction" described in the Childes Project guidelines and definitions (MacWhinney, 1991). Repetitions refer to instances where "a speaker begins to say something, stops and then repeats the earlier material without change" (p.51). Repetitions may occur on sounds, syllables, words or phrases.

e.g., Subject: I think *I think* those pills helped me.

II. Topic Analysis

Topic unit
A topic unit is defined as a sequence of utterances that address the same topic and/or an utterance that addresses a unique topic. The concepts of topic unit, on-topic, and off-topic were drawn from and based on

Keenan and Schieffelin's (1976) definition of topic, as well as definitions by Garcia and Joanette (1994) and Mentis, Whittaker, and Gramigna (1995). Several aspects of a given topic may be discussed within the scope of a single topic unit. In a topic-directed interview format topic unit boundaries typically coincided with each new question asked by the examiner, however this was not always the case.

e.g.,	Examiner:	Tell me all about your family.
Topic unit #1	Subject:	I've got an older son and he's home right now.
	Subject:	My youngest son's up in Guelph studying veterinarian medecine.
	Subject:	Good.
Topic unit #2	Examiner:	Tell me about your health right now.
	Subject:	Well it's not too bad.

Side sequence
A side sequence refers to an utterance or a series of utterances which temporarily suspend the on-going topic of discussion to address matters which arise during the course of the discussion. The definition was based on the working definition of Jefferson (1972). Two distinct types of side sequence were identified: meta-statements and external side sequences.

Meta-statement
A meta-statement is a specific type of side sequence in which the speaker suspends the on-going conversation to comment on some aspect of the topic being discussed or on some aspect of his/her performance in discussing the immediate topic. Meta-statements may occur as a single utterance or a series of utterances within the boundaries of a topic unit.

| e.g., | Examiner: | So tell me all about your family. |
| | *Subject:* | *Do you mean my personal family or my parents and stuff?* |

External side sequence
An external side sequence is an utterance or series of utterances which interrupt the on-going topic of conversation to address matters in the immediate environment or other matters which are unrelated to the immediate topic of conversation. Examples of external side sequences include:

Comments about an event which arises in the interview environment during discussion of a given topic.

| e.g., | Subject: | I started to read one of those things from a to z and it took me an hour. |
| | *Subject:* | *Is that a computer on your desk?* |

Comments about matters external to the interview situation.

e.g.,	Subject:	He had a store but that wasn't in Owen Sound it was later.
	Subject:	*We have to go home soon.*
	Examiner:	That's okay we only have a little while longer to go.

On-topic utterance
An on-topic utterance is defined as an utterance that contributes to the advancement or maintenance of the on-going topic of discussion. On-topic utterances:
* demonstrate obvious and appropriate relation to the on-going topic of discussion,
* demonstrate no obvious relation to the on-going topic in terms of semantic content, but function to permit the continuation of the topic,
* are identified as meta-statements,
* are self corrections.

Off-topic utterance
An off-topic utterance is defined as an utterance that does not contribute to the advancement or maintenance of the on-going topic of discussion and appears inappropriate within the context of the immediate topic. Off-topic utterances include:
* those which demonstrate no apparent or appropriate relation to the on-going topic of discussion,
* those which do not appear to be functioning to permit the continuation of an on-going topic,
* those which convey false or incorrect information with respect to the topic of discussion or are identified as *external* side sequences,
* perseverative or intrusive utterances.

Could not evaluate utterance
Could not evaluate utterances include those which could not be analyzed for topic measures as they lacked intelligible content. Responses identified as could not evaluate include:
* completely unintelligible utterances with no identifiable semantic content.
* utterances consisting solely of non-meaningful filler words (e.g., um, uh) or sound repetitions.

Perseverative utterance
A perseverative utterance is one which continues the preceding topic of conversation despite the introduction of a new topic by the conversational partner. It also includes exact repetition of utterances or repetition of utterances with the same propositional or non-propositional meaning.

Intrusive utterance
An intrusive utterance is one in which the speaker refers to a previously discussed topic of conversation and it is inappropriate within the context of the current topic of discussion.

Appendix B

Profile of Communicative Appropriateness

			Inapp.	Mostly Inapp.	Some App.	Mostly Inapp.	App.	CNE	Comments

R I
e n
 	
s t
 	
   e
 	
p r
 	
to l
 	
o o
 	
  c
 	
n u
 	
s t
 	
e o
 	
  r
		Request							

		Reply							
		Clarification request							
		Acknowledgment							
		Teaching Probe							
		Others							
S C e o m n a t n e t n i t c		Topic initiation							
		Topic adherence							
		Topic shift							
		Lexical choice							
		Idea completion							
		Idea sequencing							
		Others							

Profile of Communicative Appropriateness (continued)

		Inapp.	Mostly Inapp.	Some App.	Mostly App.	App.	CNE	Comments
C	Ellipsis							
o	Tense use							
h	Reference							
e	Lexical substitution forms							
s	Relative clauses							
i	Prenominal adjectives							
o	Conjunctions							
n	Others							
F	Interjections							
l	Repetitions							
u	Revisions							
e	Incomplete phrases							
n	False starts							
c	Pauses							
y	Word-finding diffic.							
	Others							
S o	Polite forms							
c S i e	Reference to interlocutor							
o n l s	Placeholders, fillers, stereotypes							
i i n t	Acknowledgment							
g i	Self correction							
u v	Comment clauses							
i i	Sarcasm/humour							
s t t y	Control of direct speech							
i	Indirect speech acts							
c	Others							

Profile of Communicative Appropriateness (continued)

		Inapp.	Mostly Inapp.	Some App.	Mostly App.	App.	CNE	Comments
C o m m N m u o n n n v i e c r a b t a i l o n	Vocal aspects: Intensity							
	Pitch							
	Rate							
	Intonation							
	Quality							
	Non-verbal aspects:							
	Facial expression							
	Head movement							
	Body posture							
	Breathing							
	Social distance							
	Gesture and pantomime							
	Others							
TOTALS								

J. Neurolinguistics, Vol. 11, Nos 1–2, p. 179–190, 1998
© 1998 Published by Elsevier Science Ltd. All rights reserved
Printed in Great Britain
0911-6044/98 $19.00 + 0.00

PII: S0911-6044(98)00012-8

 **Pergamon**

Pragmatics in the absence of verbal language: Descriptions of a severe aphasic and a language-deprived adult

Nina F. Dronkers[*†], *Carl A. Ludy*[*] *and Brenda B. Redfern*[*†]

[*]VA Northern California Health Care System; [†]University of California, Davis

Abstract—Two cases with vastly different etiologies are presented to illustrate pragmatic competence in the absence of verbal language. The first is a man with severe Broca's aphasia who lost the ability to use any propositional language after a massive left hemisphere stroke. The second is a congenially-deaf woman with no exposure to language until well into adulthood. Despite their lack of verbal skills, both cases demonstrate a full command of pragmatic abilities and function as competent social actors. This finding reinforces the view of pragmatics as a vital part of social interaction.

Introduction

Ask any linguist, psycholinguist, speech pathologist, neurolinguist, neurologist, or philosopher what pragmatics is and you may get a cautious answer. Most would venture that pragmatics has something to do with the social aspects of language, i.e., the *way* in which things are communicated rather than *what* is communicated. Yet the specifics of how the '*way*' and the '*what*' relate to each other and how they interact to achieve effective communication are much more difficult to pin down. Most people are just not sure what role pragmatics plays in communication or to what extent it relates to the other components of language.

Too frequently, 'language' is considered to consist only of syntax, semantics, morphology and phonology. Yet, we know that the inappropriate use of prosody, vocal intensity, eye contact, turn-taking, maintenance to the topic, or physical proximity to the listener can cause an utterance to be misinterpreted or cause the listener to "tune out". These errors result in poor communication just as surely as grammatical, lexical, or phonological errors cause misinterpretations. It has been known for some time that certain patients (mostly with right hemisphere disease) are shunned for their odd pragmatic behavior, even though their grammar, lexical selection, and pronunciation may be flawless. At the same time, aphasic patients may demonstrate perfect pragmatic skills and not be able to utter a word.

Even with this knowledge, pragmatics continues to be low on the totem pole of language, and its role in communication is often overlooked. In this paper, we offer some thoughts on the importance of pragmatics in communication particularly in light of its remarkable perseverance in the absence of verbal language. In our Aphasia Research Lab, we have been struck by the similarities between two patients of vastly different etiologies. One is a severe Broca's aphasic with a massive left hemisphere lesion. The other is a language-deprived adult, whose profound deafness was misdiagnosed as cognitive impairment, so that language exposure was delayed until well past the developmentally critical language learning period. Both of these individuals have severe limitations in speaking, but seem to be competent, engaged social actors. We have attempted to document their preserved pragmatic abilities and will discuss them in the context of inter-communication.

Some history and perspective

The notion of pragmatics first emerged in philosophy in the effort to relate meaning to thought and communication and to understand how the units of meaning were used in real situations. The word itself is attributed to Morris (1938) who divided 'semiotics' (the 'science of signs') into three areas: (1) semantics—the content of signs (what they refer to), (2) syntax—how signs relate to one another (how they can be combined into larger units), and (3) pragmatics—how signs are deployed by one individual for uptake by another (how they are used and interpreted). Consideration of all three realms was considered necessary for a full grasp of meaning.

Another philosopher, Austin (1962), launched these ideas to a wider public with an influential lecture series at Harvard in 1955. He sought to help grammarians and philosophers find a common ground in talking about language. The only way to approach a sentence had been to treat it as a verifiable description of a state of affairs, that is, as a true or false statement of fact. Austin showed that more could be made of language by considering it as performing deeds: people use words and sentences to do things. By looking at what they are trying to do, the meaning of the language units they are using becomes clearer.

These ideas were further popularized and brought into linguistics by Searle (1969; 1971) and Grice (e.g., 1968; 1975). With their discussions of 'speech acts', linguists could now talk about the difference between sentences and utterances. Relevance, presupposition, implication, and so on, could be handled in systematic ways. (See Levinson (1983) for description of pragmatics within linguistics.) The revolution promised by Austin was well under way.

The tools provided by studying pragmatics have been well used in applied linguistics. Practitioners in speech pathology drew directly on Austin and Searle to develop instruments to help clinicians understand how an individual uses language, with the goal of remediating such deficits in language-disordered patients (Gallagher & Prutting, 1983; Prutting & Kirchner, 1987). In neurolinguistics, it became clear that pragmatics encompassed an area of language for which right-hemisphere involvement is critical (Joanette, Goulet, & Hannequin, 1990; Molloy, Brownell, & Gardner, 1990). Patients with unilateral right-hemisphere damage usually do well with the semantic and syntactic aspects of language, but show deficits in the context-sensitive, socially appropriate uses to which language is put in everyday situations. These applications and discoveries were made possible by defining pragmatics as a language function. Today, we have nurses and other health care providers being schooled in pragmatics as "right brain communication" (Boss, 1996) and cognitive scientists using theories of utterance interpretation to discern the influence of pragmatics on people's understanding of what is communicated by other speakers (Gibbs & Moise, 1997).

Though the pragmatics revolution may be firmly launched, there is still room for growth in applying it to our paradigm of language. In particular, much of the work has focused on deficits in pragmatics. This has been useful for defining these language functions and for establishing their validity. But surely there is a positive side to pragmatic abilities, a "glass half full" vantage point for considering their communicative functions.

The tendency to treat pragmatics as an extension or attachment to the traditional functions of linguistics has obscured its independent reality in human interactions, the important role it plays in what could be called "nonverbal social communication". Even without the benefits of semantics or syntax, a great deal of social interaction is still possible. Pragmatics could be a window onto these important competencies.

The Pragmatic Protocol (Prutting & Kirchner, 1983; 1987)

For this paper, we will use the Pragmatic Protocol to describe the pragmatic abilities of our two individuals. The protocol is well recognized in the rehabilitation community as a measure of pragmatic skills in both children and adults (Sohlberg & Mateer, 1989). It is based on a 15-minute unstructured conversational sample in which the clinician can rate patients on how well they use language. The relationship between the communicative partners must be positive or neutral, so that both expect to engage in cooperative discourse. The clinician judges performance on each of 30 pragmatic behaviors, broken down into two main aspects: verbal and nonverbal. Verbal aspects of pragmatics include speech acts, topic selection and maintenance, turn taking, lexical selection, stylistic variations and paralinguistic aspects. Nonverbal aspects include kenesics and proxemics.

After watching the interaction, the clinician evaluates each of the 30 behaviors as appropriate or inappropriate over the course of the episode. The criterion is whether the behavior facilitated or detracted from the communicative exchange. Behavior is only judged inappropriate if it could be said to penalize the individual or if it makes a difference in the interaction.

Case descriptions

Patient AK (severe Broca's aphasia)

Patient AK is a 63-year-old white male who was pre-morbidly right-handed and a native English-speaker. He had some college education and had retired after a 23-year career in the Navy as a Chief Petty Officer; he was in a second career as a corporate training manager. In 1991, he suffered an extensive left middle cerebral artery infarction (see Figure 1) which left him with a dense right hemiparesis, dysarthria, apraxia of speech, and severe Broca's aphasia. His productive speech consists almost entirely of the recurring utterance /tõnõ tõnõ/ (or slight variations thereof) which permeate every attempt at verbalization, including spontaneous speech, repetition, and naming. (See Appendix A for a sample transcription[1].) His comprehension is typical of Broca's aphasia, with good single word and simple sentence understanding, breaking down on complex grammatical constructions. These behaviors are quantified in his Western Aphasia Battery (WAB) scores which can be seen in Table 1.

---

[1] A video sample of this patient's conversational skills is contained in Telerounds Program #9, "Neuroanatomical Correlates of Production Deficits in Aphasia" (Dronkers, 1993), produced by the National Center for Neurogenic Communication Disorders at the University of Arizona, Tucson.

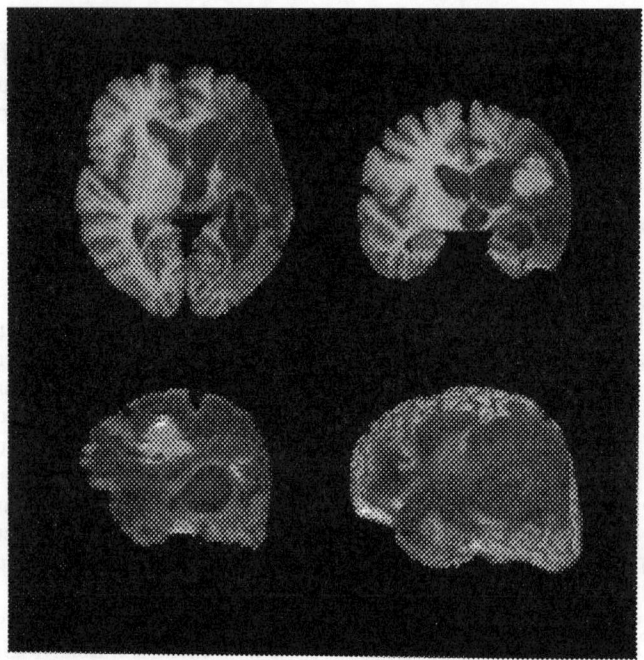

Figure 1. Three-dimensional reconstruction of Patient AK's lesion. This reconstruction
depicts the extent of the middle cerebral artery infarction in this patient with a
severe Broca's aphasia. The upper left image is a horizontal slice similar to
those seen in CT images. The upper right image depicts a coronal section, and
the lower left, a sagittal section through the left hemisphere. The lower right
template shows a 3-D reconstruction of the brain with the lesion apparent on
the lateral surface of the left hemisphere.

Date	Fluency	Aud. comp	Repetition	Naming	WAB AQ
3-92	1	5.25	.2	4	17.7
9-92	1	6.1	.4	1.4	21.8
1-97	1	7.5	0	0	17.0
max	10	10	10	10	100

Table 1. Summary scores obtained on the Aphasia Quotient of the Western Aphasia
Battery for Patient AK. The range for Broca's aphasia on the WAB is 0-4 for
fluency, 4-10 for auditory comprehension, 0-7.9 for repetition, and 0-8 for
naming.

AK's writing is limited to imitating samples and to writing his full name. His knowledge
of numbers is better preserved, and he often uses numbers traced in the air as part of his
communication (e.g., to indicate dates or specific numbers). AK's reading is severely
impaired and limited to a few single words and to letters. He also has moderate-to-severe
limb and buccofacial apraxia. His basic calculation and constructional praxis skills are

essentially intact. In spite of his limited productive speech, AK can carry a tune perfectly well, and can sing the words if he has a model to mimic. Table 2 summarizes some of the other neuropsychological test scores that reflect his language and cognitive abilities.

Peabody Picture Vocabulary Test

Date	Vocabulary age
6-97	4-1

Token Test

Date	Parts A-E (max correct=67)	Part F (max correct=96)
6-97	42	53

Wechsler Adult Intelligence Scale—Revised (Scaled Scores)

Date	Picture completion	Picture arrangement	Block design	Object assembly	Digit symbol	P IQ
6-97	5	7	8	5	3	86

Raven's Coloured Progressive Matrices

Date	Raw score (max correct=36)
3-92	22
9-92	30
1-97	32

Table 2. Sample scores obtained on other standard tests for Patient AK

AK lives with his wife and has children and grandchildren nearby. He enjoys following sports, playing computer games, and building elaborate Lego models. His social life includes visits from fellow sports enthusiasts and active involvement in several stroke support groups.

Patient AK's pragmatic abilities

Patient AK's performance on the Pragmatic Protocol is based on three ratings of a conversation with one of the authors, and is summarized in Table 3. In terms of any inappropriate pragmatic behaviors, the ratings indicate that AK only occasionally initiates directives, queries, and comments, and that the variety of his speech acts is also greatly reduced. He rarely selects or introduces the topic of discussion, and makes little effort to change it. Thus, the partner is left with the burden of carrying the conversation. These shortcomings are not surprising in light of his restricted productive speech. AK has very little verbal language with which to establish new topics and must rely on others to take that role. In other circumstances, AK can, of course, convey that he has wants or needs, but is not always successful in conveying the content of his message.

Communicative act	Appropriate	Inappropriate	N/A
Verbal aspects			
Speech acts			
1. Speech act pair analysis	Responds to directives, queries and requests; acknowledges comments	Only occasionally initiates directives, queries, comments	
2. Variety of speech acts	comments, asserts, disagrees	reduced variety	
Topic			
3. Selection		rarely selects topic	
4. Introduction		rarely introduces new topic	
5. Maintenance	contributes to maintenance		
6. Change		does not change topic	
Turn taking			
7. Initiation		partner carries burden	
8. Response	responds as a listener		
9. Repair/revision	asks for clarification		
10. Pause time	normal pauses		
11. Interruption/overlap	no interruptions or overlaps		
12. Feedback to speakers	nods, gestures appropriately		
13. Adjacency	waits turn		
14. Contingency	stays on topic		
15. Quantity/conciseness	non-verbal comments		
Lexical selection			
16. Specificity/accuracy			*
17. Cohesion			*
Stylistic variations			
18. The varying of communicative style	adjusts speech style with prosodics		
Paralinguistic aspects			
19. Intelligibility			*
20. Vocal intensity	normal intensity		
21. Vocal quality	normal vocal quality		
22. Prosody	normal prosody		
23. Fluency	recurring utterance is at normal rate and smoothness		
Nonverbal aspects			
Kenesics and Proxemics			
24. Physical proximity	normal distance		
25. Physical contacts	normal contacts		
26. Body posture	normal body posture		
27. Foot/leg hand/arm move.	normal movements		
28. Gestures	normal gestures		
29. Facial expression	normal facial expressions		
30. Eye gaze	normal eye gaze		

*not applicable/no opportunity to observe

Table 3. Summary Pragmatic Protocol ratings for Patient AK

Despite his difficulty in initiating new topics, AK does use gestures, facial expressions, and intonational variations on his recurring utterance to respond to directives, queries, and requests, and to acknowledge the comments of others. He also expresses his comments, assertions, and disagreements in this way. He contributes to maintaining the conversation by responding as a listener, and asking for clarification, again by using intonational variations, facial expression, and by nodding and gesturing. He does not interrupt and waits his turn appropriately.

AK expresses stylistic variations by adjusting his speech style with prosodics. He uses stress and intonation to modify his recurring utterance. Paralinguistic aspects include normal intensity of the recurring utterance with normal vocal quality, prosody, rate, and smoothness.

All nonverbal aspects of AK's communicative efforts are perfectly normal. He maintains a natural distance between himself and his conversational partner, with normal physical contacts. His body posture, movements, gestures, facial expressions, and eye gaze are all also normal.

Thus, AK is an example of an individual with little to no verbal output except for the recurring utterance /tōnō tōnō/ which he varies with changes in stress and intonation. This strategy works well for him in terms of maintaining near-normal pragmatics. Though he has difficulty initiating and changing topic, he exhibits excellent pragmatic abilities by using appropriate turn taking, stylistic variations, and paralinguistic and nonverbal skills that keep him part of the conversation.

The case of "Chelsea" (language-deprived)

"Chelsea" has a very different language history, although she also maintains excellent pragmatic behavior in spite of impoverished verbal language. She is a 49-year-old white female who was born the second of seven children in a rural community in Northern California. She was born with a severe to profound sensori-neural hearing loss but was misdiagnosed as mentally retarded during her childhood. Chelsea's mother knew her to be deaf and ignored professional advice to institutionalize her, raising her at home among siblings. She learned to cook and do housework and helped her mother raise the younger children. She was denied admission to local schools and a school for the deaf. As a result, Chelsea did not acquire any language or receive any formal education until the age of 32 when she was referred to a neurologist and a speech pathologist by a social worker who realized her situation. At that time, she was fitted with bilateral hearing aids and began an intensive program of oral and signed language instruction, as well as education in math and other academic subjects. Chelsea currently lives at home with her parents, and works part-time in a veterinarian's office as an assistant. Her performance on several standardized tests can be found in Table 4[2].

Peabody Picture Vocabulary Test

Date	Vocabulary age	With or without signing
7-80	2-3	without
12-80	3-2	without
4-81	3-11	without
8-81	4-3	without
8-81	5-5	with
10-81	5-3	with
3-82	6-10	with
6-82	5-11	with
1-83	5-8	with

[2] Numerous neuropsychological and language tests have been administered to Chelsea over the years, and cannot all be represented here. The first author can be contacted for further details.

Token Test

Date	Parts A-E (max correct=67)	Part F (max correct=96)
8-82	65	53
10-86	60	55
10-87	52	43
7-89	55	57

Wechsler Adult Intelligence Scale (Scaled Scores)

Date	Picture completion	Picture arrangement	Block design	Object assembly	Digit symbol	P IQ
10-80	10	4	6	10	3	77
12-81*	6	4	8	16	4	84
10-86*	12	4	8	10	4	89
10-87*	12	2	7	7	5	84

* WAIS-R

Raven's Coloured Progressive Matrices

Date	Raw score (max correct=36)
12-81	24
1-83	24
10-86	29
10-87	28

Table 4. Sample scores obtained on standard tests for Chelsea

Chelsea's case addresses many interesting questions concerning the critical age for language acquisition, particularly whether it is possible to learn language after long periods of language deprivation in childhood. Her situation is analogous to those of linguistically "feral" children, such as Genie (Curtiss, 1977) or the Wild Boy of Aveyron (Itard, 1801), who did not acquire language because of lack of exposure. For Chelsea, the lack of exposure was due to severe hearing loss, but she was otherwise raised in a normal and loving family environment, contrary to previous cases. Lenneberg (1967) believed that children who did not learn language by the age of puberty would not be able to acquire it normally, while Krashen (1973) lowered this critical age to five years. The answer offered by Chelsea's case is the same one concluded by Curtiss for Genie; the critical age for language acquisition is different for the different components of language. Both Genie and Chelsea continue to develop their vocabulary, years after beginning to learn language as adults. Their knowledge of syntax, however, remains virtually absent. Chelsea's conversational style is to string words together, with no evidence of syntax or morphology. (See Appendix B for a sample transcription.)

Chelsea's pragmatic abilities

In the realm of pragmatics, Chelsea has developed quite normally. Her social skills are most appropriate, and she, like Patient AK, is very pleasant company. This is reflected in her performance on the Pragmatic Protocol (Table 5).

Communicative act	Appropriate	Inappropriate	N/A
Verbal aspects			
Speech acts			
1. Speech act pair analysis	Responds to directives, initiates queries and comments		
2. Variety of speech acts	appropriate use and diversity		
Topic			
3. Selection			*
4. Introduction	introduces topics		
5. Maintenance	tries to maintain topic		
6. Change	makes some change in topic		
Turn taking			
7. Initiation	initiates questions		
8. Response	responds as a listener		
9. Repair/revision		rarely asks for clarification	
10. Pause time	normal pauses		
11. Interruption/overlap		some overlap when signing	
12. Feedback to speakers	nods, gestures appropriately		
13. Adjacency	waits turn		
14. Contingency	stays on topic	often repeats, doesn't add	
15. Quantity/conciseness		not always informative	
Lexical selection			
16. Specificity/accuracy	limited, but appropriate		
17. Cohesion			*
Stylistic variations			
18. The varying of communicative style	adjusts speech style		
Paralinguistic aspects			
19. Intelligibility	signs are intelligible	verbal responses are not always intelligible	
20. Vocal intensity	normal intensity		
21. Vocal quality	normal vocal quality		
22. Prosody	almost normal prosody		
23. Fluency	normal rate and smoothness		
Nonverbal aspects			
Kenesics and proxemics			
24. Physical proximity	normal distance		
25. Physical contacts	normal contacts		
26. Body posture	normal body posture		
27. Foot/leg hand/arm move.	normal movements		
28. Gestures	normal gestures		
29. Facial expression	normal facial expressions		
30. Eye gaze	normal eye gaze		

* not applicable/no opportunity to observe

Table 5. Summary Pragmatic Protocol ratings for Chelsea

The rating was based on a videotaped sample of a conversation between Chelsea and her teachers one and a half years into her training when she had considerably less language than she does now. The rating indicates that she demonstrates a variety of speech acts, responding to directives and initiating questions and comments in a normal fashion. She also introduces topics, and tries to maintain and change them. She initiates questions, responds as a listener, uses normal pauses, nods and gestures appropriately, awaits her turn in the conversation, and stays on topic. Her choice of lexical items, while limited, is appropriate, and her signs, if not her speech, are intelligible. Furthermore, she adjusts her speech style, uses normal intensity, vocal quality, rate, and smoothness, and virtually normal prosody given her hearing loss. In nonverbal aspects, Chelsea positions herself at a normal distance, with normal physical contacts, body posture, movements, gestures, facial expressions, and eye gaze.

The only area in which Chelsea could be rated with inappropriate pragmatics was in the realm of turn taking. Here, she rarely asks for clarification if she does not understand the content. Instead, she will often repeat what the other person has said, and does not add concise, new information on her own. This is not surprising considering her limited vocabulary and lack of experience with language.

Discussion

We have chosen two very different case studies to illustrate that pragmatic competence is an ability which can persist in the virtual absence of verbal language. Because both cases are largely nonverbal, this ability can be considered as independent of the verbal modality. This, we believe, is a necessary corrective to the tendency to treat pragmatics as an epiphenomenon of speech. In fact, we believe that both pragmatics and speech should be treated as phenomena of social interaction. The history of pragmatics guides us towards considering speech as social action, as accomplishing social deeds. These two cases provide further evidence that nonverbal social skills and abilities can exist in parallel to verbal language.

The contrast between our two cases is also instructive. With AK, language breakdown came after the development of full social competence. His pragmatic challenges relate to the loss of propositional communicativeness. He has had to adjust his social engagements to his decreased ability to convey meaning by speaking. Chelsea, on the other hand, did not begin developing verbal language until the age of 32. Until that time, she was only able to develop such social competence as was not dependent on language. As she continues to learn language, she must also learn the attendant conversational skills. Yet her abilities as a nonverbal social actor give her the basis for developing these skills in tandem. Ultimately, her language development may reach a limit determined by the critical learning period, while her social communicative development will continue to the full potential of her personality.

We began this paper by outlining the revolution that was wrought in philosophy and linguistics by the introduction of pragmatics as an aspect of understanding human communication. It should be clear that we believe this revolution has not finished running its course. There is still a predilection for installing speech at the apex of human interaction. Yet clinicians have had an insight for years, which has been hard to express, that there is more to communication than words and sentences. Our two cases, so far apart yet so similar, define a point in space of social connection. Other points in this space might be the development of verbal ability without pragmatics, or, preserved language with pragmatic breakdown. Only the total space defined by these points and others will reveal the full universe of human communicative potential.

Acknowledgments—We are grateful to Peter Glusker, M.D. and Catherine O'Connor, M.A. for involving the first author in Chelsea's case and for providing some of her history and test scores.

References

Austin, J. L. (1962). *How to do things with words*. Oxford: Oxford University Press.
Boss, B. J. (1996). Pragmatics: Right brain communication. *Axone, 17* (4), 81-85.
Curtiss, S. (1977). *Genie: A psycholinguistic study of a modern-day "wild child"*. New York: Academic Press.

Dronkers, N. F. (1993). *Neuroanatomical correlates of production deficits in aphasia.* Telerounds Production Nr. 9, University of Arizona, Tucson.

Gallagher, T., & Prutting, C. (Eds.) (1983). *Pragmatic assessment and intervention issues in language.* San Diego: College-Hill Press.

Gibbs Jr., R. W., & Moise, J. F. (1997). Pragmatics in understanding what is said. *Cognition, 62*, 51-74.

Grice, H. P. (1968). Utterer's meaning, sentence-meaning, and word-meaning. *Foundations of Language, 4*, 1-18.

Grice, H. P. (1975). Logic and conversation. In P. Cole and J. L. Morgan (Eds.), *Syntax and semantics 3: Speech acts* (pp. 41-58). New York: Academic Press.

Itard, J. M. G. (1801). *De l'éducation d'un homme sauvage ou des premiers développements physiques et moraux du jeune sauvage de l'Aveyron.* Paris: Gouyon.

Joanette, Y., Goulet, P., & Hannequin, D. (1990). *Right hemisphere and verbal communication.* New York: Springer-Verlag.

Krashen, S. (1973). Lateralization, language learning, and the critical period: Some new evidence. *Language Learning, 23*, 63-74.

Lenneberg, E. H. (1967). *Biological foundations of language.* New York: Wiley.

Levinson, S. C. (1983). *Pragmatics.* Cambridge: Cambridge University Press.

Molloy, R., Brownell, H. H., & Gardner, H. (1990). Discourse comprehension by right-hemisphere stroke patients: Deficits in prediction and revision. In Y. Joanette and H. H. Brownell (Eds.), *Discourse ability and brain damage: Theoretical and empirical perspectives* (pp. 113-130). New York: Springer-Verlag.

Morris, C. W. (1938). *Foundations of the theory of signs.* Chicago: University of Chicago Press.

Prutting, C. A., & Kirchner, D. M. (1983). Applied pragmatics. In T. Gallagher and C. Prutting (Eds.), *Pragmatic assessment and intervention issues in language* (pp. 29-64). San Diego: College-Hill Press.

Prutting, C. A., & Kirchner, D. M. (1987). A clinical appraisal of the pragmatic aspects of language. *Journal of Speech and Hearing Disorders, 52*, 105-119.

Searle, J. R. (1969). *Speech acts: An essay in the philosophy of language.* Cambridge: Cambridge University Press.

Searle, J. R. (Ed.) (1971). *Philosophy of Language.* Oxford: Oxford University Press.

Sohlberg, M. M., & Mateer, C. A. (1989). *Introduction to cognitive rehabilitation: Theory and practice.* New York: The Guilford Press.

Appendix A

Transcription of Patient AK's speech in conversation:

Interviewer: Do you like coming to Group? (referring to the weekly support group meeting)
Patient AK: (nodding emphatically) T̲õ-nõ tõ-nõ. Nõ-tõ. Tõ-tõ. T̲õ-nõ tõ. Tõ-tõ. Tõ-tõ.
Interviewer: It's a nice group of people, isn't it?
Patient AK: Tõ-t̲õ-tõ-tõ.
Interviewer: Does [your wife] enjoy it? [Your wife] enjoys sitting with the other...
Patient AK: (shrugs) T̲õ-tõ. T̲õ-nõ tõ-nõ. (waves hand in air) Õ̲õ, tõ-tõ tõ-nõ tõ-nõ t̲õ-nõ tõ-nõ. (makes pushing-away gesture with hand, points to self, then waves it off) Tõ-tõ t̲õ-tõ tõ-tõ.
Interviewer: Yeah.

(Underlining indicates syllabic stress.)

Appendix B

Transcription of Chelsea's speech in conversation:

(Both participants sign at the same time they are speaking.)

Interviewer: (addressing second interviewer) I've told Chelsea for the last two days that I had a gift for her.
Chelsea: Gift.
Interviewer: From Colorado.
Chelsea: Colorado.
Interviewer: I remembered! (presents gift) Do you want to open it?
Chelsea: (accepts wrapped gift, begins to untie ribbon.)
Interviewer: Ribbon.
Chelsea: Ribbon.
Interviewer: What do you think it is?
Chelsea: Think? (shakes head)
Interviewer: A book. Think it's a book?

Chelsea: Book? Don't think.
Interviewer: Is it a blouse?
Chelsea: Blouse? No. You...collar.
Interviewer: Collar? Oh, a scarf. Yes, for my birthday....
Chelsea: (continues unwrapping, still unfolding paper)
Interviewer: There's nothing....I tricked you!
Chelsea: (laughs; takes out small box) Oh! Thank you! (hugs interviewer) Jewelry!
Interviewer: That is named 'turquoise'. (finger spells 'turquoise')
Chelsea: Turquoise.

J. Neurolinguistics, Vol. 11, Nos 1–2, p. 191–206, 1998
0911-6044/98 $19.00 + 0.00

Pergamon

PII: S0911-6044(98)00013-X

The use of gestures as a compensatory strategy in adults with acquired aphasia compared to children with specific language impairment (SLI)

Bibi Fex and Ann-Christin Månsson[1]

Logopedic Unit, ENT Clinic, Helsingborg Hospital, Helsingborg, Sweden

Abstract—Gestures used in a confrontation naming task by four severely aphasic patients, two nonfluent, two fluent and four controls, were categorized according to Ekman and Friesen (1969). The findings were compared to an earlier study of gestures used by SLI children. There were similarities between the two language-impaired groups in how to compensate for the missing access to the word by using gestures as pragmatic cues.

Introduction

Communication among individuals involves a variety of modalities. The spoken word carries the burden of communication but there are also paralinguistic channels such as intonation, facial expression, eye movement and hand and arm gestures that convey meaning and contribute greatly to communication for both speaker and listener. Gestures are natural and more or less frequent in normal conversation by speakers of different languages. Gesturing is a supplementary or alternative means for communication for both children and adults. Ekman and Friesen (1969) categorized nonverbal behaviour and found the following types: emblems, illustrators, affect displays, regulators and adaptors.

Emblems have a direct verbal translation, consisting of a word or two or perhaps a phrase. Emblems are well known by all members of a group, class or culture. An emblem may repeat, substitute or contradict some part of the verbal behaviour. Illustrators are movements directly tied to speech, serving to illustrate what is being said verbally (i.e., to repeat, substitute, contradict or augment the information provided verbally); they are socially learned. There are six types of illustrators: Batons are movements which time out, accent or emphasize a particular word or phrase; Ideographs trace the itinerary of a logical journey, the direction of an idea; Deictic movements point to an object in the environment; Spatial movements depict a spatial relationship; Kinetographs are movements which depict a bodily action; Pictographs are movements which draw a picture of their referent.

Affect displays have the face as the primary site for indicating emotions such as happiness, surprise, fear, sadness, anger, disgust and interest. Regulators are related to the conversational flow, the pacing of the exchange. The most common regulators include head nods, eyebrow raises, etc. A person can perform a regulator without knowing it, but if asked s/he can easily recall it. Adaptors are movements that were first learned as part of adaptive efforts to satisfy personal needs, perform bodily actions, or manage emotions.

---

[1] Authors are listed alphabetically; both authors share equal responsibility for the paper.

Self-adaptors involve touching one's face, hair or lips to facilitate or inhibit sound production or speech. Hand-to-face adaptors are a rich source of information.

Månsson and Lundström (1996) investigated 8 specific language impaired (SLI) children and 8 normal language developed children with regard to their nonverbal communication (age 3:5 to 6:7). The children were video-recorded while they were telling the pre-school teacher about their wishes for Christmas gifts. The results showed that the SLI group used 436 gestures while the normals used 318 gestures (a statistically nonsignificant difference). There were individual differences among the groups. Emblems, adaptors and affect displays were used more often by the SLI children, while the normal children used more regulators and illustrators (statistically significant). The authors concluded that children develop gestures in the following order:

affect displays → adaptors → emblems → regulators → illustrators

The SLI children used more pictographs and deictic movements while the normal children used more batons and spatial movements (statistically significant). Månsson and Lundström (1996) suggested that these subgroups of illustrators are developed in the following order:

deictic movements → kinetographs → pictographs → batons → spatial movements.

Observing gestures in children adds knowledge that can be used diagnostically, but also prognostically. Children with more advanced gestures such as, for example, pictographs might have a better prognosis than children who do not use pictographs in their nonverbal behaviour.

Söderbergh (1980) described mother-child dialogues during play activity. Verbal, somatic and vocal language was observed. The verbal channel covered all verbal language, the vocal channel included communication that was signalled by voice, e.g., pitch, stress, etc. The somatic channel comprises communication that is signalled via the face, body and posture. The author says that verbal language is developed on the basis of body language, and that adult-child dialogue develops towards normal adult-adult dialogue. Somatic (body language) communication continues to be used by the adult speaker, not only as an accompaniment to speech, but as an independent and necessary part of dialogue. A complete model of dialogue must therefore include the somatic component. A listener must be somatically active. The listener reacts continuously to the speaker by posture and body movements and mimics and responds to the speaker.

Similarly, Vygotsky (1978) claimed that children use gestures and words at an early age to get their needs and wishes satisfied. He also stated that children use gestures to regulate others and thereby learn to regulate themselves. In that way, children take an active role in their own cognitive development.

McNeill and Levy (1982) divide nonverbal communication into three systems of gestures: iconic (pictographs according to Ekman and Friesen), metaphoric (depicting a more abstract concept) and beats (batons). They are of the opinion that icons depict whole scenes, while extranarrative statements lacking the sequentiality constraint tend to be accompanied by more formless gestures or beats. The theory traces the sensory-motor images of speakers back to the ontogenetically primitive stages of early childhood and shows how they continue to play a role in generating language for adult speakers. At age 9 or 10, deictics and iconics are well developed but beats are still quite rare.

Jancovic, Devoe and Wiener (1975) also point out that gestures become more frequent and more complicated as the child's language develops. Wilkinson and Rembold (1981) suggest that, as the child's language improves, gestures complement and enrich the verbal language. Deictics are used to give a message when verbal ability is not well enough developed and these gestures diminish with age. By contrast, pictographs expand a verbal message. Depending on age and language complexity, the use of pictographs expands. Sanmarco (1984) observed how mothers and children interacted when they tried to solve a problem in copying a model. SLI and normal language children from 3 to 11 years old were studied. It was noted that the SLI children used deictic gestures more than normal children.

Early aphasia therapy focused on re-educating linguistic skills. Dean and Skinner (1995) conclude that, when the therapist profiles the communication performance of aphasic clients in addition to results from assessments, remediation will be more balanced. Lately it has been suggested that the assessment of aphasia not only give the status of the spoken and written language, but also the person's complete communicative ability.

A series of papers have provided support for the argument that nonverbal performance should be profiled in aphasia assessment. Some studies have shown that aphasic patients can use nonverbal communication to a greater degree than spoken or written language. Duffy, Duffy and Pearson (1975) claimed that there was one central communication and symbol system in human beings and that damage to this mechanism would result in lowered performance on all symbolic tasks. Aphasic individuals would not be able to use gestures. The observation that non-verbal communication (NVC) is used for structuring verbal production, i.e., for semantic planning, and the finding that this could be seen in Wernicke's aphasics, in that the amount of gesturing increased when they had problems with semantic planning, are also highly relevant for the interpretation of NVC in aphasics, according to Delis et al. (1979). Cicone et al. (1979) studied spontaneous gestural production and spoken language in four aphasic subjects, two Broca's aphasics and two Wernicke's aphasics. In general the gestures of the patients paralleled their speech output. The total amount and nature of gesturing produced by each patient was investigated. Posterior aphasics produced more movements than anteriors. Anterior aphasics produced a higher percentage of head movements as well as movements of the whole body and the torso. No differences were revealed between the aphasic groups in measuring hand configuration and orientation. The difference in the number of movements between the groups seems unremarkable. The results of the Wernicke's aphasics more closely resembled those of the normal controls, but the aphasics' language and gestures lacked clarity. Their gestures were often elaborate but generally unclear and confusing like their speech. In Broca's aphasics, a large percentage of iconic gestures occurred; they produced simple and unelaborate units in both modalities. Their output was sparse, but their gestures were generally informative and clearly intelligible. Wernicke's aphasic patients used many more pointing gestures in their communication. These results do not support the claim that aphasics spontaneously improve their communication by gestures and the authors say that their findings can be interpreted as evidence in favour of a central organizer which controls the critical features of communication, irrespective of the modality of expression.

Spontaneous gestures used by 20 aphasic patients during a verbal confrontation naming task were studied by Helm (1979). Broca's aphasics produced at least one gesture in association with the greatest number of the 40 pictured items. The Wernicke's patients rated second to the Broca's group in the frequency of their gesturing but the quality of their gestures exceeded all other groups. The results showed that different groups of aphasic patients have differential capacities for producing self-initiated gestures. According to Helm, an evaluation of gestural skills should include both an apraxia examination and a tool for noting spontaneous use of gestures.

A study by Ahlsén (1985) clearly shows that aphasics have a much higher use of NVC than non-aphasics. Even interindividual variation was greater in the aphasic group. The results were summarized as follows: The aphasics used: (a) much more NVC illustrating factual information; (b) more NVC showing positive and negative feelings; (c) more NVC for turn keeping, turn giving and feedback eliciting; (d) less NVC for turn taking (in fact none). There were no differences between the aphasics and the controls in the amount of use of (a) NVC showing hesitation, b) feedback giving, (c) emphasis, d) emblems for "yes" and "no". The aphasics used a substantial amount of NVC, especially pointing, which was often used without verbal accompaniment. Ahlsén (1991) also studied non-verbal communication in a patient with Wernicke-type aphasia and found that he used pointing gestures and illustrators a lot more than other aphasics and controls.

Le May, David and Thomas (1988) found that hand gestures are used most by Broca's aphasics and least by non-aphasic controls. Batons, ideographs and deictic movements were associated with Broca's aphasics, while both Wernicke's and Broca's aphasics used kinetographs more than the controls. These authors raise the question of whether therapy should encourage or teach Broca's aphasics to use more batons, ideographs and deictic movements as compared to their use of pictographs. Would an increased use of pictographs increase the clarity of their communication?

Hermann, Reichle and Lucius-Hoene (1988) found that aphasic patients produced almost twice as many verbal utterances as their healthy counterparts. The patients produced repetitive utterances without value, e.g., automatisms and perseverations. The researchers calculated the percentage of nonverbal elements over the total number of verbal and nonverbal communicative actions. Aphasics used fewer speech-focused movements and significantly more codified gestures than their counterparts. There was no difference in the use of descriptive gestures. Holland (1982) points out that aphasics communicate better than they speak. There is little doubt that aphasia is often associated with nonverbal disorders according to Feyereisen (1988). The fact that some cases of aphasia are explained by general conceptual deficits and others by modality-specific disorders may suggest that an intermediate position should be taken with regard to the existence and/or role of a central organizer. Feyereisen (1983) also noted an increase in the ratio of gestures per word in aphasic patients. These observations suggest a functional role for iconic gestures in word retrieval. McNeill (1985) said that gestures dissolve along with speech in aphasia. Butterworth and Hadar (1989) point out that, if gesture is to be linked to one or more computational stages of speech production, the model must include the following stages: Stage 1: Pre-linguistic message construction; Stage 2: Determination of the grammatical form; Stage 3: Selection of the lexical items in abstract form from a semantically organized lexicon; Stage 4: Retrieval of phonological word forms on the basis of Stage 3; Stage 5: Selection of prosodic features including the location of sentence stress points; Stage 6: Phonological stage in which word forms are ordered syntactically and prosodic features marked; Stage 7: Full phonetic specification with all timing parameters specified; Stage 8: Instructions to articulators. In this model, gestures and speech are autonomous. The authors claim there is no reason why one should not occur without the other. Kendon (1986) was of the opinion that gestures can throw light upon the nature of thought. Slama-Cazacu (1976) believed in a "mixed syntax" in which gesture and speech are used in alternation, with gestures sometimes "filling in" when speech had ceased.

Glosser, Wiener and Kaplan (1986) found that moderate aphasics produce significantly fewer semantic, modifying and relational gestures and more pantomimic and deictic gestures than control subjects or mild aphasics, who did not differ from each other on these measures. The more severely impaired aphasic subjects used fewer complex communicative gestures while producing proportionally more of the nonspecific, nonconsensually shared, unclear gestures. Geschwind (1975) proposed that the left hemisphere is dominant not only for speech but also for learned movements. He said that many Broca's aphasics are unable to carry out commands with the non-paralysed left extremities. Hanlon, Brown and

Gerstmann (1990), on the other hand, found that, in their study, functional activation of the hemiplegic right arm in the production of communicative gestures favourably affects the naming capacity of nonfluent aphasics. By activating the motor system of the hemiplegic side, the production of deictic gestures had a facilitative effect on the naming performance of nonfluent aphasics but not fluent aphasics.

Duffy and Duffy (1981) stressed the active role of the left hemisphere, particularly in more complex symbolic nonverbal communication such as pantomime. For example, nonfluent subjects tend to present with a low pitch, monotonous intonation, poor intensity and quality control and a slow rate. On the whole, fluent subjects did not do this. Prosodic features, for example, intonation and rate, were thus used in a compensatory fashion on many occasions. Behrmann and Penn (1984) found that gesture, pantomime and facial expression were also frequently actively used as a compensatory strategy by nonfluent subjects. In the fluent patients, however, such actions seemed to be coincidental concomitants of the verbal message. Bosone (1977) found that, following group therapy, 4 of 17 aphasic patients self-initiated gestures for communication outside the clinical setting; 4 did not learn the gestures and 6 learned but did not transfer the use of gestures. Goodglass and Kaplan (1963) studied the gestural performance of 20 aphasics. The aphasic group not only showed impaired gestural performance but failed to improve on imitation trials of gestural tasks. The authors interpreted the gestural abnormality in aphasia as an apraxic disturbance.

The aim of this study

The aim of this study was to investigate how two nonfluent and two fluent aphasics would use different kinds of spontaneous gestures to initiate, accompany or substitute for missing words on a confrontation naming task where they were shown drawings representing nouns. Could these aphasic adults cue the target word by using gestures? Were their gestures the same or of the same kind as those used by SLI children? Would it be possible to suggest some type of gesturing as a means of therapy parallel to speech therapy by using a paralinguistic channel in order to gain access to lexical semantics?

Materials and method

Subjects

Four Swedish-speaking post-CVA aphasic adults, 3 males and 1 female, were chosen for this investigation. They were two nonfluent and two fluent aphasics. Their ages ranged between 70 and 79 years and they exhibited no gross sensory loss or dysarthria. None of them had any motor disorder except for one nonfluent aphasic who had a paralysed right arm (all of them were right-handed premorbidly) and used her left arm and hand instead. The nonfluent aphasics had anterior lesions in Broca's area, and the fluent aphasics had posterior lesions in Wernicke's area. None of the subjects had any visual deficit, but two of them had presbyacusis and wore hearing aids.

Language and speech

The two nonfluent subjects used only one-word utterances. Their speech rate was low, sentence length was short, and the melodic contour was lost. Their speech production was effortful, with more pauses than actual words. They had word finding problems but good comprehension. The nonfluent male subject had verbal apraxia in combination with his aphasia. However, he had no signs of limb apraxia, and could use both hands with different tools and when making gestures. The nonfluent female subject had a moderate tendency towards perseveration, which did not occur in the nonfluent male subject. Both subjects could read and understand short sentences, but were not capable of reading stories due to their memory deficit. The male subject was rather good at writing, but his spelling was not perfect. He did not use writing as a substitute means of communication. The nonfluent female subject did not like to write with her left hand, but used it in daily living activities.

The fluent aphasics had fluent, empty, often neologistic speech, verbal and phonemic paraphasia and poor comprehension. Verbal repetition was impaired. None of the fluent aphasics were helped by phonemic or semantic cues, but written cues such as letters and words did help. When the patient was searching for the target word, the prosodic features were unclear, but when the word was found, prosody became clear and the voice louder. These patients had perfect articulation. The speech of the fluent aphasics was more abundant than normal speech. The amount and rate of word production was higher than in normals. The melodic contour of the sentences approximated normal speech.

The four aphasic subjects were tested with the Swedish Screening Test for Aphasia, and were diagnosed as severely aphasic. They had all received language therapy in the interval since onset but no specific training in gestures. Four Swedish-speaking adults without any history of CVA were chosen as controls; their ages ranged between 64 and 78 years.

	Time since onset	Age
Nonfluent aphasic E, male	36 months	77 years
Nonfluent aphasic F, female	34 months	72 years
Fluent aphasic K, male	9 months	79 years
Fluent aphasic L, male	38 months	70 years

Method

All subjects were presented with a confrontation naming task, the SBP Skånes Benämnings Prövning by Apt (1994), which consists of 64 drawn items representing nouns. All subjects were videorecorded. There was no time limit, in order to allow the aphasics to feel relaxed and take the time they needed. The important factors were the types and number of gestures related to all 64 items, and also the number of verbal trials, not the time used for testing. The mean time for testing the controls was only two minutes.

	Time for test
Nonfluent aphasic E	22 minutes
Nonfluent aphasic F	35 minutes
Fluent aphasic K	54 minutes
Fluent aphasic L	30 minutes

The authors examined the videotapes several times frame by frame. Gestures were categorized according to Ekman and Friesen's (1969) system. The number of verbal trials, the number and type of gestures, and the intervals when the gestures were used, i.e., prior to, simultaneously or after the word, were examined.

Results

Nonfluent aphasic E

Yes or no are not adequately used. Head nodding and shaking are not adequately used either. Pictographs are usually but not always clear. The patient uses gestures to compensate and substitute for words and also to accompany word trials.

Nonfluent aphasic F

As her right arm is paretic, the patient points with her left arm and hand. She is generally helped by her pointing (and also by phonetic prompting). She uses her left index finger, holding it up in the air to point out turn keeping. When hesitating, she moves her fingers up to her lips. When she finds the correct word in her lexicon, and while pronouncing it, her voice is more intense and her prosody and intonation, more emphatic. Her gestures are not very clear; sometimes they are difficult to interpret. After she is unable to find a word she shakes her head twice in succession. Her many pointing gestures are not used to compensate or substitute for the target word but to elicit it. The two ideographs she used indicate the direction of her thought (idea).

Fluent aphasic K

All his gestures were clear and easy to understand. His few gestures were used to compensate for the word in the beginning and thereafter to accompany it. The target word is characterized by many literal paraphasias and numerous trials. Prosody is emphatic and the intensity of the voice is raised when the patient finally hits the right target word.

Fluent aphasic L

This patient's gestures are clear. It is easy to understand the referent. Prosody is emphatic and the intensity of the voice is raised when the patient hits the right target word.

The total number of gestures produced by all 4 aphasics was 412, compared to 1 gesture produced by the normals. The percentage of gestures used by all aphasics is shown in Figure 1.

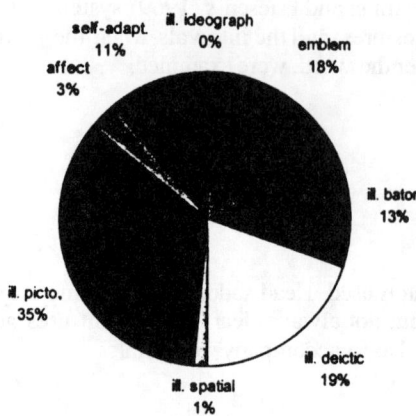

Figure 1. Gestures used by all aphasics

The nonfluent aphasics used a total of 240 gestures and the fluent group used 172, compared to the controls' 1. A statistical analysis using chi-square tests was done. The chi-square test for 2 x 4 contingency tables of the distribution of nonfluent versus fluent aphasic subject emblems, illustrators, affect displays and self-adaptors had a significance of $p < 0.05$. The chi-square test for 2 x 5 contingency tables of the distribution of nonfluent versus fluent aphasic subjects' illustrators, batons, deictics, spatials, pictographs and ideographs had a significance of $p < 0.01$.

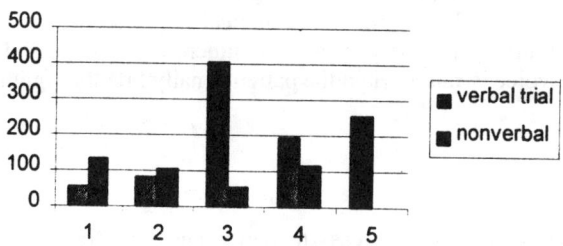

Figure 2.   Verbal and nonverbal trials in 1: Nonfluent E, 2: Nonfluent F, 3: Fluent K, 4: Fluent L, and 5: Controls

The number of verbal trials made by all aphasics was 744, compared to 256 for normals (Figure 2). The chi-square test for 3 x 2 contingency tables for the nonfluent versus fluent aphasic subjects and for the aphasics versus the controls with respect to the distribution of verbal and nonverbal trials had a significance of $p < 0.0001$.

Pictographs

Pictographs were used by the nonfluent and fluent aphasics, prior to, simultaneously and after their verbal trials.

The nonfluent aphasics produced 37 pictographs prior to verbal trials, compared to the fluent aphasics who only used 5. The fluent aphasics produced 68 simultaneous pictographs, compared to only 16 in the nonfluent aphasics.

Pictographs produced after verbal trials totalled 6 in the nonfluent aphasics and only 1 in the fluent aphasics.

Pictographs used by the nonfluent aphasics

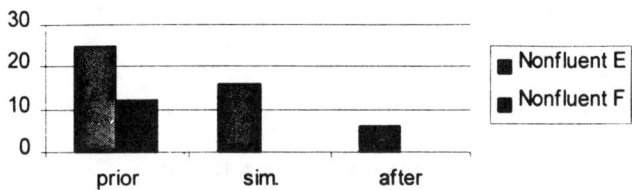

Figure 3. Pictographs used by the nonfluent aphasics

Pictographs used by fluent aphasics

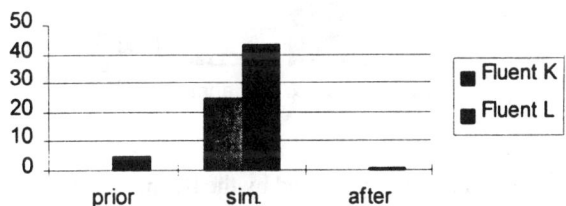

Figure 4. Pictographs used by fluent aphasics

The chi-square test for 2 x 3 contingency tables of the distribution of nonfluent versus fluent aphasic subjects' pictographs used prior to, simultaneously and after words (Figure 4) had a significance of $p < 0.0001$.

Deictics

The nonfluent aphasics produced 65 prior deictics (Figure 5) and the fluent aphasics, only 2 (Figure 6). Simultaneously used deictics amounted to 10 in the nonfluent aphasics and 20 in the fluent aphasics. Deictics used after word trials totalled 2 in the nonfluent and only 1 in the fluent aphasics.

Deictics used by the nonfluent aphasics

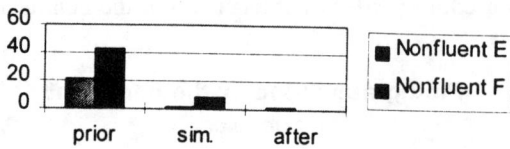

Figure 5. Deictics used by the nonfluent aphasics

Deictics used by the fluent aphasics

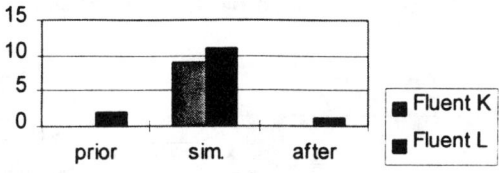

Figure 6. Deictics used by the fluent aphasics

The chi-square test for 2 x 3 contingency tables of the distribution of nonfluent versus fluent aphasic subjects' deictics used prior to, simultaneously and after words had a significance of $p < 0.0001$.

Batons

Only one baton was used prior to verbal trials by the nonfluent aphasics and the fluent aphasics did not use any. The nonfluent group used 16 simultaneous batons and the fluent group, 2. After verbal trials, 10 batons were produced by the nonfluent aphasics and only 3 by the fluent aphasics (Figures 7 and 8).

Batons used by the nonfluent aphasics

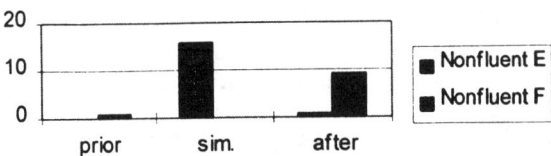

Figure 7. Batons used by the nonfluent aphasics

Batons used by fluent aphasics

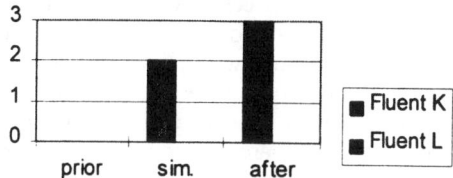

Figure 8. Batons used by fluent aphasics

The chi-square test for 2 x 3 contingency tables of the distribution of nonfluent versus fluent aphasic subjects' batons used prior to, simultaneously and after words resulted in p = 0.6032, which is not significant.

Spatials

Spatials were only used simultaneously with the verbal trials; they were produced twice by the nonfluent aphasics and 4 times by the fluent aphasics. The chi square test of 2x3 contingency tables of the distribution of nonfluent versus fluent aphasic subjects' spatials used prior to, simultaneously and after words could not be calculated because of the many results of zero obtained in the testing.

Discussion

This study found that aphasic subjects with acquired brain damage use the same type of gestures as do SLI children. This might indicate that, when people do not have access to a word, they naturally use a paralinguistic channel, in this case, gestures. Illustrators seem to be the gestures used to get access to the word. The two nonfluent aphasics used more pointing gestures than the fluent aphasics, who used more pictographs. The nonfluent aphasics used more pointing gestures and pictographs prior to uttering words indicating that these gestures play a functional role in word retrieval. The fluent aphasics use the same kinds of gestures simultaneously in order to pick out the right target word from their flow of speech. Batons are more frequently used by the nonfluent aphasics (N=27) than the fluent aphasics (N=5). Subject Nonfluent E used 16 batons simultaneously with his word trials; the gestures seemed to serve as pragmatic cueing to find the right word.

A number of authors (e.g., Duffy et al., 1975; Cicone et al. 1979; Duffy & Duffy, 1981; Glosser et al., 1986) argue in favour of a central organizer which controls both language and gesture. Thus, both verbal and gestural communication should be impaired in aphasia. Feyereisen (1988) and Christopoulou and Bonvillian (1985) argue against a central symbolic deficit, stating that the more severe the level of apraxia and aphasia, the more gestures are produced by aphasics. However, Feyereisen (1988) also shared the opinion that aphasia was associated with nonverbal disorders. He claimed that some cases of aphasia were explained by conceptual deficits and others by modality-specific disorders, thereby staking out an intermediate position in the discussion concerning a central organizer. He suggested a functional role for iconic gestures in word retrieval.

Butterworth and Hadar (1989) considered that gestures and speech are globally different and autonomous and that gesture and speech often occur one without the other. This statement contradicts McNeill's (1985) opinion that gestures dissolve together with speech in aphasia. Based on our data, our opinion with regard to the role of a central organizer favours the middle ground. The central organizer is the concept. The different paralinguistic channels, for example, one that leads to the phonological/phonetical representation and another that leads to gestural behaviour, are autonomous but seem to originate from the same centrally organized concept.

Emblems

Emblems have a direct verbal translation. In this study, the nonfluent aphasics produced 32 emblems, the fluent aphasics 11 and the controls, none. Head nods and head shakes were used more frequently by the nonfluent aphasics as a substitute for "yes" and "no", but not adequately in the case of the male nonfluent. He used this emblem to substitute for or contradict his faulty speech production. His head nods gave supplementary information, confirming to the listener what was right or wrong.

In their study of SLI children compared to normals, Månsson and Lundström (1996) found a significant difference in the distribution of the different categories of gestures. The children used the same types of gestures but interesting differences in how they used them were found. The SLI children produced more emblems, adaptors and affect displays and the normals, more regulators and illustrators. Emblems like head nods and head shakes get the listener's attention and are easy to use for the SLI children. Emblems were used parallel to or instead of verbal expressions. The authors concluded that the SLI children were at a more immature stage than the normal children in their language development and therefore used gestures that also seemed to be more immature. It appears that both SLI children and nonfluent aphasics prefer to use emblems in the absence of semantic access.

Illustrators

Illustrators are socially learned movements that are directly tied to speech. They illustrate what is said and repeat, substitute, contradict or augment the information given verbally. The controls in this study did not use any of these movements since they were administered a confrontation naming task. Normals use illustrators in ordinary conversation. The nonfluent aphasics seem to use illustrators as elicitors, while the fluent aphasics use them as a guideline in picking the right word out. Månsson and Lundström (1996) found that SLI children use more pictographs than normal children do to compensate for their language impairment. SLI children need to use illustrators socially as they lack experience and are often misunderstood when communicating; SLI children's parents are uncertain about what messages they are receiving, and thus need more knowledge about their children's NVC.

Deictics

Deictic movements are pointing gestures. This study found that deictics are used more often by nonfluent aphasics. The nonfluent female aphasic subject frequently used deictics prior to uttering a target word and continuously pointed to the object in question. This seems to be a case of pragmatic cueing, i.e., pointing in order to get access to the word. Hanlon et al. (1990) suggested that pointing with the right hemiplegic limb at the picture to be named would help nonfluent aphasics in a confrontation naming task by functionally activating the proximal motor system to reach early levels in speech production. Our nonfluent aphasic with a paralysed arm did not do this but used her left, non-paralysed, arm only. This kind of stimulation most probably needs specially designed therapy. Deictic gestures are more often used simultaneously by the fluent aphasics. They recognize and point to the picture, and thereby seem to reinforce the word and its phonetic prompting while they make several trials to find the right word. Wilkinson and Rembold (1981) showed that deictics are used when verbal ability is missing in small children. They also indicated that deictics help to develop language ability.

Pictographs

Pictographs are movements which draw a picture of their referent. In the present study, the fluent aphasic group used more simultaneous pictographs (68) in their attempts to find the target word than the nonfluent aphasics, who used only 16 pictographs, and the controls with 0. This fact is in agreement with Feyereisen's (1988) opinion of iconic gestures. Le May et al. (1988) suggest that Broca's aphasics should be encouraged to use batons, ideographs and deictics more than pictographs to clarify their communication. Our nonfluent aphasics, however, used more pictographs prior to the word (37) than did the fluent aphasics (5); this could indicate that they used the pictographs as a pragmatic cue in order to access the lexicon. In the fluent group, simultaneous pictographs seem to be used as a support and a guide to facilitate word retrieval in their copious flow of speech. By using a pictograph prior to the word and during the latency period in the word finding process, nonfluent aphasics trigger the word and thus they should be encouraged to use such gestures in therapy. Overall, pictographs were used more frequently by the fluent aphasics.

In comparison, SLI children were found to use pictographs more frequently than normal children to substitute for a missing word or to clarify matters for the listener when their phonology was incorrect or incomplete. Children who have a lexical semantic impairment are not likely to use pictographs to the same extent as children with only phonological disorders.

Batons

Batons accent or emphasize a particular word and phrase. The nonfluent aphasics used more batons than the fluent aphasics. The nonfluent and dyspraxic subject E used most of his batons simultaneously with his verbal trials while the other nonfluent subject, K, used most of her batons after the word as an accentuation feature together with a louder voice and well pronounced prosody. When the word is correctly presented, the baton stresses this fact. This can also be regarded as pragmatic cueing. The controls used none of these gestures. SLI children were not found to use batons at the same rate as normal children who had a certain knowledge about language and prosody. The SLI children, lacking this knowledge, could not use batons, which seem to be more advanced gestures.

Ideographs

Ideographs are gestures that trace the itinerary of a logical journey. Such gestures were only used twice by one of the nonfluent aphasics, and not by the fluent aphasics or the controls. The SLI children studied by Månsson and Lundström (1996) did not use these gestures either. As our test was a confrontation naming test, it did not encourage the subjects to use ideographs. Even though the nonfluent aphasic who used ideographs did not find the word she sought, she showed that she understood the logic of the process by using this gesture, which appears to be more advanced than the others.

Spatial movements

Spatial movements depict a spatial relationship. One fluent aphasic used 4 such movements simultaneously with word production and one nonfluent used 2. None of the controls used any as they had access to the word and did not need to indicate any spatial relation to it. The SLI children produced no such gestures but the normally speaking children used a few.

Affect displays

Affect displays show emotions (e.g., happiness, surprise, fear, sadness, anger, disgust and interest.) Twelve of these were produced by the fluent subjects and 8 by the nonfluent subjects. These gestures appeared when the subjects were frustrated with their word finding problems; the same was true of the SLI children.

Self-adaptors

Self-adaptors involve touching one's face, hair or lips to facilitate or inhibit speech production. According to Ekman and Friesen (1969), hand-to-face adaptors are a rich source of information. In this study, the nonfluent aphasics produced 33 self-adaptors in comparison to the fluent aphasics' 43 and the controls' 1. The self-adaptors in all three groups consisted of hand-to-mouth or -nose movements, when the subjects were searching for the target word. One of the nonfluent aphasics used the touching of her lips as a pragmatic cue. The two fluent aphasics touched the nose or ear while choosing the right word. Self-adaptors are more frequently used by SLI children than by normally speaking children. These gestures might be interpreted as a support in eliciting the word or as a sign of discomfort when one is not able to verbalize a need.

Conclusion

In this study and the one conducted by Månsson and Lundström (1996), aphasics and SLI children were found to use gestures to compensate for their language problems. In the clinical setting, when assessing developmental or acquired language disability, we strongly recommend also doing an analysis of the gestures that accompany or replace speech. The nonverbal behaviour research by Ekman and Friesen (1969) provides a good categorization of gestures. Based on this study, clinicians can describe the gestures of their patients in detail and also examine their pragmatic cueing in the absence of the spoken word. To judge from this study, it is advisable to encourage the natural use of gestures in aphasia therapy.

Clinicians working with aphasic individuals often see a discrepancy between their abilities as measured by standard assessments or tests, and the communicative possibilities revealed when the patients want to "talk" about matters important to them. More comprehensive studies are required if the relationship between verbal and nonverbal behaviour is to be fully understood.

Acknowledgment—The authors are indebted to Osamu Shiromoto, Dept. of Communication Disorders, Hiroshima Prefectural College of Health and Welfare, Japan, for his help with the statistics.

References

Ahlsén, E. (1985). Discourse patterns in aphasia. *Gothenburg Monographs in Linguistics 5.* Department of Linguistics, University of Göteborg, Göteborg, Sweden.
Ahlsén, E. (1991). Body communication as compensation for speech in a Wernicke's aphasic—A longitudinal study. *Journal of Communication Disorders, 24,* 1-12.
Apt, P. (1994). SBP Skånes Benämningsprövning. Standardized revision based on Naming. Escape. EU project.
Behrmann, M., & Penn, C. 1984. Non-verbal communication of aphasic patients. *British Journal of Disorders of Communication, 19,* 155-168.
Bosone, Z. (1977). Paper presented to the Amerind Conference, St. Lewis, cited in M. Skelly, 1979. *Amerind gestural code based on universal American Indian hand talk.* New York: Elsevier, North-Holland.
Butterworth, B., & Hadar, U. (1989). Gesture, speech, and computational stages: A reply to McNeill. *Psychological Review, 96,* 168-174.
Christopoulou, C., & Bonvillian, J. D. (1985). Sign language, pantomime, and gestural processing in aphasic persons: A review. *Journal of Communication Disorders, 18,* 1-20.
Cicone, M., Wapner, W., Foldi, N., Zurif, E., & Gardner, H. (1979). The relation between gesture and language in aphasic communication. *Brain and Language, 8,* 324-349.
Dean, A., & Skinner, C. (1995). Perceptual assessment in aphasia. In S. Wirtz (Ed.), *Perceptual Approaches to Communication Disorders* (pp. 84-99). London: Whurr Publishers.

Delis, D., Hamby, S., Gardner, H., & Zurif, E. (1979). A note on temporal relations between language and gestures. *Brain and Language, 8,* 350-354.

Duffy, R. J., & Duffy, J. R. (1981). Three studies of deficits in pantomimic expression and pantomimic recognition in aphasia. *Journal of Speech and Hearing Research, 24,* 70-84.

Duffy, R. J., Duffy, J. R., & Pearson, K. (1975). Pantomime recognition in aphasics. *Journal of Speech and Hearing Research, 18,* 115-132.

Ekman, P., & Friesen, W. B. (1969). The repertoire of nonverbal behaviour: Categories, origins, usage and coding. *Semiotica, 1,* 49-98.

Feyereisen, P. (1983). Manual activity during speaking in aphasic subjects. *International Journal of Psychology, 18,* 545-546.

Feyereisen, P. (1988). Non-verbal communication. In F. C. Rose, R. Whurr, and M. A. Wyke (Eds.). *Aphasia* (pp. 46-88). London: Whurr Publishers.

Geschwind, N. (1975). The apraxias: Neural mechanisms of learned movements. *American Scientist, 63,* 188-195.

Glosser, G., Wiener, M., & Kaplan, E. (1986). Communicative gestures in aphasia. *Brain and Language, 27,* 345-359.

Goodglass, H., & Kaplan, E. (1963). Disturbances of gesture and pantomimes in aphasia. *Brain, 86,* 703-720.

Hanlon, R. E., Brown, J. W., & Gerstman, L. J. (1990). Enhancement of naming in nonfluent aphasia through gesture. *Brain and Language, 38,* 298-314.

Helm, N. A. (1979). The gestural behavior of aphasic patients during confrontation naming. Doctoral dissertation, Boston University.

Hermann, M., Reichle, T., & Lucius-Hoene, G. (1988). Nonverbal communication as a compensatory strategy for severely nonfluent aphasics. A quantitative approach. *Brain and Language, 33,* 41-54.

Holland, A. L. (1982). Observing functional communication in aphasic adults. *Journal of Speech and Hearing Disorders, 47,* 50-56.

Jancovic, M., Devoe, S., & Wiener, M. (1975). Age-related changes in hand and arm movements as nonverbal communication. Some conceptualization and an empirical exploration. *Child Development, 46,* 922-928.

Kendon, A. (1986). Some reasons for studying gestures. *Semiotica, 62,* 3-28.

Le May, A., David, R., & Thomas, A. P. (1988). The use of spontaneous gesture by aphasic patients. *Aphasiology, 2,* 137-145.

Månsson, A.-C., & Lundström, C. (1996). En Jämförelse mellan normalspråkiga och språkstörda barns icke-verbala kommunikation. Unpublished Master's dissertation, Lund University, Institution of Logopedics and Phoniatrics, Lund, Sweden.

McNeill, D. (1985). So you think gestures are nonverbal? *Psychological Review, 92,* 350-371.

McNeill, D., & Levy, E. (1982). Conceptual representation in language activity and gesture. In R. J. Jarvella and W. Klein (Eds.), *Speech place and action: Studies in deixis and related topics* (pp. 271-295). Chichester: John Wiley.

Sanmarco, J. G. (1984). Joint problem-solving activity in mother-child dyads: A comparative study of normally achieving and language disordered preschoolers. U-M-I Dissertation Information Service.

Slama-Cazacu, T. (1976). Nonverbal components in message sequence: 'Mixed syntax'. In W. C. McCormack and S. A. Wurm (Eds.), *Language and man: Anthropological issues* (pp. 14-29). The Hague: Mouton.

Söderbergh, R. (1980). En modell for beskrivning av dialoger mellan barn och vuxna. 1 (Språkstimulering i förskoleåldern. Stockholm, Child Language Research Institute, Paper 5).

Vygotsky, L. (1978). Mind in society. *The development of higher psychological processes.* Cambridge, MA.: Harvard University Press.

Wilkinson, L. C., & Rembold, K. L. (1981). The form and function of children's gestures accompanying verbal directives. In P. S. Dale and D. Ingram (Eds.), *Child language: An international perspective* (pp. 175-190). Baltimore: University Park Press.

 **Pergamon**

J. Neurolinguistics, Vol. 11, Nos 1–2, p. 207–221, 1998
© 1998 Published by Elsevier Science Ltd. All rights reserved
Printed in Great Britain
0911-6044/98 $19.00 + 0.00

PII: S0911-6044(98)00014-1

Relationship between language impairment and pragmatic behavior in aphasic adults

Jan R. Avent*, Robert T. Wertz† and Linda L. Auther†

*California State University, Hayward, California;
†Veterans Administration Medical Center, Nashville, Tennessee

Abstract—We explored the relationship between language impairment and pragmatic performance in aphasic adults during the first 48 weeks postonset. Twenty patients, ten fluent and ten nonfluent, who had suffered a single left hemisphere stroke participated in the study. Half of the patients received 44 weeks of individual treatment and half received 44 weeks of group treatment. Language impairment—auditory comprehension, reading, oral-expressive language, and writing—was determined with the Porch Index of Communicative Ability at four weeks postonset, pretreatment, and at 15, 26, 37, and 48 weeks postonset. Pragmatic performance—verbal aspects, paralinguistic aspects, and nonverbal aspects—was assessed during conversation with a pragmatic protocol at the same points in time. The patients made significant improvement in both language impairment and pragmatic performance between 4 and 48 weeks postonset, and there were no significant differences in improvement between the fluent and nonfluent groups or between patients receiving individual or group treatment. In our fluent aphasic patients, language impairment and pragmatic performance were significantly related at four weeks postonset but not thereafter. Conversely, in our nonfluent aphasic patients, language impairment and pragmatic performance were not significantly related at four weeks postonset, but they were significantly related at 15, 26, 37, and 48 weeks postonset. The significance of the relationship between language impairment and pragmatic performance was sporadic over time in the individual and group treatment groups.

Introduction

Traditional assessment of aphasia has focused on language impairment. It is designed to determine whether a language problem exists, document specific deficits, plan treatment, and measure change in performance over time (Lomas, Pickard, Bester et al. 1989). During the past 20 years, however, assessment of aphasia has expanded to include assessment of communicative competence (Holland, 1996; Newhoff & Apel, 1997). The impact of aphasia on communication is evident, particularly, in conversational ability (Prins, Snow, & Wagenaar, 1978). However, measures of improvement in conversational ability have been ignored in most aphasia treatment studies. Because conversation is one of the most important aspects of language use (Levinson, 1983) and basic to communication in a social world (Brinton & Fujiki, 1989; Smith, 1985), its inclusion in the evaluation of treatment effects appears important.

One measure of communicative competence is pragmatics; the study of how utterances have meaning in situations (Leech, 1983). Pragmatic assessment of aphasia is appealing, because it is conducted within a social context where linguistic, paralinguistic, and nonverbal skills are observable. Pragmatic performance can be assessed by observational profiles, communicative efficiency measures, standardized testing in real or simulated life situations, family and significant others questionnaires, and composite assessments (Manochiopinig, Sheard, & Reed, 1992).

One striking observation that has resulted from the assessment of communicative competence is that individuals with aphasia, generally, display more intact pragmatic skills than they do in the more formal aspects of syntax and lexical components of language (Avent & Wertz, 1996; Holland, 1996; Roberts & Wertz, 1993). Despite the presence of more intact pragmatic abilities in individuals with aphasia, impairments in pragmatic skills are evident. In an investigation designed to test the usefulness of a descriptive taxonomy of pragmatic behaviors, Prutting and Kirshner (1987) assessed the conversational abilities of six different diagnostic groups—children with language disorders, children with articulation disorders, children with normal language, adults with aphasia, adults with right hemisphere damage, and adults with normal language. They observed distinct profiles for each diagnostic group. The aphasic subjects showed a mean of 82% appropriate pragmatic performance. The majority of the deficits noted related to linguistic constraints, including specificity and accuracy, pause time in turn taking, quantity and conciseness of the message, fluency, and the variety of speech acts produced.

Goldblum (1985) compared chronic fluent and nonfluent aphasic subjects' conversational skills using the Prutting and Kirshner Pragmatic Protocol. She found that both groups retained their social competence, but the fluent subjects had fewer inappropriate ratings than the nonfluent subjects. The majority of inappropriate behaviors for both groups were in the categories of specificity/accuracy, fluency, pause time in turntaking, and quantity/conciseness. The nonfluent group had more difficulty in pause time in turntaking and quantity/conciseness than the fluent group.

Recent evidence suggests that pragmatic skills improve with treatment during the first year postonset. Roberts and Wertz (1993) evaluated the communicative effectiveness of 20 treated aphasic subjects at one, three, six, nine, and 12 months postonset using the Pragmatic Protocol (Prutting & Kirchner, 1983; 1987) and found that pragmatic performance improved throughout the first year postonset. Mean appropriate performance was 87% at one month, 91% at three and six months, 92% at nine months, and 93% at 12 months. Thus, the Pragmatic Protocol appears useful for documenting pragmatic performance and determining change in performance over time.

Avent and Wertz (1996) investigated the influence of type of aphasia and type of treatment on aphasic patients' pragmatic abilities in conversation between 4 and 48 weeks postonset of stroke. Ten patients were fluent and ten were nonfluent. All patients received six to eight hours of treatment each week for 44 weeks. Ten patients received group treatment with no direct manipulation of language deficits, and ten patients received individual, stimulus-response treatment with direct manipulation of language deficits in all communicative modalities. Mean conversational pragmatic ability for the 20 patients was 87% appropriate at four weeks postonset. Mean performance after 44 weeks of treatment was 92% appropriate. There were no significant differences in improvement in pragmatic performance between fluent and nonfluent patients, and overall improvement in pragmatic performance was not influenced by the type of treatment, group or individual. However, group treated patients showed significant change in pragmatic abilities during the first 11 weeks of treatment, whereas the individually treated patients did not.

The results of these studies indicate that a pragmatic protocol is useful for describing differences among different disorders (Prutting & Kirchner, 1987); distinguishing differences among subtypes of aphasia (Goldblum, 1985); documenting changes over time (Roberts & Wertz, 1993); and describing treatment effects (Avent & Wertz, 1996).

However, there is no widespread use of pragmatic assessment in the management of aphasia.

Failure to assess conversational ability may result from a variety of factors. Traditional linguistic-based assessments reveal well recognized impairments in syntax, semantics, and phonology (Gallagher, 1991). Pragmatic assessments, on the other hand, capture individualistic communicative performance that is highly sensitive to interactional contexts and cultural expectations (Gallagher, 1991; Hough & Pierce, 1994). Pragmatic assessments are further hindered by the absence of a general theory encompassing spoken discourse or pragmatic aspects of language (McTear, 1985). This lack of an accepted theoretical framework has resulted in confusing terminology and no defined procedures for assessment (Gallagher, 1991; Penn, 1993; McTear, 1985).

Another factor which may influence the scarcity of conversational assessment is historical reliance on standardized testing. Typically, treatment outcome is determined with standardized tests. However, the sole use of standardized tests raises questions about the purpose of documenting treatment effectiveness. While standardized tests have played an important role in establishing the efficacy of aphasia treatment, from a clinical perspective, discrepancies exist between standardized language test performance and observations of aphasic individuals' conversational skills (Aten, Caligiuri, & Holland, 1982; Binder, 1984; Blomert, 1990; Holland, 1980; Newhoff & Apel, 1997; Penn, 1985; Sarno & Levita, 1971; Ulatowska, Haynes, Hildebrand, & Richardson, 1977). Standardized tests, generally, are based on a linguistic framework that does not provide a clear understanding of how an aphasic individual uses his or her language (Penn, 1993; Chapman & Ulatowska, 1992; Gurland, Chwat, & Wollner, 1982; Holland, 1980; Ulatowska et al., 1977). Therefore, to demonstrate that treatment creates a change in communicative competence, other measures may be more appropriate, for example, a pragmatic analysis (Lund & Duchan, 1993; Green, 1984).

Sarno (1993) suggests the goal of rehabilitation is to "restore the person's role as a communicator, regardless of whether certain symptoms have been eradicated or particular linguistic skills have improved" (p. 325). Conversational performance provides an opportunity to study "the communicator role" of individuals with aphasia. Moreover, Holland (1982) observed aphasic people communicate better than they talk, implying a difference between pragmatic performance and language impairment. Because a measure of conversational ability was not used in the majority of aphasia treatment studies, it has not been determined whether aphasic patients' pragmatic abilities are related to their linguistic performance on standardized tests. The purpose of this study was to determine whether there is a relationship between pragmatic performance and severity of language impairment in aphasia subsequent to stroke. We attempted to answer the following questions: What is the relationship between pragmatic performance and language impairment in aphasic people during the first year postonset; does the relationship between pragmatic performance and language impairment differ between fluent and nonfluent aphasic people during the first year postonset; and is the relationship between pragmatic performance and language impairment influenced by the type of treatment, individual or group, aphasic people receive?

Methods

This investigation was a retrospective study of change in aphasic people who were treated during the first year postonset. The data come from a standardized language measure and pragmatic analysis of videotaped conversational samples by subjects who participated in the first Veterans Administration Cooperative Study on aphasia (Wertz, Collins, Weiss et al., 1981). Prior to receiving treatment, four weeks postonset, each patient conversed with his clinician about events during the previous week. These conversations were repeated after each 11-week treatment period at 15, 26, 37, and 48 weeks postonset. Similarly, a standardizedlanguage measure was administered at the same points in time.

Subjects

Study patients were aphasic subsequent to a first, left hemisphere thromboembolic infarct; four weeks postonset at entry; under 80 years of age; premorbidly literate in English; and had no present or previous other neurological disease (Wertz et al., 1981). Three criteria were used to select subjects from the pool of 34 subjects who completed the VA Cooperative study. First, videotapes containing a conversational sample had to be available for evaluations conducted at 4, 15, 26, 37, and 48 weeks postonset. Second, patients were selected to include an equal number of fluent and nonfluent individuals. And, third, patients were selected to include an equal number of individuals in each of the two treatments, individual and group. Performance by 20 subjects who satisfied these criteria was analyzed. Table 1 shows age, education, and Porch Index of Communicative Ability (PICA) (Porch, 1967) overall percentile performance at four weeks postonset, pretreatment; type of treatment received; and type of aphasia for all subjects.

Subject	Age (years)	Education (years)	PICA overall percentile	Treatment	Aphasia
1	73	16	43	Individual	Nonfluent
2	63	9	72	Group	Fluent
3	58	6	50	Group	Fluent
4	65	8	42	Individual	Nonfluent
5	45	12	15	Group	Nonfluent
6	57	12.5	16	Individual	Nonfluent
7	63	8	60	Individual	Fluent
8	78	8	66	Group	Fluent
9	41	12	73	Individual	Fluent
10	51	12	44	Group	Fluent
11	68	10	49	Group	Fluent
12	65	8	55	Individual	Nonfluent
13	62	9	58	Individual	Fluent
14	50	12	49	Group	Fluent
15	64	6	35	Individual	Nonfluent
16	55	7	39	Group	Nonfluent
17	78	3	26	Group	Fluent
18	52	13	47	Individual	Nonfluent
19	52	12	15	Individual	Nonfluent
20	61	12	19	Group	Nonfluent

Table 1. Patient descriptive data at four weeks postonset for age, education, language severity, type of treatment, and type of aphasia.

Table 2 shows descriptive data for the subjects classified by type of aphasia, fluent and nonfluent, and Table 3 shows descriptive data for the subjects classified by the type of treatment received, individual or group. The two groups, classified by type of aphasia, did not differ significantly (p < .05) in chronological age or in years of education. However, the groups differed significantly in language severity on the PICA. Nonfluent aphasic subjects scored significantly lower (p < .003) than the fluent aphasic subjects. The two groups, classified by type of treatment received, did not differ significantly (p < .05) in age, education, or language severity on the PICA.

Variable	Type of aphasia	
	Fluent (n = 10)	Nonfluent (n = 10)
Age in years		
Mean	61.20	58.90
Standard Deviation	11.25	7.82
Range	41-78	45-73
Education in years		
Mean	8.90	10.65
Standard Deviation	2.74	3.03
Range	3-12	6-16
PICA Overall Percentile		
Mean	54.70	32.60
Standard Deviation	14.24	15.03
Range	26-73	15-55

Table 2. Descriptive data at four weeks postonset for age, education, and language severity for subjects classified by type of aphasia.

Variable	Type of treatment	
	Individual (n = 10)	Group (n = 10)
Age in years		
Mean	59.40	60.70
Standard Deviation	8.62	10.73
Range	41-73	45-78
Education in years		
Mean	10.45	9.10
Standard Deviation	2.94	2.95
Range	6-16	3-12
PICA Overall Percentile		
Mean	42.90	44.40
Standard Deviation	18.73	18.66
Range	15-73	15-72

Table 3. Descriptive data at four weeks postonset for age, education, and language severity for subjects classified by type of treatment received.

Treatment.

Each subject received six to eight hours of treatment each week for 44 weeks. Individual treatment was conducted by a speech-language pathologist and consisted of four hours of stimulus-response type of treatment designed to improve language in all communicative modalities—auditory comprehension, reading, oral-expressive language, and writing—and two to four hours of machine-assisted treatment. Three fluent and seven nonfluent aphasic patients received individual treatment. Group treatment was conducted by a speech-language pathologist in groups of three to seven patients for four hours each week, and an additional two to four hours of group recreational activity was provided. Group treatment was designed to facilitate language use in a social setting, but there was no direct manipulation of speech or language deficits. Seven fluent and three nonfluent aphasic patients received group treatment.

Pragmatic analysis

The pragmatic aspects of language were analyzed for each patient with the Pragmatic Protocol (Prutting and Kirchner, 1983; 1987). The Pragmatic Protocol assesses verbal, paralinguistic, and nonverbal aspects of language use and is designed to screen overall communicative abilities. It consists of 30 pragmatic aspects of language that were extrapolated from normative data on children and adults. For the purposes of this study, the category of stylistic variations was omitted from the ratings, because it required observations not available in the videotape samples of conversation. Pragmatic aspects assessed in the 29-item protocol we employed included verbal aspects, for example, variety of speech acts, topic selection, specificity/accuracy, etc.; paralinguistic aspects, for example, intelligibility, vocal intensity, prosody, etc.; and nonverbal aspects, for example, physical proximity, body posture, gestures, etc.

According to Prutting and Kirchner (1987), the Pragmatic Protocol rating should be completed after viewing an unstructured, spontaneous conversational interaction between the subject and his or her partner. Each of the protocol behaviors is scored as appropriate, inappropriate, or no opportunity to observe. All conversational samples were between the patient and a speech-language pathologist who conducted the periodic evaluations.

Procedures

Prior to the beginning of the study, the 100 subject videotapes, five for each of the 20 subjects—4, 15, 26, 37, and 48 weeks postonset, were randomized and coded by number to control for bias from knowing the time postonset. A pragmatic protocol for each conversation was completed for each of the 100 conversations. Five percent of the videotapes were selected randomly to determine point-to-point reliability. Interjudge reliability between two judges for the pragmatic ratings was 94 percent.

PICA performance at 4, 15, 26, 37, and 48 weeks postonset utilized the data from the VA Cooperative Study. Interjudge reliability between two judges for PICA overall percentile performance, obtained in the VA Cooperative Study, was 98 percent.

Results

Language impairment and pragmatic performance

Table 4 shows language impairment, PICA overall percentile, and pragmatic performance, percent appropriate behaviors, for all subjects at entry, four weeks postonset, and at 15, 26, 37, and 48 weeks postonset. Language impairment improved significantly over time, 44th percentile at four weeks postonset and 69th percentile at 48 weeks postonset. Similarly, pragmatic performance improved significantly over time, 87 percent appropriate at four weeks postonset and 93 percent appropriate at 48 weeks postonset. At all points in time, pragmatic performance was better than language performance.

| Time | Measures | | | |
| | PICA %ile | | Pragmatic % App | |
	Mean	Range	Mean	Range
4 Weeks	43.65	15-73	87.45	72-97
15 Weeks	62.00	15-90	91.15	76-97
26 Weeks	64.70	13-90	91.35	79-97
37 Weeks	67.60	20-93	92.25	79-100
48 Weeks	68.70	32-95	92.90	83-100

Table 4.   Mean and range in the PICA overall percentile and percent appropriate pragmatic performance at 4, 15, 26, 37, and 48 weeks postonset for all subjects (n = 20).

As shown in Table 5, pragmatic performance was more inappropriate in verbal aspects than in paralinguistic or nonverbal aspects at all points in time from 4 to 48 weeks postonset. Subjects' specificity/accuracy was the least appropriate of all verbal aspects, and it remained least appropriate during the 44-week treatment trial. Intelligibility and fluency were the least appropriate paralinguistic aspects. Nonverbal aspects were generally appropriate, and none emerged as more inappropriate than the others. The total number of inappropriate behaviors in the 20 subjects declined over time from 73 at four weeks postonset to 42 at 48 weeks postonset. However, at 48 weeks postonset, half of the 20 patents remained inappropriate in specificity/accuracy.

Behavior	Weeks postonset				
	4	15	26	37	48
Verbal Aspects					
Speech act pair analysis					
Variety of speech acts	2				
Topic Selection	1			1	
Topic Introduction	1			1	
Topic Maintenance	4		1		1
Topic Change		1			1
Turntaking Initiation	5	3	5	1	4
Turntaking Response					
Turntaking Repair/Revision				1	
Turntaking Pause Time	1	1			1
Turntaking Interpretation/Overlap	3		1	1	1
Turntaking Feedback to Speaker					
Turntaking Adjacency					
Turntaking Contingency	6	5	7	6	4
Turntaking Quantity/Conciseness	5	2		3	
Specificity/Accuracy	18	16	14	15	10
Cohesion	6	4	2	3	6
Subtotal	53	32	30	32	28
Paralinguistic Aspects					
Intelligibility	9	9	8	4	6
Vocal Intensity	1	1	2	1	1
Prosody	2	1	2	1	1
Fluency	6	7	6	6	4
Subtotal	18	18	18	12	12
Nonverbal Aspects					
Physical Proximity					
Physical Contacts					
Body Posture					
Foot/Leg and Hand/Arm Movement		1		1	1
Gestures	1	1	1		1
Facial Expression					
Eye Gaze	1		1	1	
Subtotal	2	2	2	2	2
Total	73	52	50	46	42

Table 5. Number of subjects exhibiting specific, inappropriate verbal, paralinguistic, and nonverbal pragmatic behaviors at 4, 15, 26, 37, and 48 weeks postonset.

Fluent and nonfluent

Table 6 indicates fluent patients displayed less language impairment, indicated by the PICA overall percentile, than nonfluent patients at all points in time. Differences ranged from 23.4 percentile units at 15 weeks postonset to 13.6 percentile units at 48 weeks postonset. Similarly, pragmatic performance was better in fluent patients than in nonfluent patients at all points in time. Differences ranged from 3.1 percent at 26 weeks postonset to 5.0 percent at 48 weeks postonset.

While fluent patients performed significantly better on the PICA than nonfluent patients, there was no significant difference between fluent and nonfluent patients in the amount of improvement attained—21 percentile units in the fluent group and 29 percentile units in the nonfluent group—during the 44-week treatment trial. Pragmatic performance did not differ significantly between groups at any point in time, and there was no significant different between groups in the amount of improvement attained—5.7 percent in the fluent group and 5.2 percent in the nonfluent group—during the 44-week treatment trial.

Time		PICA %ile	Comparison		Pragmatic %	
	F	NF	D	F	NF	D
4 Weeks	54.7	32.6	+22.1	89.7	85.2	+4.5
15 Weeks	73.7	50.3	+23.4	93.2	89.1	+4.1
26 Weeks	73.8	55.6	+18.2	92.9	89.8	+3.1
37 Weeks	75.0	60.2	+14.8	94.4	90.1	+4.3
48 Weeks	75.5	61.9	+13.6	95.4	90.4	+5.0

Table 6. Mean and mean difference (D) in the PICA overall percentile and percent appropriate pragmatic performance between the fluent (F) and nonfluent (NF) aphasia groups at 4, 15, 26, 37, and 48 weeks postonset.

Individual and group treatment

Ten patients, three fluent and seven nonfluent, received individual treatment, and ten patients, seven fluent and three nonfluent, received group treatment between 4 and 48 weeks postonset. Table 7 shows comparisons between the individual and group treatment groups at 4, 15, 26, 37, and 48 weeks postonset.

There were no significant differences between groups in PICA overall percentile performance or pragmatic performance at any point in time. Both groups attained a similar amount of improvement on the PICA—27 percentile units for the individual treatment group and 23 percentile units for the group treatment group—during the 44-week treatment trial. And, there was no significant difference between groups in the amount of improvement attained. Both groups displayed similar improvement in pragmatic performance—4.2 percent in the individual treatment group and 6.7 percent in the group treatment group—during the 44-week treatment trial. And, there was no significant difference between groups in the amount of improvement attained. However, a significant time by type of treatment interaction emerged. This resulted from significant improvement in group treated patients, 5.3 percent, between 4 and 15 weeks postonset, and nonsignificant improvement in individually treated patients, 2.1 percent, during the same time period.

| Time | Comparison | | | | | |
| | PICA %ile | | | Pragmatic % | | |
	I	G	D	I	G	D
4 Weeks	46.0	41.3	+4.7	88.0	86.9	+1.1
15 Weeks	65.5	58.5	+7.0	90.1	92.2	-2.1
26 Weeks	67.7	61.7	+6.0	91.5	91.2	+ 0.3
37 Weeks	71.3	63.9	+7.4	92.8	91.7	+1.1
48 Weeks	72.7	64.7	+8.0	92.2	93.6	-1.4

Table 7.   Mean and mean difference(D) in PICA overall percentile and percent appropriate pragmatic performance between individual (I) and group (G) treated patients.

Relationship between language impairment and pragmatic performance

As shown in Table 8, the PICA overall percentile was significantly related with pragmatic performance at all points in time for the total sample of 20 patients. However, when the sample was divided into fluent and nonfluent patients, PICA performance was significantly related with pragmatic performance in the fluent group at four weeks postonset but not thereafter. Conversely, in nonfluent patients, PICA and pragmatic performance was not significantly related at four weeks postonset, but it was significantly related at 15, 26, 37, and 48 weeks postonset. Comparison of PICA and pragmatic performance in individual and group treated patients indicated a significant relationship for individual treated patients at all points in time except 37 weeks postonset. Conversely, in group treated patients, PICA and pragmatic performance was significantly related at only 37 weeks postonset.

In addition, we examined the relationship between total improvement, between 4 and 48 weeks postonset, in PICA and pragmatic performance. As shown in Table 8, no significant relationships emerged for all subjects combined, fluent patients, nonfluent patients, individually treated patients, or group treated patients.

| Time | Cohorts | | | | |
	A	F	NF	I	G
4 weeks	.68*	.66**	.59	.86*	.31
15 weeks	.63*	.33	.66**	.84*	.59
26 weeks	.54**	.01	.66**	.69**	.42
37 weeks	.58*	.19	.69**	.57	.66**
48 weeks	.57*	.02	.81*	.69**	.56
Change between 4 and 48 weeks	.36	.49	.31	.52	.33

* Significant at $p < .01$
** Significant at $p < .05$

Table 8.   Relationship (r) between PICA overall percentile and percent appropriate pragmatic performance for all subjects (A), fluent (F) and nonfluent (NF) groups, and individual (I) and group (G) treated patients at 4, 15, 26, 37, and 48 weeks postonset and change in PICA and pragmatic performance between 4 and 48 weeks postonset.

Discussion

If pragmatic performance in conversation is considered to be a measure of communication and performance on a standardized test of aphasia, for example, the PICA is a measure of language impairment, our results support Holland's (1982) observation. Our aphasic people communicated better than they talked. However, we found no differences in the severity or amount of improvement in pragmatic performance between fluent and nonfluent aphasic people. Moreover, the type of treatment received, individual or group, appeared to have no influence on the severity of or amount of improvement in pragmatic performance. The relationship between pragmatic performance and language impairment is inconsistent over time, and the point in time when this relationship exists appears to differ between fluent and nonfluent aphasic patients. And, while the severity of pragmatic performance and the severity of language impairment appears related in a combined group of fluent and nonfluent aphasic people, improvement in pragmatic performance is not related with improvement in language impairment.

Pragmatic performance and language impairment

Generally, our sample of aphasic people displayed better pragmatic performance in conversation than language ability on a standardized test for aphasia. This was the case at four weeks postonset, pretreatment—72 to 97 percent pragmatically appropriate compared with 15th to 73rd PICA overall percentile—and at 48 weeks postonset, post-treatment—83 to 100 percent pragmatically appropriate compared with 32nd to 95th PICA overall percentile. Performance in both domains improved significantly over time, 4 to 48 weeks postonset. Mean change in the PICA overall percentile was greater, 25 percentile units, than in percent appropriate pragmatic performance, 4.45 percent. Probably, this results from a "ceiling effect" on the pragmatic protocol. Appropriate pragmatic behaviors were high early postonset, 87 percent, and had little room to improve over time.

Pragmatic behaviors that were most impaired at four weeks postonset—specificity/ accuracy, cohesion, intelligibility—remained the most impaired at 48 weeks postonset. Prutting and Kirchner (1987) reported 100 percent of their aphasic sample were inappropriate in specificity/accuracy. Ninety percent of our sample demonstrated inappropriate specificity/accuracy at four weeks postonset. Fifty percent continued to display inappropriate specificity/accuracy at 48 weeks postonset. This aspect of pragmatic behavior, along with cohesion and intelligibility, may be most similar to the semantic, syntactic, and phonologic measures contained in standardized measures of language impairment, for example, the PICA.

Fluent and nonfluent

Difference between pragmatic performance and language impairment became apparent when fluent and nonfluent aphasic people were compared. Our fluent aphasic patients displayed significantly less impairment on the PICA than our nonfluent aphasic patients. However, we found no significant differences in pragmatic performance between our fluent and nonfluent groups. The difference between groups in language impairment is probably an artifact of the dichotomy—fluent and nonfluent. The nonfluent group included the most severely aphasic people, global, and the fluent group included the most mildly aphasic people, anomic. Thus, measures of language impairment typically result in group differences when aphasic patients are classified into fluent and nonfluent groups. However, pragmatic analysis indicates fluent and nonfluent aphasic people do not differ significantly in their percent of appropriate pragmatic behaviors. This was our observation, and it is in line with Goldblum's (1985) results.

When improvement in pragmatic performance and language impairment is compared in fluent and nonfluent groups, the groups appear to change similarly. Both of ours made essentially the same amount of improvement of both measures during a 44-week treatment trial, and there were no significant differences between the fluent and nonfluent groups in the amount of improvement attained on either measure. Thus, the response to time and treatment appears to be similar in fluent and nonfluent aphasic patients.

Individual and group treatment

The influence of the type of therapy appeared minimal on both pragmatic performance and language impairment. Individually treated and group treated patients made essentially the same amount of improvement on both measures, and there were no significant differences between groups on either measure at any point in time. The only significant difference between treatments was significant improvement in pragmatic performance in group treated patients during the first 11 weeks of treatment and the absence of significant improvement in pragmatic performance during the same period in individually treated patients.

Failure to find a treatment group effect is somewhat surprising. Individually treated patients received stimulus-response treatment designed to improve language impairment in all communicative modalities—auditory comprehension, reading, oral-expressive language, and writing. Group treated patients received methods designed to improve communication in a social setting and no direct manipulation of specific language deficits. One might predict that these different treatments would result in individually treated patients making more improvement in language impairment and group treated patients making more improvement in pragmatic performance. This did not occur.

Aten, Caligiuri, and Holland (1982) reported that group treatment with chronic aphasic patients that was designed to improve functional communication resulted in significant pre- to post-treatment improvement on the Communicative Abilities in Daily Living (CADL) (Holland, 1980) but not on the PICA. We found group treatment designed to improve communication in a social setting did not result in significantly better pragmatic performance than individual, traditional, stimulus-response treatment designed to improve language impairment. Our patients entered treatment at four weeks postonset and continued until 48 weeks postonset. Aten et al.'s patients were chronically aphasic when treatment began. Moreover, similarities between what we call functional communication and pragmatic aspects of language are not clear. We observed that pragmatic performance was generally better than performance on a standardized test for aphasia. However, the relationship between pragmatic performance and functional communication remains to be demonstrated.

Relationship between language impairment and pragmatic performance

In our total sample of 20 aphasic patients, pragmatic performance and language impairment were significantly related pretreatment, four weeks postonset, and at 15, 26, 37, and 48 weeks postonset. However, when nonfluent aphasic patients were compared with fluent aphasic patients, pragmatic performance and language impairment were significantly related in the fluent group only at four weeks postonset. Conversely, in the nonfluent group, pragmatic performance and language impairment were not significantly related at four weeks postonset, but they were significantly related at 15, 26, 37, and 48 weeks postonset. And, when the relationship was examined according to the type of treatment received, pragmatic performance and language impairment in individually treated patients were significantly related at all points in time except at 37 weeks postonset. Conversely, in group treated patients, pragmatic performance and language impairment were significantly related only at 37 weeks postonset. Finally, when improvement in pragmatic performance and improvement in language impairment were compared, we found no significant relationship for all patients combined, fluent patients, nonfluent patients, individually treated patients, or group treated patients.

Some (Aten et al., 1982; Binder, 1984; Blomert, 1990; Holland, 1980; Newhoff and Apel, 1997; Penn, 1985; Sarno and Levita, 1971; Ulatowska et al., 1977) suggest discrepancies exist between standardized language test performance and observations of aphasic patients' conversational skills. While discrepancies exist, our results indicate pragmatic performance during conversation and performance on a standardized aphasia test are significantly related in a sample of aphasic people. Thus, performance on one measure predicts performance on the other. However, the strength of the relationship, correlations ranging from +.57 to +.68, is not overwhelming.

The relationship between the two measures changes when aphasic people are subdivided into fluent and nonfluent groups. At four weeks postonset, pragmatic performance and language impairment was significantly related in fluent aphasic patients but not in nonfluent aphasic patients. Failure to find a significant relationship in nonfluent patients resulted from good pragmatic performance, 80% appropriate, coexisting with severe language impairment, PICA performance below the 20th percentile, in 30 percent of the nonfluent sample. Conversely, the two measures were significantly related in nonfluent patients between 15 and 48 weeks postonset but not in fluent patients. Scatter plots indicate that nonfluent patients' language impairment, by 15 weeks postonset, had improved sufficiently to coincide with their generally intact pragmatic performance at four weeks postonset. By 15 weeks postonset, both pragmatic performance and language impairment in fluent aphasic patients had improved, and there was a narrow range of severity in both behaviors among members of the nonfluent group.

The sporadic relationship over time between pragmatic performance and language impairment in the different treatment groups probably results from unequal assignment of fluent and nonfluent patients to the different treatments. Seven nonfluent and three fluent patients received individual treatment, and three nonfluent and seven fluent patients received group treatment.

Finally, while we observed a significant relationship between pragmatic performance and language impairment at all points in time between 4 and 48 weeks postonset for our entire sample, no significant relationship was observed between improvement in pragmatic performance and improvement in language impairment between 4 and 48 weeks postonset. Thus, while performance on one measure appears to predict severity on the other measure, improvement on one measure does not predict improvement on the other measure.

The work to be done is exploration of aphasic patients' pragmatic performance and its relationship to their communicative competence. While Leech (1983) suggests pragmatic performance is a measure of communicative competence, it is not clear how pragmatic performance in aphasia relates to functional communication. Previous reports (Avent and Wertz, 1996; Binder, 1984; Goldblum, 1985; Prutting and Kirchner, 1987; Roberts and Wertz, 1993) have compared pragmatic performance and language impairment in aphasia. Similarly, our results indicate pragmatically appropriate performance can coexist with significant language impairment. We may want to ask what pragmatic appropriateness, as indicated by a pragmatic protocol, implies about an aphasic person's functional communication.

References

Aten, J. L., Caligiuri, M. P., & Holland, A. L. (1982). The efficacy of functional communication therapy for chronic aphasic patients. *Journal of Speech and Hearing Disorders, 47*, 93-96.

Avent, J. R., & Wertz, R. T. (1996). Influence of type of aphasia and type of treatment on aphasic patients' pragmatic performance. *Aphasiology, 10*, 253-265.

Binder, G. M. (1984). A societal and clinical appraisal of pragmatic and linguistic behaviors. Unpublished master's thesis, University of California, Santa Barbara, CA.

Blomert, L. (1990). What functional assessment can contribute to setting goals for aphasia therapy. *Aphasiology, 4*, 307-320.

Brinton, B., & Fujiki, M. (1989). Conversational management with language-impaired children. Rockville: Aspen Publishers, Inc.

Chapman, S. B., & Ulatowska, H. (1992). Methodology for discourse management in the treatment of aphasia. *Clinical Communication Disorders, 2*, 64-81.

Gallagher, T. M. (1991). *Pragmatics of language: Clinical practice issues.* San Diego: Singular Publishing Group, Inc.

Goldblum, G. (1985). Aphasia: A societal and clinical appraisal of pragmatic and linguistic behaviors. *South African Journal of Communication Disorders, 32*, 11-18.

Green, G. (1984). Communication in aphasia therapy: Some of the procedures and issues involved. *British Journal of Disordered Communication, 19*, 35-46.

Gurland, G. B., Chwat, S. E., & Wollner, S. G. (1982). Establishing a communication profile in adult aphasia: Analysis of communicative acts and conversational sequences. In R. H. Brookshire (Ed.), *Clinical Aphasiology Conference Proceedings* (pp. 18-27). Minneapolis: BRK Publishers.

Holland, A. L. (1980). *Communicative abilities in daily living.* Baltimore: University Park Press.

Holland, A. L. (1982). Observing functional communication of aphasic adults. *Journal of Speech and Hearing Disorders, 47*, 50-56.

Holland, A. L. (1996). Pragmatic assessment and treatment for aphasia. In G. L. Wallace (Ed.), *Adult aphasia rehabilitation* (pp. 161-173). Boston: Butterworth-Heinemann.

Hough, M. S., & Pierce, R. S. (1994). Pragmatics and treatment. In R. Chapey (Ed.), *Language intervention strategies in adult aphasia* (third ed.) (pp. 246-268). Baltimore: Williams and Wilkins.

Leech, G. N. (1983). *Principles of pragmatics.* London: Longman.

Levinson, S. C. (1983). *Pragmatics.* Cambridge: Cambridge University Press.

Lomas, J., Pickard, L., Bester, S., Elbard, H., Finlayson, A., & Zoghaib, C. (1989). The Communicative Effectiveness Index: Development and psychometric evaluation of a functional communication measure for adult aphasia. *Journal of Speech and Hearing Disorders, 54*, 113-124.

Lund, N. J., & Duchan, J. F. (1993). *Assessing children's language in naturalistic contexts* (third ed.) Englewood Cliffs, N.J.: Prentice Hall.

McTear, M. F. (1985). Pragmatic disorders: A question of direction. *British Journal of Disorders of Communication, 20*, 119-127.

Manochiopinig, S., Sheard, C., & Reed, V. A. (1992). Pragmatic assessment in adult aphasia: A clinical review. *Aphasiology, 6*, 519-533.

Newhoff, M., & Apel, K. (1997). Impairments in pragmatics. In L. L. LaPointe (Ed.), *Aphasia and related neurogenic language disorders*, second ed. (pp. 250-264). New York: Thieme Medical Publishers, Inc.

Penn, C. (1985). The profile of communicative appropriateness: A clinical tool for the assessment of pragmatics. *South African Journal of Communication Disorders, 32*, 18-23.

Penn, C. (1993). Aphasia therapy in South Africa: Some pragmatic and personal perspectives. In A. L. Holland and M. M. Forbes (Eds.), *Aphasia treatment: World perspectives* (pp. 25-53). San Diego: Singular Publishing Group, Inc.

Porch, B. E. (1967). *Porch index of communicative ability.* Palo Alto, CA: Consulting Psychologists Press.

Prins, R. S., Snow, C. E., & Wagenaar, R. (1978). Recovery from aphasia: Spontaneous speech versus language comprehension. *Brain and Language, 6*, 192-211.

Prutting, C. A., & Kirchner, D. M. (1983). Applied Pragmatics. In T. M. Gallagher and C. A. Prutting (Eds.), *Pragmatic assessment and intervention issues in language.* San Diego: College-Hill Press, Inc.

Prutting, C. A., & Kirchner, D. M. (1987). A clinical appraisal of the pragmatic aspects of language. *Journal of Speech and Hearing Disorders, 52,* 105-119.

Roberts, J. A., & Wertz, R. T. (1993). Communicative effectiveness in treated aphasic patients during the first post onset year. In M. L. Lemme (Ed.), *Clinical Aphasiology, 21,* 291-298.

Sarno, M. T. (1993). Aphasia rehabilitation: Psychosocial and ethical considerations. *Aphasiology, 4,* 321-334.

Sarno, M. T., & Levita, E. (1971). Natural course of recovery in severe aphasia. *Archives of Physical Medicine and Rehabilitation, 52,* 175-179.

Smith, L. (1985). Communicative activities of dysphasic adults: A survey. *British Journal of Disordered Communication, 20,* 31-44.

Ulatowska, H. K., Haynes, S. M., Hildebrand, B. H., & Richardson, S. M. (1977). The aphasic individual: A speaker and a listener, not a patient. In R. H. Brookshire (Ed.), *Clinical Aphasiology Conference Proceedings* (pp. 198-213). Minneapolis: BRK Publishers.

Wertz, R. T., Collins, M. J., Weiss, D., Kurtzke, J. F., Friden, T., Brookshire, R. H., Pierce, J., Holtzapple, P., Hubbard, D. J., Porch, B. E., West, J. A., Davis, L., Matovitch, V., Morley, G. K., & Resurreccion, E. (1981). Veterans Administration cooperative study on aphasia: A comparison of individual and group treatment. *Journal of Speech and Hearing Research, 24,* 580-594.

**Pergamon**

J. Neurolinguistics, Vol. 11, Nos 1–2, p. 223–241, 1998
© 1998 Published by Elsevier Science Ltd. All rights reserved
Printed in Great Britain
0911-6044/98 $19.00 + 0.00

PII: S0911-6044(98)00015-3

A cross-language analysis of conversation in a trilingual speaker with aphasia

Luise Springer*, Nick Miller[†] and Frauke Bürk[†]

*RWTH Aachen, Germany; [†]University of Newcastle upon Tyne.

Abstract—This study explores the relationship between formal and functional language performance of a trilingual speaker with aphasia, focusing on her monolingual use of German and English in free conversations and semistructured interviews with different interlocutors. We use a conversation analysis framework to highlight the contrasting pictures between her formal test scores and sources of trouble across different conversations. In the discussion possible reasons for divergences across settings and languages are examined.

Introduction

In aphasiology, as in other fields of language enquiry, it has been recognised for some time now that an account of successes and problems in communication is incomplete unless the full, collaborative nature of conversational exchanges is encompassed in any descriptive or analytic procedure. Achieving mutual understanding is as much a task of the listener as it is of the speaker. Several approaches have been developed aimed at characterising conversational interaction.

One of the approaches, developed from ethnomethodological perspectives on language behaviour, is conversation analysis (Schegloff, Jefferson & Sacks, 1977; Heritage, 1989). Essentially, this is a data driven, bottom-up approach to analysing the sequential, collaborative construction of a conversation, that attempts as far as possible to embark on the analytic task with as few prior assumptions as possible regarding the nature of what might cause trouble for participants and how repairs of any trouble spots in conversation might be negotiated.

Conversation analysis (CA) and derivatives have been applied to the description and analysis of talk in an increasing range of communication settings—including the field of speech-language pathology. CA has been employed to gain further insights into the nature of communication in conversations between children with semantic-pragmatic disorders and adults (Brinton & Fujiki, 1989; Willcox & Mogford-Bevan, 1995), speakers with aphasia and their partners (Ferguson, 1994; 1996; Milroy & Perkins, 1992; Perkins, 1995), aphasic speakers with each other (Klippi, 1995), speakers with dementia and their partners (Hamilton, 1994; Orange, Lubinski & Higginbotham, 1996), and speakers with head injury (Friedland & Miller, in press). Researchers have examined the nature of breakdowns, the properties of self repairs (Laakso, 1997), and variations across settings, partners and topics (Ferguson, 1994). To our knowledge there are no accounts yet applying CA principles to the comparison of performances across different languages (though see Friedland & Miller, 1997) in bilingual speakers with aphasia, the topic of this report.

One claimed advantage of CA over formal language testing is that it does not look at the results of isolated levels of linguistic analysis divorced from communication. Such isolated results may tell one about the potential strengths and weaknesses of a speaker and underlying impairments they have to face. However, they may give little indication of the speaker's true communicative competence. Aphasia test results also ignore the fact that in a person's conversations there may be formal 'errors' that cause no breakdown in communication and there may be well-formed utterances that do. CA, by contrast, concentrates in assessment on what a speaker is actually able to achieve with her/his (limited) formal language skills, and, just as importantly, what the speaker can achieve in tandem with another speaker.

CA lends itself, too, to addressing issues of communication (breakdown) in bilingual settings in general, and in individual speakers. For instance, if a pragmatic disorder is evident while using one language, is it also present and of the same type and degree while speaking the other language? Are differences in overall severity of impairment between the languages concerned reflected in the comparative severity of conversational breakdowns across languages? Do the trouble spots for the conversational partners in the two languages reflect common underlying deficits in language processing, or do language specific trouble spots arise? For speakers who habitually maintain strict language separation, does code switching or mixing become a problem in aphasia and what kind of consequences does this have for their communication with monolingual and bilingual speakers?

We examined some of these issues in a German-English-Italian trilingual speaker with agrammatic aphasia (although the analyses below pertain only to her German and English interactions). Our enquiries were directed at an examination of the possible relationships between the speaker's (HK) formal language profiles in German and English and the trouble sources and repair strategies used in her monolingual conversations in German versus English. For this we employed elements of CA methodology.

More specifically we wished to establish whether HK's success at communicating in German and English, in comparable settings, was equal across languages or whether success mirrored, or was influenced by, the significant differences established on formal language assessments. We examined whether the trouble sources in conversation could be directly attributed to the identified formal language impairments or whether for one or both languages there existed a pattern of troubles independent of the formal language profile. We analysed the nature of the trouble and repair instances to see whether a similar pattern was observable across languages or whether HK employed strategies available to her in one language but not the other. A further target of our enquiries concerned the effects of different conversational types (informal versus semistructured) and partners (Ferguson, 1994) on the quantity and quality of trouble spots and repairs.

We made the following predictions and hypotheses:
1) Given that HK is a right-handed speaker with a lesion in the left hemisphere, we hypothesized that there would be no pragmatic disorder in either of her languages, i.e., turn taking, topic maintenance, self-monitoring and collaboration with the interlocutor would be normal.
2) From a descriptive point of view, we predicted that results on formal language tests would show a quantitative score in favour of German, but that the profile of the types of impairment would be qualitatively similar for German and English.
3) Consequent on the predicted qualitative similarity, we hypothesized that the sources of trouble in both languages, with all speakers and in all settings would be identical.
4) However, the differences in quantitative scores across languages would lead to a greater number of incomplete or fragmentary utterances in English, a more unfavourable type-token relationship for closed as well as open class words, a greater presence of interjections and pauses/hesitations and in turn a raised number of trouble sources in English. Furthermore, it was hypothesized that these depressed English scores, in

comparison to German, would lead to longer and more complex repair trajectories. On the basis of the quantitative dissimilarity, it was also predicted that there would be more intrusions of German into the English conversations than vice versa.

Methods and materials

This section gives the medical and language background of HK, details the assessments we carried out with her and explains the measures and analyses we conducted on the data gathered.

Case history

Background

At the time of this study HK was 34 years old, married with three children. She was born in Germany, but at the age of five moved with her family to Italy. There she attended a German school (where she had some instruction in Italian and learned some English) for 7 years before moving back to Germany. After a four month stay in Germany the family moved, with her father's job, to Britain. In London HK went to an English school and passed her school leaving exams (A-levels). HK reported that she had many English friends and spoke English outside of the home. Her command of the English language was very good. At the age of 20 HK and her family returned permanently to Germany where she then passed the German 'Abitur' (equivalent to the English A-levels). Subsequently she trained and worked as a horse-riding instructor. According to HK, before her subarachnoid haemorrhage (SAH) her language skills were equal in German and English whereas her Italian was not of the same standard. Focus in this study was on HK's dominant languages, German and English.

Medical history

At the age of 33 HK suffered a SAH due to an aneurysm in the left posterior communicating artery (diagnosed by CT-scan). The episode left her with a right hemiplegia and a marked expressive as well as receptive dysphasia with comprehension relatively better preserved. She had a speech apraxia. On the German version of the Wechsler Adult Intelligence Test HK achieved a performance IQ of 98.

Initially HK received regular speech therapy in German in hospital and after discharge further treatment in the community. Eight months after her haemorrhage she came for a 7-week stay at the specialist Aphasia Ward, at the Aachen University Hospital in Germany, for an intensive rehabilitation programme with daily speech and language therapy.

Data gathering

At the end of her treatment in Aachen HK was assessed using the Aachen Aphasia Test (AAT) in German and the English adaptation of the AAT (Miller, De Bleser & Willmes, 1997). The AAT commences with a semistructured interview with the therapist, with predetermined questions (about the person's illness, their family, work and hobbies), designed to gather a spontaneous language sample. The directions for the interview instruct the therapist to give maximum encouragement and time to the speaker but preferably with minimum intervention on the part of the therapist. The German and English versions of this interview (referred to below as the semistructured interview) were video recorded.

In addition, two naturally occurring, general conversations without predetermined topics were videotaped, one in each language (i.e. German and English), again with both speakers visible on the screen. These are the conversations referred to below.

Three of the conversation partners, who were strangers to HK, were introduced as monolingual speakers. For the English AAT interview HK knew the interlocutor as someone she had received therapy from in German, although the therapist had a native-like command of English.

Transcription, data coding and analysis of conversation

The conversations were transcribed orthographically from the video recordings by two of the authors working independently. Their versions were compared and any disagreements reviewed on the videos until consensus was reached. Unintelligible utterances were transcribed in a broad phonemic script or recorded as the number of syllables spoken if a clear transcription was not possible (e.g. due to simultaneous talk or laughter by the interlocutor). No utterances, including apparently meaningless vocalisations, laughter, pauses, attempts at written communication and silences were omitted. Notation followed Lesser and Milroy (1993).

The transcripts were divided into speaker turns. A turn was taken to be a conversational contribution by one speaker followed by either silence or the start of another speaker's contribution. For the purpose of analysis the length of conversations was matched by taking only those turns in the longer of each conversational pair (German and English interview; German and English free conversation) up to the length of the number of turns contained in the shorter of the pair. Turns were further subdivided into utterances ranging from complete well-formed phrases to interjections.

Following Orange et al. (1996), sources of trouble (TS) were divided into phonological, morphological-syntactic, semantic, discourse and 'other' trouble sources. The categories for describing repair types and resolution patterns were also taken from Orange et al. (1996) — viz. repair by repetition, elaboration (interlocutor expands on possible meaning of speaker's utterance), reduction (elliptical responses, confirmation and denial responses, indications of not knowing the answer to a question), substitution (use of alternate but equivalent form of meaning or altered syntax but preserved meaning), and unrelated (interlocutor does not respond to a repair initiator or gives a reply unrelated to the trouble source). Repair resolution was registered as 'most successful' (single trouble source, initiator and repair), 'successful' (more than one repair initiator and repair used to resolve a single trouble source), or 'unsuccessful' (trouble source not repaired). Repair complexity was coded as simple (single trouble source) or complex (primary TS with embedded TS's). Full definitions are available in Orange et al. (1996).

Repair initiation was counted as either self (HK) or other (conversation partner) initiated and repair resolution as self, other or collaboratively achieved. This gave a taxonomy of self-initiated self repairs (SI-SR), self-initiated other repairs (SI-OR), self initiated collaborative repairs (SI-Col R), other initiated self repairs (OI-SR), other initiated other repairs (OI-OR), and other initiated collaborative repairs (OI-Col R). Examples are given in the appendix.

We analysed the pattern of topic initiation, maintenance and shift from the point of view of presence and appropriateness. We observed the ability of the speakers to take turns, to appropriately relinquish and resume them and their ability to acknowledge trouble in the conversation and to work together to repair it.

Coding and counts were done by hand by two authors independently. Disagreements were analysed again until consensus was reached. In practice, disagreements ranged from 5-10 percent of the counts.

Control speakers

Our prime interest concerned HK's own performance in her different languages and across settings, rather than a comparison with 'healthy' (monolingual) speakers of German and English. As such HK acted as her own control. However, we did monitor the communicative behaviour of her conversation partners. Apart from some instances of abrupt topic shifts (discussed later) that were trouble sources for HK, there were no occasions where the phonological, morpho-syntactic or lexical-semantic content of the interlocutor's talk acted as a TS for HK or led to the need for any repairs.

Data processing and analysis

Three sets of comparisons were made to address the hypotheses regarding performance across the different languages, conversation types and partners. Firstly we compared the formal AAT results in the two languages; the number and distribution of linguistic units across languages and settings; and the number, length and distribution of interjections and pauses. The second set of comparisons concerned the number and nature of TS, and the pattern of repair initiation sequences. The third set of comparisons examined the repair strategies and the length, complexity and success of repair attempts.

Comparisons were made using appropriate exact non-parametric statistical procedures. Fisher's exact test was used for 2 x 2 tables and its generalisation for general two-way contingency tables employing the StatXact program-package (1989). In addition, permutation tests for two-sample problems (Edgington, 1995) were used whenever counts per turn were analysed, again making use of the StatXact program.

Results

This section reports first the results of formal tests in German and English and descriptive statistics for the different conversations in terms of turn totals, time taken and distribution of linguistic units and pauses. We go on to examine the sources of trouble, the nature of repair initiation and the strategies used in repair negotiation. Finally we look at the length and success of repair sequences across speakers and settings.

Comparison of the two versions of the AAT

According to the AAT diagnostic criteria, aphasia was classified as Broca's aphasia in both languages. Spontaneous speech was characterised by agrammatism (see syntax rating), word retrieval problems (see semantic rating and naming scores), speech apraxia (see articulation ratings and repetition scores). Only in English was there some reliance on automatised language (e.g. idioms, formulaic language). The general trend shows English scores to be more depressed, both in spontaneous speech and on the formal subtests.

	AAT-Spontaneous speech ratings						AAT-subtests (percentile ranks)				
	COM	ART	AUT	SEM	PHO	SYN	TT	REP	WRIT	NAME	COMP
German	3	4	5	3	4	2	17 (70)	119 (58)	48 (51)	99 (79)	112 (98)
English	2	3	3*	3	3	1	22 (60)	103* (44)	33* (39)	46* (35)	100 (85)

* denotes significant differences between English and German.

Table 1. AAT Performances.
Spontaneous speech is rated 0 (absent or unscorable) to 5 (normal range). The rating scales are: COM: communicative abilities; ART: articulation and prosody; AUT: automatized speech; SEM: semantic structure; PHO: phonemic structure; SYN: syntactic structure. The AAT-subtests are: TT: Token Test; REP: repetition; WRIT: written language; NAME: confrontation naming; COMP: auditory and written comprehension.

Since the (preliminary) norms for the English AAT (Miller et al., 1997) are very similar to the original German test, as are the reliability (consistency) coefficients, psychometric single case analysis (Willmes, 1985) was employed to compare performances for the five AAT subtests for both languages. Subtests 'naming', 'repetition' and 'written language' were significantly better in German. The profiles of impairment are approximately parallel in the two languages with the exception of 'naming' which is significantly worse in English. Generally the degree of severity was moderate in German and moderate to severe in English.

Table 2 presents descriptive information on the two German and two English conversations: namely conversation time, total client turns, the percentage of trouble-source-repair-sequences (TSR) and the mean length of trouble-source-repair-units. We extracted the first 212 turns from the German conversation to match the turn-length of the shorter English conversation, while the interviews were matched at the 175 turns of the shorter German interview. Analyses below are based on the 106 and 87 turns respectively that constituted HK's part in these conversations.

Conversation times in the informal conversations were shorter than in the interviews, in both languages. In contrast, the number of turns is distinctly higher in the former.

Numerically, most of the trouble-sources and repairs appeared in the German interview, i.e. in 34.5% of HK's conversational turns we identified signals where HK or her partner initiated repairs, guesses, repetitions and comments about the troubles. There are, however no significant differences in the relative frequency of TSR (Fisher's exact test, all p > .10, two-tailed, for conversation vs. interviews in each language respectively, and German vs. English for each conversation type). Nevertheless, the mean number of turns to repair troubles is generally short. The highest mean length was required to repair trouble-sources in the English interview. Employing the randomisation test for the independent two-samples problem (Edgington, 1995) a significant difference was found in the mean length of the TSR when comparing the interview in German vs. English (p = .003, two-tailed), and conversation vs. interview in English (p = . 0207, two-tailed).

	German conversation	German interview	English conversation	English interview
Conversation time (min.:sec.)	19:12	23:15	18:22	23:23
Total no. of client turns	106	87	106	87
Percentage of Trouble-Source-Repairs (TSR)	23.58% 25	34.48% 30	25.47% 27	25.28% 22
Mean Length of TSR-units* Range	2.6 1-10	1.9 1-5	2.8 1-9	4.0 1-18

Table 2. Conversation time, number of turns, percentage and mean length of trouble-source-units.

*Mean Length of TSR is calculated by averaging the number of consecutive unsuccessful turns that comprise each Trouble-Source-Repair sequence.

	German conversation n=106 turns	German interview n=87 turns	English conversation n=106 turns	English interview n=87 turns
Total no. of interjections mean/turn:	187 1.76	319 3.66	286 2.69	381 4.37
Percentage of client turns with just yes/no answers	44.33% 47	16.09% 14	60.37% 64	26.43% 23
Total length of pauses (sec.) mean/turn:	142 1.33	146 1.67	249 2.6	176 2.02
Percentage of 1 sec. pauses	49.05% 52	52.87% 46	39.62% 42	57.47% 50
Percentage of 2-4 sec. pauses	30.18% 32	21.83% 19	58.49% 62	45.97% 40
Percentage of 5-6 sec. pauses	1.88% 2	0	7.54% 8	6.89% 6
Micropauses	88	59	54	17

Table 3. Number and percentages of interjections and pauses.

Mean duration of pauses was longer in the English conversation and interview (Monte-Carlo version of independent two-samples permutation test p = <.001 and <.002 respectively). Long, 5-6 second, pauses are almost exclusive to English.

The number of interjections, yes/no answers and pauses and the percentages per turn are given in Table 3. HK used significantly more interjections in the German and English interview in comparison to the informal conversations. That this difference is not due to just yes/no interjections appearing in elliptic answers is demonstrated in the percentage of yes/no answers. As can be seen, fewer yes/no answers were present in the interviews in contrast to the conversations (Fisher's exact test, p < .0001 for German as well as English).

Since we were interested in the consequences of the linguistic impairments, we analysed performances at phrase and word level in the four different conversations. Table 4 presents the mean length of utterances (words), and the number of complete and deviant sentences and the number of one-word utterances.

	German conversation	German interview	English conversation	English interview
Mean length of utterances (MLU) (words/phrase)				
mean:	3.52	3.78	2.92	2.16
range:	1-8	1-9	1-9	1-7
Total no. of one-word phrases (without interjections)	11	5	12	27
Total no. of sentences (without one-word phrases)	72	201	40	61
Percentage of complete sentences (including complete elliptic answers)	22.22% 16	33.83% 68	10% 4	9.83% 6
Percentage of deviant sentences	77.77% 56	66.16% 133	90% 36	90.16% 55

Table 4.   Mean length of utterances (MLU) and number and percentage of complete, deviant sentences and one-word utterances.

The mean length of utterances and the total number of sentences is higher in the German conversations. The ratio of one-word : complete sentences was significantly different between the German conversation and the German interview (p = .0007) and between the German and English interview (p < .0001). The comparison of complete and deviant sentences showed significant differences between the conversation and interview in English (p = .0003). The same tendency was found in German (marginally significant p = .0750). The type of conversation influenced sentence production, with HK producing more sentences in the interviews, particularly for German.

	German conversation n=106 turns	German interview n=87 turns	English conversation n=106 turns	English interview n=87 turns
Total no. of open class words	190	371	83	118
Intrusion of other language	3	0	11	14
No. of different words	119	219	57	66
Type-token-ratio:	0.62	0.59	0.68	0.55
Total no. of nouns	82	161	39	38
Different nouns	53	96	28	20
Type-token-ratio:	0.64	0.59	0.71	0.52
Total no. of verbs	47	114	13	38
Different verbs	37	82	10	20
Type-token-ratio:	0.78	0.71	0.72	0.52
Total no. of adjectives	58	96	20	25
Different adjectives	27	41	11	18
Type-token-ratio:	0.46	0.42	0.55	0.72
Total no. of closed class words	146	125	54	88
Intrusion of other language	4	0	7	14
No. of different words	37	75	34	32
Type-token-ratio:	0.25	0.60	0.62	0.36

Table 5. Number and type-token-ratio of open and closed class words.

Table 5 gives the number and type-token-ratio of open and closed class words in the four conversations. Significant differences were found in the distribution of open class : closed class words between the German conversation and German interview (Fisher's Exact Text p < .0001 two-sided). There were no differences between the two English settings.

It is clear from these data that despite the shorter turn totals, more content words are produced in the interviews. The type-token ratio for closed class words in the German interview also indicates that a greater range of function words was produced. A comparison with the English interview illustrates that this effect is not simply due to the longer conversation time available. Despite comparable time scales the ratio of open to closed class words is significantly different between the German and English interviews (Fisher's Exact Test p < .0001).

As predicted, the direction of intrusions from one language into another was almost wholly unidirectional, from German into English. The only instance of English within an otherwise clearly German utterance was when HK was talking about the problems of keeping her languages separated and she stuck on the word 'und' (and).

Table 6 gives the number and percentage of trouble-sources (TS), differentiated according to self-initiated-self repair versus interactive repairs.

As HK's spontaneous speech is characterised by agrammatism we would expect that morphological-syntactic TS's would feature prominently. Despite the low ratings on the AAT for syntax, the majority of trouble-sources were lexical-semantic in all four settings. There were only marginally significant differences in the distribution of types of trouble-sources between the German conversation and the German interview (p = .051). No comparison between any other conversation pairs reached significance.

Trouble-Source (TS)	German conversation n=106 turns		German interview n=87 turns		English conversation n=106 turns		English interview n=87 turns	
	SI-SR	Other repairs	SI-SR	Other repairs	SI-SR	Other repairs	SI-SR	Other repairs
Phonological	0	1 0.94%	2 2.29%	0	2 1.88%	2 1.88%	0	2 2.29%
Morphological-syntactic	3 2.83%	3 2.83%	1 1.14%	0	1 0.94%	0	0	0
Lexical-semantic	5 4.71%	7 6.6%	9 10.34%	19 20.68%	8 7.54%	11 10.37%	1 1.14%	16 18.39%
Discourse	0	5 4.71%	0	0	0	2 1.88%	0	0
Code-switching	0	1 0.94%	0	0	0	1 0.94%	1 1.14%	2 2.29%
Total no. of Trouble-Source (TS)	8 7.54%	17 16.03%	12 13.79%	19 20.68%	11 10.37%	16 15.09%	2 2.29%	20 22.98%

Table 6. Number and percentage of Trouble-Sources (TS) in self-initiated and self-repair sequences and in interactive repairs.
Other repairs = interactive TSR: OI-SR, SI-OR, SI-Col R, OI-Col R

Table 7 summarises the general trouble-source-repair-sequences, also broken down according to self initiated self repairs versus interactive repairs. Although TSR totals appear small, the proportion still represents more than one in five turns over a period of 23 minutes.

We compared the trouble-source-sequences SI-SR and OI-SR which are most common in normal conversation with other and collaborative repairs (OI-OR, CoR) which are common in clinical settings. There was a marginally significant difference between the conversation and interview in English (p-value = .056) and the German and English interview (p = .062), otherwise no other comparisons between settings reached significance.

Trouble-Source-Repair (TSR) Sequences	German conversation n=106	German interview n=87	English conversation n=106	English interview n=87
SI - SR	8 7.54%	12 13.79%	11 10.37%	2 2.29%
OI - SR	2 1.88%	0	1 0.94%	1 1.14%
SI - OR	6 5.66%	7 8.04%	8 (-2) 7.54%	9 (-1) 10.34%
OI - OR	5 4.71%	5 5.74%	2 1.88%	0
SI - Col R	2 1.88%	5 5.74%	2 (-2) 1.88%	7 8.04%
OI - Col R	2 1.88%	1 1.14%	3 2.83%	3 3.44%
Total no. of TSR sequences	25 23.58%	30 34.48%	27 (-4) 25.47%	22 (-1) 25.28%

Table 7. Number and percentage of Trouble-Source-Repair (TSR)-sequences.
SI: self-initiated   SR: self-repair – failed repair;   OI: other initiated;   OR: other repair;   ColR: collaborative repair

Table 8 shows the types, number and percentage of repair strategies used by HK's conversational partners. The distribution of repair strategies was significantly different between the English conversation and English interview (exact p = .0019). Subsequent comparisons per category revealed that unrelated repairs were significantly more frequent in the English conversation and substitutions were significantly more frequent in the English interview. Unrelated sequences were evidenced solely in the English conversation (mostly topic shifts). Failed repairs were also exclusive to English.

Partner Repair Strategy (RP)	German conversation	German interview	English conversation	English interview
repetition	5 22.72%	6 30%	5 15.62%	4 11.76%
unrelated	0	0	9 (-4) 28.12%	1 (-1) 2.94%
elaboration	14 63.63%	12 60%	13 40.62%	20 58.82%
reduction	1 9.09%	0	3 9.37%	0
substitution	2 18.18%	2 10%	2 6.25%	9 26.47%
Total no. of RP* (including recursive turns)	22	20	32	34

* interactive repair-sequences (without SI-SR)

Table 8.   Number and Percentage of Partner Repair Strategies (RP).

As shown in Table 9 (cf. also Table 2) the English interview had, as predicted, the longest repair sequences. The shortest and most successful resolutions occurred in German.

Resolution (RS)	German conversation n=25 TSR		German interview n=30 TSR		English conversation n=27 TSR		English interview n=22 TSR	
	SI-SR	Other repairs	SI-SR	Other repairs	SI-SR	Other repairs	SI-SR	Other repairs
most successful and simple	8 32%	12 48%	12 40%	12 40%	11 40.74%	4 14.81%	2 0.9%	8 36.36%
successful and simple	0	2 8%	0	4 13.33%	0	2 7.4%	0	1 4.45%
successful and complex	0	3 12%	0	2 6.66%	0	6 22.22%	0	10 45.45%
unsuccessful and simple	0	0	0	0	0	2 7.4%	0	1 4.45%
unsuccessful and complex	0	0	0	0	0	2 7.4%	0	0
Total no. of RS percentage/client turns	8 7.54%	17 16.03%	12 13.79%	18 20.68%	11 10.37%	16 15.09%	2 2.29%	20 22.98%

Table 9. Success and Complexity of Self and Other Repair Resolution.

Discussion

We have presented a speaker who, despite having attained high levels of proficiency (to senior school leaving level and beyond) in two of her three languages, nevertheless showed some significant differences in performance in these languages after her SAH. In the following we discuss these differences and similarities in relation to our hypotheses and to possible explanations for the pattern of performances found.

We hypothesized that following a circumscribed lesion affecting functions of the left anterior cerebral hemisphere HK would not evidence a pragmatic disorder. This indeed seemed to be the case. HK was able to turn-take appropriately, she maintained topics both within and across turns, she relinquished her turn appropriately and took up talk again when the interlocutor signalled a turn end. Even where HK did not or could not signal her turn-taking intentions verbally, it was clear from the video that she appropriately used eye-contact, body posture, hand gestures and prosodic cues (loudness, speed and pitch changes) to hold open her turn or signal her wish to take over a turn. When she experienced problems continuing her turn because of word finding or sentence structuring problems, she was able to hold her turn open through the use of interjections or the nonverbal gestures. In this way she showed intact control of local management systems (Levinson, 1983, Glindemann, 1990)

The number of self-initiated repairs and her ability to join in collaborative repairs indicated that she had intact self and other monitoring skills. Her ability to recognise and attempt to repair inappropriate intrusions of one language into the other also testifies to the intactness of her pragmatic skills. The instances where 'discourse' featured as a trouble source (Table 6) actually involved discourse disruptions on the part of her conversation partners. For instance, in the English conversation, some of the failed repairs stem from the interlocutor's failure to recognise that HK was trying to hold open her turn, was engaged in (word or syntax) searching behaviour, or was appealing for assistance in repair. The conversation partner brought about a breakdown by interrupting HK's turn, failing to recognise that HK had more to add; by responding erroneously, based on information from just part of HK's turn; or, because she had failed to understand HK and was unable/unwilling to engage in repair negotiation, inappropriately changed the topic.

The profile of impairment across languages was, as predicted, similar (Table 1), though lexical-semantic problems in English did deviate away from the otherwise parallel impairment pattern in other areas. Such a profile might have been expected from the close structural relationship of German and English. It also fits an interpretation of aphasia as being an impairment of central language processes (Bates, Wulfeck & MacWhinney, 1991; Menn, O'Connor, Obler & Holland, 1995), which may nevertheless be affected to differing degrees individually. It also concurs with arguments for the non-separation of languages in the brain, at least as regards neurological localisation (Zatorre, 1989; Paradis, 1992), and the view that if there is apparent differential impairment of languages in an aphasic speaker these relate more to control and resource issues or aspects of language specific structures than to differential impairment of central language processes.

The greater facility in German was apparent in the descriptive details of HK's utterances (Tables 4 and 5). In both interactional settings HK produced in German significantly more complete sentences and failed attempts at sentences (deviant sentences). This is reflected in the significant differences in mean length of utterances for the matched settings, though HK does show herself capable on occasions of producing structures in English of equal length to the German sentences. In the analysis of word level units HK's German performance again outranks the English. There are more words and more different words (see type-token relationships) produced in both the German settings compared to the equivalent English ones. If one accepts the presence of pauses and non-specific interjections to indicate searching and planning behaviour, then the comparison of pause times and number of interjections per turn (Table 3) also reveals the greater ease in speaking German.

The quantitative scores in favour of German may be interpreted to arise from several sources. German was acquired long before English. It has been shown that, at least up to the age of puberty, there is a positive relationship between earlier acquisition and greater proficiency, or resistance to attrition, in a language (Harley & Wang, 1997). Furthermore, HK had not used English for the 13 years before her SAH. Having just completed an intensive course of German speech-language therapy may also have exercised an influence on the relative scores of English and German.

We hypothesised that consequent on the comparable pattern of impairments across languages, the profile of trouble sources in both languages would be closely related. This was true, with TS's restricted almost exclusively to lexical-semantic causes (Table 6). Despite the fact that HK was clearly agrammatic, with many instances of morpho-syntactic deviations and heavy reliance on single word utterances (see Tables 1 and 4), these did not feature as TS. This illustrates well the advantages of a conversation analysis approach, which focuses on communicative success and trouble from whatever source rather than making (as it would turn out here) false predictions solely on the basis of formal language scores, structural analyses, or preconceived notions of categories of discourse trouble, divorced from the communicative context and partners. The lack of morpho-syntactic TS's may also reflect the finding (Levinson, 1983; Brinton and Fujiki, 1989) that it is generally not considered appropriate for adults in a symmetrical relationship to correct the other's grammar.

Regarding the number of TS's arising in the matched settings, our hypothesis that there would be a significantly greater total in English was not upheld (Tables 6 and 7). Nevertheless, if one examines the length of repair trajectories required to solve the troubles (Tables 2 and 9), it is obvious that repair sequences for the English TS's are both significantly longer and more complex. This may be interpreted as a manifestation of HK's poorer formal scores in English. That this is not simply a reflection of less experience at speaking with an aphasic partner on the part of the English interlocutors is seen from a comparison of the English conversation and interview. The partner with less experience (conversation) has the less complex repair trajectories. In this case the greater number of failed repairs in the conversation is more likely to be a manifestation of inexperience.

The range of repair interactions (Table 7) was similar across languages, as predicted. There was only one (marginally) significant difference between self initiated self repairs versus interactive repairs, between the English conversation and interview. Three possible accounts for the fewer SI-SR in the interview are possible. It may be a reflection of the greater alertness and support available from the more experienced conversation partner in the interview. It could stem from HK's familiarity with the interlocutor (from whom she had received therapy) and her resultant feeling that either she did not need to monitor herself and/or the knowledge that she would receive support for any TS's. A third possibility may be that HK was less able to self-monitor in the linguistically more demanding setting (see below for arguments concerning this).

The range of partner repair strategies used in the two languages was, as hypothesized, similar. Elaboration was the most commonly employed strategy. However, there were significant differences. In particular the 'unrelated' category arose only in English. There were significantly fewer of these in the English interview, while at the same time the number of 'substitutions' employed there was significantly higher.

We interpret this pattern of more unrelated and failed repairs in the English conversation as a reflection of the pragmatically less skilled interlocutor. The AAT overall language behaviour ratings show that there was considerably more burden placed on the interlocutor in the English settings than the German ones. The English conversation partner was clearly overburdened at times. By contrast, in the English interview the TS totals point to a similar task facing the interlocutor. On this occasion, though, the partner is able to assist constructively in the negotiation of repairs. The size of the task (compared to in German) this time is discernible in the longer and more complex repair trajectories.

Contrasts in turn totals and formal linguistic accuracy between the free conversation and semi-structured interview settings are worthy of consideration. The interviews lasted longer, yet were associated with a decreased number of turns. The interview turns were characterised by more interjections, less turns consisting of just yes/no, and more turns with sentences or attempts at sentences. A possible explanation for this stems from the nature of the setting and is interpretable within the framework of adaptation theory (Kolk & Heeschen, 1990; 1996).

Kolk et al. claim that agrammatic aphasic speakers suffer from an underlying computational processing deficit, causing them to have difficulty in processing language in real time. In the fast exchanges of a normal conversation agrammatic speakers are able to compensate for this by reducing computational load in the planning stage, by for instance producing reduced syntactic structures (in German these are elliptical structures similar to context ellipsis in everyday language) or relying on minimal turns. They thereby increase their conversational speed and achieve better maintenance of their place in a conversation. Heeschen and Kolk found within-subject stylistic variations like this in approximately half of the agrammatic speakers they studied. HK fits this pattern in as far as she is able to change speech styles according to situational/task demands.

The compensatory aspect of adaptation theory would suggest that HK would produce more minimal turns, elliptical responses and fragmentary structures in the conversation, which demands faster alternation of turns and only partial control (if at all) of the conversation by HK. By contrast, in the interviews, the directions in the AAT manual for gathering the speech sample encourage the therapist to permit the speaker with aphasia as much time as possible, giving verbal and non-verbal support to carry on, but not intervening as soon as a silence or a TS occurs. There is therefore not the communicative pressure on the speaker to exchange turns rapidly as in normal conversational settings, leading to less demands on processing speed. The aphasic speaker is granted time to formulate and does not have to spend time and attention on holding or regaining her turn. Although there is no explicit instruction for the speaker to aim for formal accuracy over functional success, the very fact that the interview is the prelude to a formal test may engender this attitude.

Between the German conversation and interview (Table 3-5) there is a fall in minimal yes/no turns and one word utterances and the percentage of deviant sentences. There is a rise in the number of complete sentences and range of lexical items. To this extent the picture supports Kolk and Heeschen's contentions. However, there are still apparent costs to HK. There is a concomitant rise in the proportion of interjections and mean length of pauses per turn. These higher percentages of interjections within the interviews could be taken as surface symptoms for the underlying lexical and syntactical problems and the increased aim for formal accuracy. Concurrently there is also a rise in TSR's in the German interview. Thus, although HK, given time, can attempt and does produce more words and more complete structures, the setting only serves to expose more her underlying impairments. As before, however, the prevailing TS's are lexical-semantic, even when longer and more complete utterances occur that potentially could unmask her syntactic problems.

The other observation when contrasting the conversations and the interviews, is that a significant improvement is seen between the German settings which does not materialise for English. A possible account for this failure to improve in the English interview is offered below.

As regards the intrusions of one language into another, several observations might be made. The only intrusions of English into German are when HK was actually talking about the topic of switching and mixing. The occurrence of German words and phrases in the English exchanges was significantly higher. The number of German intrusions into English was highest in the AAT interview in what was meant to be a monolingual English conversation. The fact that HK knew the therapist was able to speak German may have had some bearing on the number of intrusions. Nevertheless HK was aware of the intrusions and either self corrected or appealed for help with finding the word or phrase. As far as accounts for this profile of intrusions go, two related ideas may be considered.

As an agrammatic speaker with an anterior lesion, we may assume that a major problem for HK is that there is a raised threshold of activation to produce utterances (Bates, Wulfeck & MacWhinney, 1991), both syntactic frames, morphological modifications and lexical items. Hence the predominance of canonical forms, fragmentary utterances, high frequency words and high type-token relationship, in both languages, but particularly for English. As a language that had been learned first and used exclusively over the preceding 13 years German would have had the advantage in activation both from assumed higher automaticity (Bates et al., 1991) and from being a more high frequency language, and consequently being more primed for activation, or having a higher resting level in preparation for activation.

Further, there could be another cost, in terms of brain resources (von Studnitz & Green, 1997), to activating English over German. Not only would English be expected to cost more in resources to activate the lower frequency, less 'automatised' language, but more resources would be needed to keep the more accessible German suppressed.

This may help to explain further significant differences between languages and/or between conversation types. There was a significant improvement in performance in the

German AAT interview over the conversation, which was not paralleled in the English exchanges. The greater success in the German interview was interpreted above to stem from the added time available, less conversational pressure and the focus on achieving formal accuracy. In English this may not have been achieved because the continued need to devote resources to suppressing German deprived HK of the advantages that may have accrued from the greater time available. Also, given HK knew the interlocutor could speak German, we may surmise that German remained relatively activated (Grosjean, 1989) for HK, hence also the greater susceptibility to intrusions.

HK did not feel she was using German as a conscious or unconscious strategy to activate English words. If it was this, then it was unsuccessful, and, if the supposed monolingual English speakers had not been able to understand German, then it would actually have led to more breakdowns. Another pointer to the problem being one of suppression of German is the observation that the majority of intrusions concern either highly automatised phrases (e.g. oh *Mensch* - oh goodness) or closely related words–e.g. *Musik* for music, *Vater* for father, *mein* for my.

Conclusions

We set out to explore the contrasts in performance in HK's two main languages, in particular within the context of the relationship between her formal test scores and her success in conversations. The latter was examined in the framework of conversation analysis, which was able to highlight aspects of communication that were not immediately apparent from HK's AAT results. A main observation was that formal test results, especially in German, pointed towards syntax as being a prime area of impairment, and one that might be beneficially tackled in therapy. By contrast the conversation analyses implicated lexical-semantic sources of trouble as the main causes of communication difficulties, even in German where the formal naming scores were relatively (to English and to syntax) good.

Despite overall similar patterns of impairment, differences did emerge between settings and languages which were not accounted for merely by the divergence in apparent severity of impairment in German and English. We interpreted these to have arisen in part from varying opportunities afforded by the different settings (e.g. time and greater control of the conversation for HK in the interviews; willingness/ability of the interlocutor to engage in negotiative repairs) and the extent to which HK was able to exploit the potential advantages or cope with the possible drawbacks of the situation. These contrasts emphasize that identical or like (formal) language impairments have different consequences for communication according to the conditions and participants present in a setting and according to how interlocutors handle the situation.

This further underscores the contention that assessment of communication in aphasia is incomplete without a consideration of the effects of the conversational partner and setting and that intervention directed at partner behaviour is as important as any remediation targeting the underlying impairments. Indeed, as shown here, it is only in live communicative settings that one can accurately gauge which formal impairments have the greatest consequences for mutual (mis)understanding.

For HK a management programme would ideally have included work directed at her lexical difficulties. Simultaneously though, and integrated with this, much would be gained from therapy addressing the communicative style of significant others in her family and social circle (Miller, 1989; Lesser & Algar, 1995). The conversation analysis framework also provided several parameters that could be used as indicators of progress in therapy. These included the proportion of negative (e.g. failed and unrelated) repairs compared to successful repair types; the length and complexity of repair trajectories; the proportion of self-initiated and self-repairs to ones dependent on others. A simple 'time taken to communicate' measure was not predictive of the level of formal and functional success.

Further aspects of HK's language picture were discussed from a neurolinguistic perspective. Without having to posit separate anatomical localisations for her different languages or different functional circuits within a psycholinguistic model of language functioning, it was possible to account for the variations (and similarities) seen between German and English. Specifically, the poorer formal scores in English (both on the AAT and structural description), the absence of an improvement in the interview setting parallel to across German settings and the predominantly unidirectional intrusion of German into English could be accounted for in terms of processing capacity and costs. The greater costs for English in turn could be led back to discrepant ages of acquisition, the length of time English had remained unused and the effects of recent concentration on German in therapy.

Acknowledgment—We are grateful to Klaus Willmes for guidance in conducting the statistical analyses for this study.

References

Bates, E., Wulfeck B., & MacWhinney (1991). Cross linguistic research in aphasia, *Brain and Language, 41*, 123-148.

Brinton, B., & Fujiki, M. (1989). *Conversational management with language-impaired children: pragmatic assessment and intervention*. Rockville, Maryland: Aspen Publishers.

Edgington, E. S. (1995). *Randomization tests* (3rd ed.). New York: Marcel Dekar.

Ferguson, A. (1994). The influence of aphasia, familiarity and activity on conversational repair, *Clinical Linguistics and Phonetics, 8*, 143-157.

Ferguson, A. (1996). Describing competence in aphasic/normal conversation, *Clinical Linguistics and Phonetics, 10*, 55-63.

Friedland D., & Miller, N. (1997). Conversation analysis of language choice and separation in bilingual speakers with dementia, *Disorder and Order in Talk, Conference Proceedings* (p.15). London: University College.

Friedland, D., & Miller, N. (in press). Conversation analysis of communication breakdown after closed head injury. *Brain Injury*.

Glindemann, R. (1990). Welche Probleme haben Aphasiker beim turn-taking? In R. Mellies, F. Ostermann, and A. Winnecken (Eds.), *Beiträge zur interdisziplinären Aphasieforschung* (pp. 1-29). Tübingen: Narr.

Grosjean, F. (1989). Neurolinguists beware: The bilingual is not two monolinguals in one brain, *Brain and Language, 36*, 3-15.

Hamilton, H. (1994). Requests for clarification as evidence of pragmatic comprehension difficulty: The case of Alzheimer's Disease. In R. I. Bloom, L. K. Obler, S. De Santi, and J. S. Ehrlich (Eds), *Discourse analysis and applications: Studies in adult clinical populations* (pp. 185-199). Hillsdale, NJ: LEA.

Harley, B., & Wang, W. (1997). The Critical Period Hypothesis: Where are we now? In: A. De Groot and J. Kroll (Eds). *Tutorials in bilingualism* (pp. 19-51). Mahwah, NJ: Lawrence Erlbaum Associates.

Heritage, J. (1989). Current developments in conversation analysis, In D. Roger et al. (Eds.), *Conversation: An interdisciplinary perspective* (pp. 24-47). Clevedon: Multilingual Matters.

Klippi, A. (1996). Conversation as an achievement in aphasics, *Studia Fennica Linguistica 6*, Helsinki: Suomalaisen Kirjallisuuden Seura.

Kolk, H. H. J., & Heeschen, C. (1990). Adaptation symptoms and impairment symptoms in Broca's aphasia. *Aphasiology, 4*, 221-231.

Kolk, H. H. J., & Heeschen, C. (1996). The malleability of agrammatic symptoms: A reply to Hesketh and Bishop. *Aphasiology, 1*, 81-96.

Laakso, M. (1997). Self-initiated repair by fluent aphasic speakers in conversation, *Studia Fennica Linguistica*, 8, Helsinki: Kirjallisuuden Seura.

Lesser, R., & Algar, L. (1995). Towards combining the cognitive neuropsychological and the pragmatic in aphasia therapy. *Neuropsychological Rehabilitation, 5*, 67-92.

Lesser, R., Milroy, L. (1993). *Linguistics and aphasia: Psycholinguistic and pragmatic aspects of intervention*. London: Longman.

Levinson, S. (1983). *Pragmatics*. Cambridge: Cambridge University Press.

Menn, L., O'Connor, M., Obler L., & Holland, A. (1995). *Non-fluent aphasia in a multilingual world.* Amsterdam: Benjamins.

Miller, N. (1989). Strategies of language use in assessment and therapy for acquired aphasia. In P. Grunwell et al. (Eds), *The functional evaluation of language disorders* (pp. 97-124). London: Croom Helm.

Miller, N., De Bleser R., & Willmes K. (1997). The English language version of the Aachen Aphasia Test. In W. Ziegler and K. Deger (Eds.), *Clinical Phonetics and Linguistics* (pp. 257-265). London: Whurr.

Milroy, L., & Perkins, L. (1992). Repair strategies in aphasic discourse Towards a collaborative model. *Clinical Linguistics and Phonetics, 6,* 27-40.

Orange, J., Lubinski, R., & Higginbotham, D. (1996). Conversational repair by individuals with dementia of the Alzheimer's type. *Journal of Speech and Hearing Research, 39,* 881-895.

Paradis, M. (1992). The Loch Ness Monster approach to language lateralisation. *Brain and Language, 43,* 534-537.

Perkins, L. (1995). Applying conversation analysis to aphasia: Clinical implications and analytic issues. *European Journal of Disorders of Communication, 30,* 372-383.

Schegloff, F., Jefferson, G., & Sacks, H. (1977). The preference for self correction in the organisation of repair in conversation. *Language, 53,* 361-382.

StatXact (1989). Statistical software for exact nonparametric inference. Cambridge: Cytel Software Corporation (Version 2.0, 1991).

von Studnitz, R., & Green, D. (1997). Lexical decision and language switching. *International Journal of Bilingualism, 1,* 3-24.

Wilcox, A., & Mogford-Bevan, B. K. (1995). Assessing conversational disability. *Clinical Linguistics and Phonetics, 9,* 235-254.

Willmes, K. (1985). An approach to analysing a single subject's scores obtained in a standardized test with application to the Aachen Aphasia Test (AAT). *Journal of Clinical and Experimental Neuropsychology, 7,* 331-352.

Zatorre, R. (1989). On the representation of multiple languages in the brain. *Brain and Language, 36,* 127-147.

Appendix

Examples of Trouble-Source-Repair (TSR) Sequences:

1. Self-Initiated-Self Repair Sequences (SI-SR)

Example (English conversation):
67 HK: er (1.0) riding er instructor um ah in-struct-ress
108 HK: um um my father has um (5.0 *writes down*) in London

Example (German conversation):
14 HK: (*laughs*) Apotheke (.) ja genau (*laughs*) danke () Apotheke gefahren (.) und mein Sohn hat hingele NEIN ICH mich hingesetzt und mein Sohn hat äh vorne äh gewartet

2. Other-Initiated - Self-Repair Sequences (OI-SR)

Example (English interview):
168 HK: oh er (1.0) erm (2.0) erm er I now (.) is erm (1.0) erm erm (*writes down*) dress-age and um (*simultans.*)
169 FB: sorry? (*simultans.*)
170 HK: dressage
171 FB: umu

3. Self-Initiated - Other-Repair Sequences (SI-OR)

Example (English conversation):
4   HK: and mein (*German*) eh (2.0) father and mother and (.) sister and (2.0) me has erm (1.0) erm (2.0) erm
5   CH: you've moved on ?
6   HK: yeh yes

Example (English interview):
32 HK: bath and um (2.0) [is] (1.0) erm
33 FB: you went to the bathroom or (simultans.
34 HK: yes (simultans.)

4. Other-Initiated - Other-Repair Sequences (OI-OR)

Example (English conversation):
```
55  HK:   yeh erm (1.0) um (2.0) er (4.0) n..now [a] er all forgotten
56  CH:   all your Italian or?
```

Example (German interview):
```
11  HK:   also äh ähm ähm äh (2.0) Harmonie und Smartie habe ich zur Weide gebracht und äh konnt ich mir
          schwindelig oder äh (1.0) mein Sohn hat gesagt äh setz dich hin und mach das (1.0) meine Mutter is
          krank äh ähm vier fünf sechs Jahre alt
12  OB:   sechs Jahre is das her?
13  HK:   ja
```

5. Self-Initiated - Collaborative Repair Sequence (SI-ColR):

Example (English interview)
```
138 HK:   oh er born er no er no (3.0) I. was. born (2.0) I don't know (laughs)
141 FB:   umu
142 HK:   I erm
143 FB:   you don' t know? (laughs)
144 HK:   erm er erm um I ne.. ne (German) er be nein (German) er erm ts (1.0) erm ich nein ich weiß nicht
          (German) (3 sylls. unintelligible) e r
145 FB:   you're not (.) er tell me about the city where you were born erm
146 HK:   erm
147 FB:   is that right?
148 HK:   erm doch (German) ah
149 FB:   umu which country were you born?
150 HK:   um (writes down <GER>) Germany
151 FB:   oh Germany
```

6. Other-Initiated - Collaborative Repair Sequence (OI- ColR):

Example (English conversation):
```
82  HK:   erm (3.0) um (2.0) my father has erm (.) um (1.0) tube
83  CH:   tube?
84  HK:   yeh er: (1.0) er er so um (.) [fater] (gloss: German father) erm (2.0) um mein (German) ah father
          has erm (1.0) own er (1.0) um (2.0) own um (2.0) erm (4.0) oh Mann (German) (2.0) um tube e r
          (writes/ draws) [bazer]
85  CH:   umu? (2.0) he he was a salesman of that?
86  HK:   er yeh je ja (German)
```

Example (English interview):
```
121 FB:   and then you stayed for how long in hospital?
122 HK:   (3.0) er no
123 FB:   a week or longer?
124 HK:   no no no no no (1.0) two months
125 FB:   oh really in that hospital or
126 HK:   yeh no no (simultans.) no er I er er (1.0) erm
127 FB:   umu (simultans.) in the first hospital you stayed
128 HK:   yeh erm (1.0) this es (writes down) e r
129 FB:   day?
130 HK:   yeh
131 FB:   one day
132 HK:   yeh yeh
```

Examples of Partner Repair Strategies (RP)

1. Repetition (Example: English interview):

```
96  HK:   no um no um oxy-gen (writes down)
97  FB:   um oxygen
```

2. Unrelated repairs (Example: English conversation):

```
122 HK:   um mover (1.0) um (1.0) um further er down and (.) er (.) I'm (.) [ts] whats this called um
123 CH:   do you think you could better write in English than speak in English?
```

3. Elaborations (Example: English interview):

40 HK: um um (1.0) I have um (.) er (2.0) erm (.) taxi erm
41 FB: umu (.) you called a taxi?

4. Reductions (Example: English conversation):

190 HK: yeh well (.) [ei] er (2.0) erm (3.0) NO (2.0) (*unintelligible 1 syll.*) er (2.0) I (.) could (.) can't (.) wor (.) diese (*German*) ehm words er (1.0) wordsn..er (1.0) words erm (2.0) erm
191 CH: which words?

5. Substitutions (Example: English conversation):

130 HK: um (1.) um (3.0) um (1.0) the german is genauso (*German*) (*laughs*)
131 CH: yeh (1.0) the same (3.0) um

Examples of German intrusions into English:

Example (English conversation):

45 CH: how old was you then?
46 HK: vier (*German*) eh (2.0) um five nein (*German*) er six nein oder (*German*) erm five (1.0) years was I

Example (English interview):

24 HK: ah ah ich (*German*) yeh er doctor has erm erm (2.0) got me ah tablets [ants]um (.) was is (*German*) er erm (1.0) ts doctor said er er Migräne (*German*)
43 HK: je er mein (*German*) er little girl um ts erm (2.0) um (.) erm um

Examples of English intrusions into German:

Example (German conversation):

120 HK: na sicher (.) NOW IS eh ja ähm NOW IS (3.0) ähm (1.0) MIST (.) UP (*gloss: mixed up*)
150 HK: ja no (.) Schrand oder (.) ähm (3.0) es rensiert ein nicht AND zwei äh (.) Wochen ich (*laughs*) nachholen ne?

 **Pergamon**

J. Neurolinguistics, Vol. 11, Nos 1–2, p. 243–251, 1998
Published by Elsevier Science Ltd. Printed in Great Britain
0911-6044/98 $19.00 + 0.00

Author Index

— N —

— O —

Pergamon

J. Neurolinguistics, Vol. 11, Nos 1–2, p. 253–257, 1998
Published by Elsevier Science Ltd. Printed in Great Britain
0911-6044/98 $19.00 + 0.00

Subject Index